W0042078

This comprehensive biography traces the life and works of Robert Maillart, one of the most important engineers and designers of the twentieth century. His career developed around a central issue of modern technological society: the debate between two antithetical views of engineering opposing applied science, which relied on general mathematical theories for understanding structures, against design, which Maillart championed. Maillart considered structures not merely works of utility, but also as works of art. As utilitarian objects, he created a series of innovations of lasting significance, including the concrete hollow box, the concrete flat-slab floor, the concrete deck-stiffened arch, and the concept of the shear center. Aesthetically, Maillart shaped his three innovations in concrete to create surprising and often stunning new forms. Providing an analysis of these innovations, this biography also connects Maillart's aesthetic ideas with the private and professional context in which he worked.

Robert Maillart

Marie-Claire and Robert Maillart in Orselino, 1912.

Robert Maillart

BUILDER, DESIGNER, AND ARTIST

David P. Billington
Princeton University

CAMBRIDGE
UNIVERSITY PRESS

CAMBRIDGE UNIVERSITY PRESS
Cambridge, New York, Melbourne, Madrid, Cape Town, Singapore, São Paulo

Cambridge University Press
The Edinburgh Building, Cambridge CB2 8RU, UK

Published in the United States of America by Cambridge University Press, New York

www.cambridge.org
Information on this title: www.cambridge.org/9780521571326

© Cambridge University Press 1997

This publication is in copyright. Subject to statutory exception
and to the provisions of relevant collective licensing agreements,
no reproduction of any part may take place without the written
permission of Cambridge University Press.

First published 1997
This digitally printed version 2008

A catalogue record for this publication is available from the British Library

Library of Congress Cataloguing in Publication data
Billington, David P.
Robert Maillart : builder, designer, and artist / David P. Billington.
p.   cm.
Includes bibliographical references (p.   ).
ISBN 0–521–57132–4 (hardback)
1. Maillart, Robert, 1872–1940.   2. Civil engineers – Switzerland –
Biography.   3. Bridges, Concrete – Switzerland – Design and construc-
tion.   4. Architecture – Switzerland – Aesthetics.   5. Structural design.   I. Title.
TG140.M3B57   1997
624'.092 – dc20
[B]   96-23133
CIP

ISBN 978-0-521-57132-6 hardback
ISBN 978-0-521-05742-4 paperback

This book originated with the Architectural History Foundation.

This book is dedicated to

Marie-Claire Blumer-Maillart

A Loyal Daughter
A Joyful Collaborator
A Gracious Friend

Contents

Illustrations

Preface

This biography represents a summary not just of Maillart's 46-year career, but also of my own exactly equal-length professional life. Discovering Maillart was fortuitous because research into his life and work seemed natural thanks to four stages in my own development beginning with undergraduate education at Princeton in the late 1940s. I followed a program invented by the Dean, Kenneth Condit, called ambiguously Basic Engineering, that brushed us through all fields of engineering but still allowed time to take more liberal arts electives than many liberal arts majors. My favorite courses were in music, literature, and art history, but I had enough engineering to win a Fulbright Fellowship for study of bridge reconstruction in Belgium. There I studied structural engineering at Louvain University, visited numerous new and partly built concrete bridges, and courted a talented and beautiful pianist. After marriage, we returned to Belgium where I studied the major new idea in structures, prestressed concrete, with one of its pioneers, Gustav Magnel, at Ghent University.

After academic studies, I entered a second stage, returning to the United States and capitalizing on this new idea, to work for 8 years as a structural designer in New York City. There I met Otto Gruenwald, who taught me analysis, and Anton Tedesko, who inspired me to learn about another new idea, thin-shell concrete structures. Our firm did design of difficult structures, ones that most other engineers avoided. This type of design was an important part of Maillart's practice as well.

All the while my motivation to write and to speak in public began to overtake my urge to de-

sign, and thus when offered a position at Princeton, after some hesitation, I accepted and entered a third stage in preparation for Maillart. At Princeton, my mentor, Norman J. Sollenberger, gently eased me into the academic world and recommended that I work with Robert Mark, a civil engineer then running an experimental program in stress analysis at Princeton's Plasma Physics Laboratory. Mark and I wrote proposals to the National Science Foundation; but not having doctor's degrees and having no track record in such research, we were consistently unsuccessful until the program officer, Michael Gaus, recognizing our plight, simply overrode the mediocre peer reviews and awarded us a small grant. Thus began a fruitful collaboration during which Mark and I learned about the disciplines of academic research and the necessity for publication in refereed journals.

It was then in the mid-1960s that the architects began to exert their influence and I entered a fourth stage. In 1966, I received a National Science Foundation science faculty fellowship to study thin-shell theory with Warner Koiter, a world leader, at the Technical University in Delft, the Netherlands. Prior to leaving for Europe, Kenneth Frampton, a distinguished architectural historian then at Princeton, gave a series of lectures, two of which had a profound influence on me. One was on Gustav Eiffel and the cultural context for his bridges and tower; the second and of more immediate interest was on the Dutch modern movement in art, architecture, and engineering call *de Stijl*. Frampton opened my eyes to the connection among Dutch culture, modern art, and the engineering of an artificial landscape at just the time we were preparing to spend half a year in that waterlogged environment.

So it was that I returned to the lowlands to study not only an abstract mathematical theory, but also an abstract artistic culture in which the ideas of beauty and utility were intimately connected.

Without Frampton's stimulus I would never have seen the whole of the Netherlands and on returning, I was more than ever eager to find some way to express this wholeness in scholarship that would be somehow academically respectable in the fragmented, specialized universe of the research university. It was then that Maillart seemed to be a natural.

Having studied German and French in school, having lived in small European nations, having practiced design in concrete structures, and having found colleagues and students willing to collaborate and to learn, I was prepared to take on the complexity of a multilingual Swiss engineer whose structures had already been canonized by an exhibition at the Museum of Modern Art (1947), by a long chapter in Sigfried Giedion's hugely influential *Space, Time and Architecture* (1941), and whose works were elegantly detailed in a trilingual book by Max Bill (1949).

This biography culminates 28 years of studying the life of Robert Maillart, a name I had never heard of during my own college education and my 8 years in practice. Only when I began teaching did the architecture students bring me pictures of Maillart's bridges and buildings and ask me to teach them about such visually striking forms. At that time, in the early 1960s, there was no technical writing on Maillart in English and no writing of any type on him in the contemporary engineering literature.

Maillart's forms were so far out of the ordinary that I could find no way to incorporate them into a standard course on structural engineering either to architects or engineers. Here was a dilemma: the standard courses in structures were devoid of powerful visual images, whereas a small number of structural engineers following Maillart and others

were creating such images solely as engineers not architects. It was as if literature were being taught using texts taken from the front pages of newspapers while Hemingway, Fitzgerald, and Faulkner were creating their celebrated works.

This dilemma inspired me and my colleague Robert Mark, who faced a similar quandary, to talk to Whitney Oates, head of Princeton's Council of the Humanities, who told us to see Princeton's President Robert Goheen. Those two humanists sent us to Washington to visit the newly created National Endowment for the Humanities, an institution both men had helped to bring into being. There we met Herbert McArthur, a program officer, who was immediately sympathetic, and soon we had a small grant plus a larger amount that required matching funds. Oates and Goheen saw to it that we got those funds from the Ford and Rockefeller foundations.

Thanks to that beginning, I went with my son David to Zurich in the summer of 1970 where Marie-Claire Blumer-Maillart graciously received us and there began the most fruitful and unusual biographical collaboration that I know of. The following summer, with my daughter Elizabeth, I visited her again, this time at St. Jean Cap Ferrat where we discussed her father's centennial in 1972. Robert Mark and I organized a colloquium at Princeton in Maillart's honor and Madame Blumer-Maillart came with her husband Eduard Blumer. At that time, in October 1972, I began serious discussions with her about a biography and she responded warmly. The following year my wife Phyllis and I came to Switzerland and there I began to go through Maillart's papers and allied documents that had been collected and organized by Madame Blumer-Maillart and her husband.

In 1974, as a visitor to the Institute for Advanced Study, I began to write a biography. There was little material on Maillart's private life, so I focused on his works and was more than half way finished when Madame Blumer-Maillart and her daughter found in an attic trunk a mass of letters written by Maillart dating from his youth to his last months. Clearly, I needed to read these, but what to do with the writing then nearly completed? Marshall Claggett, the historian at the Institute who sponsored my visit, gave me the needed advice – finish the book as an essay on Maillart's major work and then write a proper biography. This sound advice led to *Robert Maillart's Bridges: The Art of Engineering*, published by the Princeton University Press in 1979.

Meanwhile I read through the letters with Madame Blumer-Maillart. Together we relived much of Maillart's life, primarily from the construction of his first major work, the 1901 Zuoz Bridge, to the destruction in 1940 of his last great design, the Zurich Cement Hall. She would interpret words, names, places, and events that I could never have gotten on my own. It was an indispensable experience, especially reading those letters written while the Blumers were in Indonesia from 1929 to 1940. We read letters in Zurich, we read letters during another stay at the Institute for Advanced Study, and we read more in the Princeton Maillart Archive, founded in 1974 and installed permanently at the Engineering School in 1980, thanks to the support of our departmental chairman, Ahmet Cakmak.

I began writing the biography in 1978, once the first book was in press, but it proved more difficult than I had anticipated. As I struggled with the large mass of documentation, Martin Kessler, President of Basic Books, showed up one day in my office and stimulated me to write a book for him on the idea of art in modern structures. My brother, James H. Billington, had urged Kessler to see me and the result was *The Tower and the Bridge: The New Art of Structural Engineering*, published in 1983 with Maillart as the central figure in a story that began with Thomas Telford, Gustav Eiffel, and John Roebling, and ended with the legacy of Maillart as seen in the late twentieth-century structural artists Felix Candela, Heinz Isler, Christian Menn, and Fazlur Khan.

That book grew out of a course that I had begun teaching at Princeton in 1974 entitled "Structures and the Urban Environment," which in turn had come from my studies on Maillart. I wanted to show that Maillart was part of a modern tradition begun with the Industrial Revolution and flourishing today. Following the publication of *The Tower and the Bridge* I returned to the biography, rewriting and extending the story. By 1988, I was once again nearly finished when I got a call from Edgar Kaufman stating that he wanted me to write a heavily illustrated book on Maillart.

This call was not my first contact with Kaufman. Four years earlier he had called after receiving a letter from George Collins, Professor of Art History at Columbia University and a good friend of mine since 1969. Collins had suggested that the Architectural History Foundation publish a book of color photos as a sort of companion to *The Tower and the Bridge*. Collins believed that the theme of structural art needed a book with elegant pictures much as is frequently done for architects. That project appeared too daunting in 1984, but Kaufman's more focused idea of 1988 became a reality with the 1990 bilingual book, *Robert Maillart and the Art of Reinforced Concrete*, published by the Architectural History Foundation in New York and the Verlag für Architektur Artemis in Zurich and Munich.

I told Mr. Kaufman about the biography, which he asked to see and then enthusiastically recommended to the Architectural History Foundation. The foundation's president, Victoria Newhouse, accepted it and following publication of the 1990 book, she began to work closely with me in revising what had become an unwieldy manuscript. Thanks to her fine editorial eye, the work began to take shape but only after she had helped me reorganize it rather drastically. We spent many hours together reading over sections, reworking parts, and discussing content. When she decided in 1994 to close down the foundation, she sent the manuscript to Beatrice Rehl of Cambridge University Press through whose encouragement and help the manuscript went through one more revision before it was finally accepted in August 1995.

This long journey from the 1960s to the 1990s, therefore, has been completed thanks to the sustaining help of many people. But there is another story that has evolved through this 28 years of writing. At the outset, I was intrigued by the works and, as a structural engineer, I wanted to study them technically. Did their aesthetic power derive from a pure engineering design or did they acquire elegance by means other than engineering, that is, decoration or nonstructural form? Put the other way, was Maillart a technically inferior engineer by stressing appearance over cost and performance?

My first studies revealed that Maillart was far ahead of his profession technically, so I needed to tell that story to a nonengineering audience; but then came the next question: To what extent could his superior engineering and his fine aesthetic talent be connected to his private life? This is an old question in biography, especially for biographies of artists. The more I learned about Maillart as a person, as a family man, and as a colleague, the more difficult it became to answer that question. It was only after I had amassed the documentation and written a long manuscript that I could begin with the help of reviewers and editors to see where the private life would illuminate the professional life and where it would not.

The private story began during the Victorian certainty of a stable Europe in the comfortable bourgeois world of little Switzerland. It then met the convulsion of World War I, which led into a wildly swinging interwar Europe and culminated in the outbreak of the second half of the Great War. In a curious and unusual way Maillart lived through this tumult. Outwardly a Victorian with conventional attitudes and a routine life, he adjusted to a partly monastic existence after losing his wife and his wealth during the war. He never lost his settled attitudes and his love of routine, but his motivation did change from that of large-scale builder to small-scale designer. It was a change enforced by penury, but it was one that fit his personality.

During his 15-year marriage, he strove to create a thriving business as his family grew, but following the war, he had no home and reverted to the 7-year bachelor life he had had before marriage. The last quarter century of his life, living mostly alone, he

created the structures on which his fame rests. His sometimes weekly letters give the sense of a routine life and rarely do they digress into the professional controversies so characteristic of his work. Exceptions occur in letters to his daughter in Indonesia from 1929 to 1940. These documents themselves have a remarkable history. Sent to the Dutch colony, they were saved by Marie-Claire Blumer-Maillart; and then during her desperate escape from the invading Japanese and her incredible 4-year survival in the jungles of Java, she kept them in a small briefcase so that finally after circling the globe they returned to Switzerland intact at the war's end.

In these letters, written rapidly and unselfconsciously, we get a glimpse of Maillart's reflections and responses to weekly trials. His sense of humor never left him, his optimism remained in the face of depression, and his loneliness became more apparent as the 1930s came to a close. He even reveals at times his delight in a new little bridge that he was given because of its difficulty. These factors of personality do illuminate the works or at least they seem consistent with them. Yet the direct connection between works and events or public designs and private thoughts remains elusive. Perhaps that is always so when utilitarian objects rise to the level of great art whether portraits painted for prephotographic records or architecture created out of cost-constricted programs.

To take the example that keeps recurring in these studies, the Salginatobel Bridge, I find its design to be some kind of a mystery. I can explain it structurally and yet its visual impression is always more complex without ever leaving the realm of pure structure. In his simplicity, Maillart reminds me of two contemporaneous painters, the Dutchman Piet Mondrian and German Swiss Paul Klee. Both artists created many works of deceptive simplicity and both lived lives often apart from mainstream society. Maillart himself was half Dutch and Belgian as well as half German Swiss. The simplicity of his lines and even of his patterns of steel reinforcement within concrete reminded Max Bill of Mondrian; and Maillart's playfulness with form is reminiscent of some Klee paintings. If one could explain once and for all the meaning of Mondrian's *New York Boogie Woogie,* Klee's *Doublezelt,* and Maillart's *Salginatobel,* then they would cease to be art.

———

The theme of structure as art first became public with the 1947 Museum of Modern Art exhibition. Following that event, Professor Mark and I proposed a Maillart exhibition in the Princeton University Art Museum to go together with the Centennial Symposium in October 1972. David Steadman, acting director of the museum, organized the exhibition and J. Wayman Williams put it together out of original Maillart documents, recent photographs, and models. Four years later, in honor of Madame Blumer-Maillart's visit, we put together a new photographic exhibition again with the help of Mr. Williams and the director of the art museum, Peter Bunnell. In 1978, with director Fred Licht and Wayman Williams, we mounted an exhibition entitled "The Bridges of Christian Menn," and in 1980, with the new director Alan Rosenbaum and again Wayman Williams one on "Heinz Isler, Structural Artist," both showing the legacy of Maillart on present-day Swiss engineering designers of bridges and buildings.

All the while I was bringing copies of Maillart documents to the Princeton Maillart Archive, Madame Blumer-Maillart was establishing in Zurich a Maillart Archive at the Federal Institute of Technology. Most of the original drawings and calculations from Maillart's post-1920 offices in Geneva and Bern are now in the Zurich Archive, assembled and catalogued by Dr. Beat Glaus and Mr. Clemente Regassi. Then in 1990, the Museum für Gestaltung in Zurich held a large exhibition on Maillart that thereafter traveled throughout Europe.

One climax of these Maillart events came on August 21, 1991, when James E. Sawyer, President of the American Society of Civil Engineers, and Madame Blumer-Maillart unveiled a plaque designating the Salginatobel Bridge an International Historic Civil Engineering Landmark. Part of the Swiss celebration for the 700th anniversary of the founding of their confederation, this Maillart Day event placed his wilderness work alongside only twelve other such landmarks, which include the Eiffel Tower, the Firth of Forth Bridge in Scotland, and the Panama Canal. We made a pilgrimage that day up from the Landquart River Valley along the single-lane roadway far into the Alps; it is a pilgrimage that should be made by anyone who wishes to contemplate the theme of Maillart's life and work: Design, in this fragmented world, when carried out by a highly rational, deeply educated engineer, can integrate utility and beauty within the constraints of responsible economy with public funds.

Acknowledgments

As the preface makes clear, Marie-Claire Blumer-Maillart has been central to this biography, and her husband Eduard Blumer played a major role in collecting documents, in organizing records, and in escorting me to archives. His enthusiasm for Maillart, his unfailing sense of humor, and his ability to make friends quickly in almost any European language made him a joy to be with and an invaluable colleague. Their daughter Marie-Claire and her husband John Cuniberti have also been of considerable help especially in hosting us in Geneva and putting me in contact with Maillart works and documents there.

In 1974, while searching for more Maillart works with my daughter Jane, I had the good fortune to visit René Maillart in Marseilles; he graciously agreed to the construction of his recollections of his father. René's grandson Laurent Maillart spent 3 months in Princeton in 1993 and did a study of the Salginatobel Bridge, which added new insight into his great grandfather's most famous design. I am also much indebted to the son of Maillart's eldest child, Robert Maillart, for documents concerning his grandfather. In all, the family has been open, unfailingly helpful, and a great pleasure to know.

Along with the family, the support of Christian Menn has been crucial to this research. Switzerland's foremost living bridge designer and after Maillart the greatest master in concrete bridges of this century, Menn has opened doors to the offices of bridge engineers and cantonal archives all over his small but decentralized country. But of far greater importance is my coming to know him, his

work, and his ideas that get me closer to the ideals of Maillart. Professor Menn also secured substantial financial support for me in Switzerland to cover expenses and to allow me to be a visiting professor at the ETH (Federal Technical Institute in Zurich).

Heinz Isler, the other major Swiss structural artist practicing today, was a direct help with people and documents in the Canton of Bern and an indirect help by explaining his own works, which are exemplars of the same intrinsic interest as those of Maillart and Menn. His wife Maria Isler was a gracious host and another colleague in my Swiss adventures.

Four other Swiss engineers, all with connections to Maillart, gave me invaluable insights into his works and professional life. The late Ernst Stettler was Maillart's chief engineer in Bern from the late 1920s until his death when the younger man took over the practice. Stettler's published reflection provides a firsthand account of Maillart's design activity and of his life in Bern during the 1930s. Likewise, Marcel Fornerod's reflections give a picture of Maillart's Zurich activities during the early 1930s. I had known Fornerod long before I knew about Maillart; already a highly respected engineer in the United States, Fornerod wrote me after our 1972 symposium to tell me of his work in Switzerland with Maillart 40 years earlier. Pierre Tremblet, heading the successor office in Geneva, gave me access to his archive and kindly had copies of it all sent to the Princeton Maillart Archive. The originals have since been deposited in the ETH Archive in Zurich. Finally, Hanspeter Bernet sent me copies of his Bern Archive and hosted us elegantly in his

city. He had taken over the Maillart office when
Stettler retired. Dr. Hans Eichenberger, executive
officer of the Swiss Society of Cement, Chalk and
Gypsum Manufacturers provided financial support
for expenses and with his assistant, Kurt Müller,
was an invaluable guide to Swiss concrete structures.

In the middle 1970s, I sent letters to all the other
former employees of Maillart's that I could identify
and to my surprise Karl Lehr, the only one alive
who had worked with him in Russia, responded
from the New Jersey shore only an hour's drive
from Princeton. I immediately went to see him, conducted a long and friendly interview with an old
and feeble man, and thus secured an invaluable recollection of a relatively early period in Maillart's
life. Also I had the good fortune to find Hans
Kruck, the only architect to work for Maillart, still
living in the late 1970s in Zurich where we had
several lengthy meetings. Thanks to his daughter, I
secured another significant recollection. Similarly,
Professor Fritz Stüssi, an adversary of Maillart's
during the 1920s, received us warmly at his home
in Bäch and provided me with a different perspective on Maillart as well as sending me a full set of
lecture notes taken by an ETH student, Hans Mis-
bach, from Maillart's lectures there before World
War I.

The noted Swiss architect, Alfred Roth, gener-
ously shared his recollections of Maillart with me,
and Vinzenz Losinger sent me many photos and
materials from his firm's archives. I am deeply indebted also to the late Max Bill, for his pioneering
writing of Maillart and for his inspiring keynote
addresses both at our 1972 symposium in Princeton
and at the opening of the 1990 Maillart exhibit in
Zurich. Anyone writing about Maillart also must
be grateful to Sigfried Giedion, whose early recognition of Maillart as an artist showed great insight.
He also helped organize the 1947 exhibition on
Maillart at the Museum of Modern Art.

A large number of Swiss have aided my research
beginning with Alvin E. Jaeggli, archivist at the
ETH in Zurich. His successor Dr. Beat Glaus has
been indispensable and has gone out of his way to

provide me with documents and advice as well as
friendship. His assistant, Clemente Rigassi, produced a masterful catalogue for the Zurich Maillart
Archive and was unfailingly helpful to me. A number of engineers throughout Switzerland opened
their archives and provided me with valuable documents: Engineers Letta and Bosch of St. Gallen,
Engineers Stampf and Tschudin of Chur, Engineer
Schlumpf of the Rhätische Bahn, Engineer Hirt and
Mr. Stalden of Zurich, and Engineer E. Woywod
of the Aargau who escorted me to archive sites at
Aarburg, at Rheinfelden, at Laufenburg, and at
Aarau. I benefited from Professor R. Favre's discussions about the Aarburg and Zuoz bridges for
which he did the rehabilitation design. Others who
helped me were B. Jotteraud of the Bière-Apple-
Morges Railway, M. Masshardt of the Losinger
Company, E. Gruner who wrote me a recollection
of the Laufenburg Bridge, and Margaret Siegrist of
the ETH alumni association. Also of help were
Walter Meierhans of the Zurich Warehouse Company, Ulrich Bähler of the St. Gallen City Public
Works Department, and Andreas von Waldkirch of
the Bern Meliorationsamt.

Of special assistance was Prof. Hans Hauri, formerly Rector of the ETH, who in addition to general advice, set up and taped an interview that I
carried out with Prof. Karl Hofacker, formerly an
assistant of Prof. Max Ritter, Maillart's bitterest detractor at the ETH. Prof. Hofacker gave me another
perspective on Maillart. Tom Peters, formerly at the
ETH and now a professor at Lehigh University, has
been a helpful colleague in my Maillart work. With
Madame Blumer-Maillart, I met Miss Oechslin
(Mucci) who had known Maillart well in the 1930s.

The principal financial support for this book has
come from the Division of Research Programs of
the National Endowment for the Humanities
headed by Harold Cannon and from the Program
for the History and Philosophy of Science of the
National Science Foundation headed by Ronald
Overmann. Robert Mark and I have had grants
from the National Endowment for the Humanities,
from the Ford and Rockefeller Foundations, as well
as from the Andrew W. Mellon and Alfred P. Sloan

Foundations, all of which have helped the studies leading to this book. Of particular benefit to this book were two grants from the National Endowment of the Arts, and especially the encouragement of Thomas Cain. Of great help also were a series of grants from the Alfred P. Sloan Foundation in its program, The New Liberal Arts, aimed at bringing engineering to liberal arts students. I am especially grateful to its former president, the late Albert Rees, and to its program officer, Samuel Goldberg.

At Princeton, a number of colleagues contributed to this research, especially my academic mentor Norman J. Sollenberger and my close colleague Robert Mark. John Abel, now at Cornell, played a major role during the first years of this project especially for the Maillart symposium and its publications for which he served as editor and coeditor. Ahmet Cakmak, who succeeded Sollenberger as chairman of Civil Engineering, strongly supported this effort, encouraged me to publish the early research results, and urged me to give a course to include this material. The late Donald Egbert, professor of Architectural History, served as an advisor to me during a leave in 1969 and taught me about writing history. Also of help in that endeavor were François Bucher, who first collected Swiss documents for us in 1968, and Kenneth Frampton, who showed me how connections could be made among art, architecture, and engineering. The late William Shellman was a continual help in acquainting me with the relationship among art, architecture and engineering; he also made elegant drawings of Maillart's bridges for our art museum exhibitions.

The late George Collins, professor at Columbia University, wrote an article at our suggestion on Maillart and Modern Art, which I believe to be the most perceptive such writing ever done. He was always a guide and friend. Carl Condit, the pioneering historian of civil engineering in the United States, was an inspiration for our work, and Edwin Layton, another distinguished historian of technology, now at the University of Minnesota, has been a longtime teacher to us engineers. A third major American historian of technology, Merritt Roe Smith of MIT, read an early version of this biography and gave cogent advice and continual encouragement.

The late Myron Goldsmith, both an architect and an engineer, has never ceased to stimulate me; his immense enthusiasm for Maillart helped keep me at work. His colleague at Skidmore, Owings and Merrill, the late Fazlur Khan, was a Maillart enthusiast, as well and Khan's own work, like Goldsmith's, served to give me a sense of Maillart's legacy. Like Khan and Menn, Felix Candela, the gifted structural artist from Mexico, wrote a moving essay on Maillart's influence and helped me formulate ideas on structure through his lively discussions.

No one has sustained my efforts professionally more than J. Wayman Williams, whose critical judgment and engineering good sense go together with a talent for organizing all our exhibitions, for making many of the slides for my teaching, and for putting together publications of all sorts. His wife Patti typed one version of this book as well. Jean Carlucci, Thelma Keith, Susan Cleary-Diaz, and Lilya Lorrin also typed sections. Kathy Posnett, my present secretary, typed the final version and serves in many other ways to keep things organized and running smoothly. The former engineering librarian at Princeton, Dee Hoelle, was a continual help.

During this entire Maillart project I have been fortunate to have a series of excellent students serve as research assistants, starting with Peter Cole who helped greatly with the 1972 symposium, David Lamb, Ellen Lieng, Mark Herron, Kent Smith, Robert Shulock, Michael Hein, Neil Hauck, Paul Gauvreau, Scott Hunter, Ronald Wakefield, Christopher Peck, Rosemary Secoda, John Matteo, Karen Mielich, Roger Haight, Nicholas Edwards, Susan Lyons and Eric Hines who also checked the notes, especially for correct German. Byron Pipes first called my attention to Maillart's shear center discovery. Mark Reed made many of the fine drawings for this book as did Clark Fernon and Colin Ripley. Of central importance was James Chiu, who volunteered to organize the copies of the Maillart Geneva Archive and was thereby responsible for the founding of the Princeton Maillart Archive.

Edward Tenner, editor for *Robert Maillart's*

Bridges, helped me during the early stages of this work. My sister-in-law, Lynn Billington provided me with fine editing and careful retyping at a crucial stage in this biography. Then from 1990 to 1994, Victoria Newhouse became my editor and gave the essential reworking that finally put the book in its present shape. When some issues still remained unresolved, she urged me to contact Stanford Anderson, professor of Architectural History at MIT, who gave me significant assistance in late 1994. Since then Beatrice Rehl at Cambridge University Press has taken over as editor and provided the necessary advice and actions to put the book in production. Ernie Haim has deisgned and organized the final form of this book with great skill. I am also grateful for the splendid photos taken by Bruno Mancia and Franziska Bodmer Mancia.

Finally, and of the greatest importance, my family has not just been supportive, but also actively engaged in this research and writing. My brother Jim has long served as a model both of academic integrity and of historical scholarship. He read numerous versions and drafts and always had incisive suggestions often of major themes. Each of my children has traveled with me to visit Maillart sites, often to locate documents, and always to provide critical reactions and collegial companionship. Sarah, the youngest, has become herself a fine structural engineer and has already collaborated on two writing projects with me. She studied at the ETH with Christian Menn and developed a close relationship also with Madame Blumer-Maillart. My two sons, Stephen and Philip, traveled through Switzerland with my wife and me in 1987; they are both artists though not engineers and thus provided sharp reactions to the works and an often comic sense of the wilderness Swiss context replete with cows and multilingual little valleys. My daughter Jane, a teacher now, ferreted out many little bridges in 1974, and my eldest daughter, Elizabeth, also an artist along with my eldest son David, an historian, searched archives in 1975 with Eduard Blumer. Both Jane and Elizabeth have always exacted a steadying influence on me during the stress of foreign travels especially in rural settings.

At the very end of this project when acceptance was still in doubt, David read the entire manuscript and saw a central theme that had eluded me. He then drafted an introduction, which is largely the present version, and then I was able to make the final revisions that led to the book as it now stands. His clear insight and his historical imagination have been my greatest support. David has also done the index with great skill.

This biography is properly dedicated to Marie-Claire Blumer-Maillart. Without her collaboration, I would not have attempted this kind of scholarship. Although in the end, the judgments are mine with all their limitations, her dedication to her father and to my efforts has sustained this research and writing for over a quarter of a century.

In a deep and general sense, all my work is dedicated to my wife, Phyllis. Her piano playing fills the house as I write and her management of a large and complex family not excluding a disorganized professor has kept that operation together with a model of beauty and utility.

David P. Billington
Princeton, N.J.
September 12, 1996

Robert Maillart

Introduction

The life and works of Robert Maillart developed around one central issue of twentieth-century technological society: the debate between the applied science view and the design view of engineering, between the ideal of "one best way" and the belief in many possible best ways, between reliance on general mathematical theories for understanding engineering works and the inspiration of specific works already in service.

Maillart's life was a constant battle for the design view of engineering against the applied science view. The two competing perspectives gave rise to very different approaches to education, research, and practice. These differences in turn illustrate a wider division over modern technological society itself: between a view that favors centralized rational planning and central authority in society, on the grounds that these are the ways technology itself works, and a view that technology and a modern technological society can also favor decentralized, local, and diffused responsibility and a high degree of choice. It is no accident that Maillart was able to accomplish in Switzerland works that would have been difficult or impossible in the larger and more centralized nations of Europe.

No other structural engineer during the first half of the twentieth century so fully engaged in this debate over the real meaning of modern technology as Maillart. He taught and publicly attacked the applied science education he saw emerging in engineering schools such as his own in Zurich. He wrote heatedly against applied science research and he pioneered innovative research and design in direct opposition to authorities and peer groups that

had been seduced by the applied science view.

In his own field of structural engineering, proponents of the applied science view saw education as the teaching of fundamental principles, general theories, and basic mathematical methods illustrated with abstract diagrams and algebraic formulations. Maillart himself had a quite different education that formed his view that principles ought to be illustrated by specific examples of actual structures and that methods of analysis should be visual. He later gave courses along these lines.

In research, structural engineers taking the applied science approach sought to discover general principles by the systematic study of the elements of structures (e.g., beams, columns, joints) under controlled laboratory conditions. Their goal was to write codes to regulate practice in accordance with research results. The measure of performance in concrete structures was taken to be stress (force per unit area) determined by complex mathematical theories. Maillart reacted strongly against this kind of abstract reductionism. Instead, he did his own research on large-scale models of real structures or on actual structures in service. His goal was to understand specific structural forms under actual service conditions so as to develop simplified but true calculations of their safety.

Maillart had an even deeper objection to the applied science view of practice: It discouraged real innovation. He found that innovation, especially in bridge design, came not from laboratory work and mathematical theories, but from design offices and construction sites. Numbers play an essential role in engineering. But innovation in bridge design was the product of a visual-geometric imagination, not the outcome of abstract numerical studies or deduction from general theories.

Maillart took an integrated view of structure that started with a geometric idea and then developed the calculations necessary to make it work. To design a bridge, the engineer must imagine how the weight will transfer from the bridge span to the supports, how the structure will be built, and what it will look like in service. These three aspects are independent and make the designer's task complex. Most engineers focus solely on the transfer of weight by accepting a standard frequently built form that is known to be safe. By contrast, the best structural designers keep all three criteria in mind. To do this, they rely on simplified calculations. These are not less rigorous or less accurate for being simpler; rather, they are another way to look at the structure, one that is visual as well as numerical, geometric rather than entirely algebraic, and one that suggests new forms.

Simplified calculations might be better called conceptual computations appropriate to the form being studied and not necessarily a special case from a general theory. But conceptual computations are also not merely preliminary ones; instead, the best designers use them to direct the design and often they are the only ones made. The crucial decision in structural design is always the chosen form and how this choice is made marks off two types of engineers: one who accepts traditional forms and relies on general theories and the other who questions these forms and looks for specific conceptual computational procedures.

Finally, in the applied science view, engineering practice was merely a working out of an optimum solution found by theory. The objective was to find the least expensive structure that satisfied the restrictions of a building code. Aesthetic appearance was either neglected or treated by the view that true economy would produce elegance. If more beauty was needed, then an architect could be employed to provide decoration or conceal the ugliness. To Maillart, such an approach to appearance was anathema. Appearance was to him the responsibility of the engineer and one of the tests of a good design.

In the late 1920s, a few writers popularized the view that engineering works could be works of art in the manner of architecture and sculpture. Writers such as Sigfried Giedion "discovered" Maillart in the early 1930s and began to publish photographs of his structures in modernist journals. These writers recognized that culture was being fundamentally changed by modern technology and they were searching for examples of modern engineering that combined both modern design and elegant appearance.

Maillart's design ideas owed nothing to such writings even though he certainly read some of them. Maillart's engineering vision had an entirely different origin: in the essence of structural engineering itself, in the geometric patterns of forces in reinforced concrete forms as they transmitted weight to the ground. A significant number of structural engineers have since recognized, as Maillart did, that elegant appearance could arise from the patterns traced by these forces. Elegance arose from the structure itself and not from an extraneous idea of beauty.

Maillart went beyond the structural engineers of his time in striving wherever possible for striking visual expression. This aesthetic motivation makes it possible to call him a structural artist. Yet he did not think of himself as an artist. He never used the term and he had little interest in contemporary artistic trends. He was not an artist in any sense that modernists understood the term. He was a modern figure with an aesthetic vision rooted in that defining characteristic of the modern world, its engineering.

CHAPTER ONE

Student and Designer

1872–1901

The Synthesis of Cultures
BERN AND ZURICH

BELGIAN AND GERMAN-SWISS

In the second half of the nineteenth century, the ideal Swiss city in which to raise a future bridge engineer would certainly have been Bern. The setting stimulates a bridge designer's imagination at once. Unlike Geneva, Basel, and Zurich, Bern was not in a strategic location for transportation and trade. Rather, its location high above a sharp bend in the Aare River made it an ideal fortress for the last Duke of Zähringen, who founded it in 1191.[1] Up until 1844, this natural, isolated peninsula was still almost completely unabridged. Even today, the old city of Bern gives the feeling of a stronghold. It is there that Robert Maillart was born on February 6, 1872.

His mother, Bertha Küpfer Maillart (1842–1932; Fig. 1), was German-Swiss and his father, Edmond Maillart (1834–74; Fig. 2), was Belgian. Robert was the fifth of six children, the first three of whom were born in French-speaking Geneva and the last three in German-speaking Bern. Thus, the Maillarts were culturally mixed: part of a native Swiss family (the Küpfers), and part of a displaced Belgian family (the Maillarts). Robert became a Swiss citizen in 1886. Maillart's great grandfather, Phillipe Joseph Maillart (1764–1856), was a well-known engraver and a painter.[2]

Both the Küpfer and the Maillart families were Calvinists, and this strict Protestant background may well have played a part in Robert Maillart's extraordinary self-discipline. Brought up in French Switzerland, as a youth, Robert's father began to study theology, presumably with the intention of becoming a minister, but a speech impediment deterred him and he entered the business of banking instead. On May 21, 1864, he married Bertha Küpfer of Bern and the couple settled in Geneva. Shortly after the birth of their third child the family moved to Bern where the other three, including Robert, were born. A mere 7 months after the birth of Maximilian, Edmond died suddenly (April 24, 1874), leaving his widow with five small children (one girl having died in 1873) and little money. Robert was only 2 years old when his father died, and he therefore had no memory of him.

The Küpfers were an old and well-known bourgeois Bernese family; Bertha's lineage can be traced back as far as Johannes Küpfer, a member of the Bern Grand Council in 1550.[3] Many of her relatives were ministers, many were in business, and one of her father's cousins, Ludwig (1803–79), was a Bern cantonal architect. Although she had only one sister, Bertha grew up within a large family of aunts, uncles, and cousins. Indeed, when her fifth child, Robert, was christened, the three godparents were all from her family, and when her husband died and left her practically destitute, it was one of her aunts who helped her financially.[4]

Robert Maillart's father spoke French as his native tongue and his mother German; together they conversed in French. But Robert's native language was German, which he always spoke with his

Figure 1. Bertha Maillart (1842–1932), mother of Robert Maillart. (*Source:* Madame M.-C. Blumer-Maillart)

Figure 2. Edmond Maillart (1834–74), father of Robert Maillart. (*Source:* Madame M.-C. Blumer-Maillart)

mother. She was his confidante, and from early childhood he developed a respect for her that sharply curtailed any kind of boyish misbehavior. He was so upset by the sorrow caused Bertha by his older brother Paul's misconduct that he resolved at an early age never to risk a similar affront.

Until the eighteenth century Bern was an imperialist capital whose leaders had conquered surrounding territories by force. Although modern Switzerland has no hereditary nobility, the conquering families of Bern took the aristocratic prefix "von." Their exclusivity had spurred the Küpfers to support an abortive revolt in 1749.[5] The Maillart family name is mentioned as far back as 988 in records of the region of Liège. In 1017, Jean Coley, dit Maillart, commanded a victorious army for the

Bishop of Liège but was blinded in battle. This event is presumed to have been the origin of the French version of Blind Man's Bluff, known in French as "Colin-Maillard." With neither social position nor family wealth, Maillart grew up feeling that he was on the fringe of Swiss society. He and his brothers made their careers on their own merit and each married non-Swiss women.

Whereas it is clear that Robert Maillart inherited his mother's emotional stability and self-reliance, little is known about his father. Undocumented family legend has it that Edmond Maillart's banking partner, by some questionable practices, led Edmond into bankruptcy, an event that may have contributed to his early death.

Figure 3. Maillart Family, circa 1884, (*left to right*) Alfred, Bertha, Rosa, Paul, Max, and Robert. (*Source:* Madame M.-C. Blumer-Maillart)

DISCIPLINE AND PLAY

Robert Maillart would develop into a man focused, almost to the point of obsession, on work and family. He was indifferent to the performing or visual arts, read little, except for literature related to his profession, and had little interest in travel or hobbies. The one deviation from this single mindedness was his love of parlor games, which soon was focused on cards. What appealed to Maillart in these games was their combination of discipline – represented by the rules – and "play" or the freedom allowed by these rules.

In his book, *Homo Ludens,* the Dutch historian, Johan Huizinga (1872–1944), elevates playing man to a level of cultural significance equal to knowing (*Homo Sapiens*) and making (*Homo Faber*). Hui-

zinga's description of play fits Maillart's attitude closely: "it creates order, *is* order. The least deviation from it 'spoils the game,' robs it of its character and makes it worthless. The profound affinity between play and order is perhaps the reason why play . . . seems to be to such a large extent in the field of aesthetics. Play has a tendency to be beautiful. It may be that this aesthetic factor is identical with the impulse to create orderly form, which animates play in all its aspects."[6] Maillart's career would center on structures that had to obey the rules of nature but within those rules allowed for the creation of new forms that grew from an aesthetic sensitivity.

Maillart's disciplined love of play appears in his early interest in chess. The seriousness with which he took the game – a seriousness that marked all his pursuits – is illustrated in the elegant and precise little journal he published in 1888–9 reporting on a teenage chess club of which he was secretary (also it would seem president, treasurer, editor, and chronicler). The names of thirteen young men appear in neat records of their wins and losses with Robert consistently the leading winner. In the journal, he laments that most of the club members did not take the game seriously enough.[7] But he also displayed a sense of humor in some of the writings. In an early family photo, only he exhibits the slight smile that would later characterize a quiet joi de vivre (right in Fig. 3).

The chess club membership, including young men from old established Bernese families (von May and von Tscharner), demonstrates something of the society in which Maillart grew up. He also had some contact with the city's artistic and intellectual elite through his close chess-playing friend, Fritz Widmann (1869–1937), who later became a painter and friend of Herman Hesse and Paul Klee.[8] But Maillart never felt at ease with socially or artistically prominent people and rarely saw Widmann after their boyhood friendship in Bern.[9]

Between 1885 and 1889, Maillart attended the Bern gymnasium, where he showed talent in mathematics and drawing. He excelled in both freehand sketching and technical drawing. In 1889, he passed the State examinations with 4.8 out of 6.[10] His grades qualified him for admission to the Federal Polytechnical Institute in Zurich (ETH), where he would study engineering. But being too young to enter, he went to Geneva for a year to study mechanics in watchmaking school.[11] This Geneva experience foreshadowed a critical aspect of his talent: He earned a 4 (out of 6) in practical work (watchmaking) and a 6 in design and in mechanics. His aptitude was obviously not for making things with his own hands, but rather for design and for the analysis of how things work.

With his stint in Geneva complete, in the fall of 1890, Maillart moved to Zurich to enter the 35-year-old ETH. The young man carried with him a strong bond to his family, and an independence of mind despite his restrictive upbringing. He had re-

flected on and developed many ideas on career, virtue, and religion, ideas that because of his natural reserve had not been tested in discussion.

ENGINEERING IN ZURICH: THE SWISS SYNTHESIS

Maillart entered the Federal Polytechnical Institute with a good grounding in mathematics and the sciences; he graduated 4 years later with the finest possible education in structural engineering. By the 1890s, the Institute was world famous; with many foreign students, it also educated more Swiss than the small country could support. Since its founding in 1855, the Institute had maintained the high standard established by two remarkable teachers, Carl Culmann (1812–81) and Karl Wilhelm Ritter (1847–1906). From the latter, Maillart learned to approach structures with a design view rather than with the view of an applied scientist.

Among the Institute's first appointments, civil engineer Carl Culmann (Fig. 4) was one of the four who were perhaps the best academics in their fields: the others were architect Gottfried Semper (1803–79), physicist Rudolf Clausius (1822–88), and historian Jacob Burckhardt (1818–97).[12] The contrast between Semper and Culmann characterizes the contrast between the world of architecture and that of engineering as the latter grew to prominence in nineteenth-century higher education. Semper (Fig. 5) practiced architectural design – he designed the first main building of the Institute (Fig. 6) built between 1858 and 1864 – he wrote widely on art and architecture, and he was politically active. His participation in the 1849 revolution in Dresden, where he was director of the Bauschule of the Royal Academy, forced him into exile in London. It was his Dresden friend and corevolutionary, Richard Wagner, exiled in Zurich, who recommended him for the position there, which he accepted in 1855.

Culmann, by contrast, did not practice engineering design once he came to Zurich; rather he served as a consultant on technical problems ranging from flooding mountain torrents to bridge building. He wrote not on accessible subjects like art, but in the

Figure 4. Carl Culmann (1821–81), founding head of the Civil Engineering Department at the Swiss Federal Polytechnical Institute in Zurich. (*Source:* Eidgenossisches Polytechnikum, Festschrift I)

Figure 5. (*upper right*) Gottfried Semper (1803–79), founding head of the Architecture Department at the Swiss Federal Polytechnical Institute in Zurich. (*Source:* Eidgenossisches Polytechnikum, Festschrift I)

Figure 6. (*right*) Federal Polytechnical Institute in Zurich. (*Source:* Eidgenossisches Polytechnikum, Festschrift II, p. 328)

specialized field of structural analysis, and he had little interest in politics. The formal portrait of Culmann in the ETH 50-year celebration book shows a serious middle-aged man in black suit and formal tie, whereas Semper's portrait has him sporting an ermine-collared cloak: the staid, conservative engineer and the fashionable, radical architect.

What was to become the highly influential modern Swiss bridge tradition began with Culmann. Born in Bavaria, Culmann studied French graphical analysis in Metz and then received a German engineering diploma from the Karlsruhe Polytechnic Institute in 1841, after which he worked for the Bavarian state railways until invited to Zurich in

1855.[13] In addition to gaining firsthand field experience with railroad bridges built during the early days of the German rail boom, Culmann took an interest in recently completed structures elsewhere, and in 1849–50, he made a 2-year trip to Britain and the United States to study their more advanced bridges and railways. Britain was still the leading industrial nation and its bridges represented the best designs built up to mid-century. The United States, however, was rapidly catching up and would soon dominate bridge innovation. Many of the best American bridge engineers were German-trained but had rejected German dogmatism. John Roebling, the most famous one, had been quite explicit about this and much like those of the Swiss, his great works combined ideas coming from Germany (his education in Berlin) and from France (parallel-wire cables for suspension bridges). The result of Culmann's trip was a widely read report.[14] He was a pioneer, possibly the first major European professor of structures to have made an extended study tour to specifically broaden his vision, rather than to develop foreign business connections.

Culmann also began detailed studies of structural engineering methods. Early in his teaching career, he started to systematize these studies, and, in 1866, he published them in *Graphic Statics,* probably the single most influential book in its field at the time.[15] His basic idea was to demonstrate structural behavior through geometric diagrams rather than through algebraic formulas. "Drawing is the language of the Engineer" he used to say, "the geometric way of thinking is a view of the thing itself and is, therefore, the most natural way, while with analytic method, as elegant as that may also be, the subject hides behind unfamiliar symbols."[16] Culmann imparted a sense of the three-dimensional reality of man-made structures set within the natural environment.

In Zurich, he founded an approach to design based on firsthand field experience, an international outlook, and a visual-geometric approach, all in sharp contrast to the algebraic approach taken by leading engineering teachers at the time. Because of his unorthodox methods, Culmann's influence

might not have extended into the twentieth century had it not been for Karl Wilhelm Ritter, chosen to succeed him in 1882.

RITTER AND THE BRIDGE: MAILLART AND AMMANN

Wilhelm Ritter (he dropped the first name very early) had been Culmann's best student. Culmann's lectures, while inspiring, were often unclear, and it was to Ritter (Fig. 7) that his fellow students looked for help in understanding their professor. After graduation, Ritter spent 4 years as Culmann's assistant, proving to be not only a gifted scholar, but also a brilliant teacher. When in 1873, the new Polytechnical School in Riga, Latvia (founded 1862), asked Culmann to recommend someone for the chair in structural engineering, he urged them to take Ritter, who became a full professor there at the age of 26. Following Culmann's death, Ritter was called back to Zurich, where in 1882 he took over the chair of structural engineering and bridge design.[17] Ritter revised and simplified Culmann's approach in a series of books on graphic statics and articles and lectures on design.

Ritter was a teacher whose personality and professional ideas had a profound influence on a generation of Swiss engineers. He was not an aggressive self-promoter. Sensitive to others, he was never known to talk against anyone behind their back. None of his students benefited more than Robert Maillart, and on none of them did Ritter make a more lasting impression. His influence came in two ways: first, through a series of courses taught to juniors and seniors in civil engineering, and, second, through his role as a bridge consultant to three public agencies for which Maillart was to design the early works in which he incorporated his first revolutionary ideas.

As an educator, nurturing students for only a fraction of their lives, Ritter provided an essential link between his student's inborn talents and their future careers. As a teacher, he communicated the recent tradition of his country and of his profession. As a scholar, he took scientific formulations and

Figure 7. Wilhelm Ritter (1847–1906), Professor of Civil Engineering at the Swiss Federal Polytechnical Institute in Zurich; teacher of Robert Maillart. (*Source:* Eidgenossisches Polytechnikum, Festschrift I)

shaped them into clear ideas to reveal better their design potential.

He stood aside and let the students pass on. He did not become in any way competitive by having a design office or domineering by having a design ideology. He was an interpreter of technical events: to his students through his lectures, to the profession through his writings, and to public officials through his detailed consultations that led to Swiss codes for metal structures and works of reinforced concrete.[18]

A strong argument can be made for the judgment that the two greatest bridge designers of the twentieth century were Maillart, using concrete, and Othmar Ammann (1879–1965), using steel (for New York's largest bridges: the George Washington, the Bayonne Arch, the Bronx-Whitestone, and the Verazzano among others). Both had the same professor for bridge design: Wilhelm Ritter. Ritter's name is nearly forgotten, but he touched students of whom Maillart and Ammann are only the most spectacular examples.

RITTER AND THE DESIGN VIEW

To understand the basis for Maillart's education, we need some insight into Ritter's ideas, particularly as contrasted with current ideas in Germany. In 1892, Ritter responded to the German professor, Franz Engesser, who had argued against the full-scale load test for small bridges (driving heavy trucks over newly built bridges to observe their performance) on the grounds that it was more economical and just as reliable to calculate the results mathematically.[19] Ritter's detailed defense of what had become the common Swiss practice of load testing contrasted a Swiss position with a German one: the full-scale field test versus purely mathematical studies. In a broader sense, Ritter reflected a pragmatic, design-oriented attitude as opposed to a more theoretical and applied scientist approach. The Swiss tended to be more open to visual demonstrations of performance.

Maillart accepted Ritter's viewpoint that public works set in a difficult environment are always built with uncertainty. There was no way in the late nineteenth century to predict mathematically the full response of a public structure to its loads. In spite of many new mathematical theories, detailed textbooks, and immense computer power, the same condition exists in the late twentieth century. The validity of any work rests, Ritter emphasized, with "the probing expert" who must give "a reliable judgment"; in short, it always rests finally on the judgment of a person about an existing structure and not on the solution to an equation.

Unlike Culmann, Ritter did not have extensive field experience; his early brilliance led him too soon into a professor's chair for that. But he put high value on the experience he gained as a consultant in the use of load tests, which would play a central role in Maillart's career. Ritter's defense of such field experiences, against German objections, allowed structures to be built that could not be done in Germany in the early twentieth century and for which complex and so-called more rigorous analyses would have obscured the design potential. As a consultant, Ritter could not provide a satisfactory mathematical analysis for Maillart's first major design, the Zuoz Bridge of 1901. But Ritter did direct and interpret the full-scale load test, which demonstrated the validity of Maillart's simple calculations for that structure.

Ritter continued Culmann's effort to widen the engineer's horizons. In a report and in lectures arising from a trip to the United States in 1893, Ritter broadly reviewed forms and new details, and gave a relatively complete technical picture of a few selected bridges.[20] This wide visual variety of solutions to what is essentially the same set of problems is just the combination of insights that most intrigues the design-oriented, as opposed to the analytically oriented, student. Unlike the vast majority of engineers at this time, Ritter did not hesitate to introduce aesthetic judgments as, for example, in describing the high bridge built in 1888–9 over the Mississippi River at St. Paul. "The structure, in spite of its extraordinary size, makes a rather insipid impression; it leads us to realize that aesthetic ideas must have been completely neglected in favor of some utilitarian principles."[21]

As every bridge designer knows, however, good overall form means nothing if the details are bad: All the pieces must fit together and none must be structurally weak. To the watchmaking Swiss, details are a critical part of design. Ritter devoted about one-third of his report on the trip overseas to a detailed review of joints, connections, eyebars, and rivets. Many of his drawings of details are refined and Ritter did not hesitate to criticize other details when he found them to be ugly. As illustrated in a pioneering article on bridge design, Ritter taught structural theory in a simple, elegant, and practical way.[22] He used a minimum of algebraic analysis, which he presented within the context of its design implications partly drawn from his experience in the United States. Ritter had seen in the United States the creative results of a design tradition based on extensive field experience and sound academic training.

Maillart's class notes of 1893 and 1894 are full of drawings and diagrams illustrating Ritter's ideas. They also show Maillart's growing enthusiasm for

structural design in general and for bridges in particular. On one page of his notes, he drew a stiffened wooden-arch bridge, a form also featured in Ritter's report on American bridges; underneath it he wrote "a marvelous bridge."[23]

At the time Maillart graduated from the ETH in March 1894, he was a serious-looking 22-year-old: about 5'9", slightly built, with a light mustache, prominent nose, and a high forehead, indicating an already receding hairline (Fig. 8). His myopic eyes were deep-set and intense. The pince-nez he began to wear shortly after finishing school signaled his conservative taste. Maillart's first job was working for a railroad designer in Bern. His education gave him a solid scientific basis for structural engineering, but his close contact with Wilhelm Ritter had made him suspicious of dogmatic theory and provided a strong impetus for design innovation.

Stone Versus Concrete
Zurich, 1894–1899

THE NEW MATERIAL AND THE TRADITION OF MASS

During the first 5 years of Maillart's career, reinforced concrete came from being a little-used novelty to a major building material throughout the western world. This evolution was due largely to three pioneers: a French gardener-inventor Joseph Monier (1823–1906), a French builder-designer François Hennebique (1843–1921), and a German engineer G. A. Wayss (1851–1917).

Monier took out numerous patents on iron-bar and wire-mesh, reinforced-concrete shells, beams, and columns. In the late 1880s, Wayss bought these patents, founded a business based on designing and building reinforced-concrete structures, and began a testing program that lead to standardized formulas as a basis for design. The Wayss and Freytag firm became a leader in concrete structures through-

Figure 8. Robert Maillart (1872–1940): (a) in 1887, (b) circa 1890, and (c) in 1894. (*Source:* Madame M.-C. Blumer-Maillart)

out Europe. In 1890, it built a concrete-arch bridge in Wildegg, Switzerland, that was well publicized and impressed the young Maillart.[24] When Maillart began to work, the subject of reinforced concrete was still largely in the hands of these pioneers. In the United States, a young Austrian, Fritz von Emperger (1864–1942) was working doggedly and without much success to introduce the new material there.[25] For the young Swiss engineer, therefore, 1894 was a good year to strike out onto this new terrain.

Once home in Bern, Maillart found that his high ideals collided with the dull realities of an engineering apprenticeship. There is a major difference between the "marvelous bridges" illustrated by a brilliant teacher and the mundane structures on which most newly graduated engineers have to work. From the pinnacle of education, young graduates often drop to the footings of ordinary practice, which can easily discourage and make the dramatic forms glimpsed in academia seem elusive indeed. The best structural engineers like Maillart have faced this problem and, in part, their greatness has been marked by an ability to learn from the ordinary in order to create something extraordinary.

For a young man whose greatest excitement lay in his work, Maillart's first project in practice could

not have been more disappointing. On March 14, 1894, he began work with the firm of Pümpin and Herzog on the layout of a small water tunnel for the little town of Stolzenmühle in the hills about 15 kilometers south of Bern.[26] Although Maillart complained to one of his classmates about supervising this drab assignment, it gave him responsibility for an installation critical to the local village.[27] He began to gain experience with water tunnels that 25 years later would establish him as a leader in this field. Moreover, he saw at firsthand the lost wilderness of Canton Bern, so close to the Swiss Capital yet so removed from city life. He was beginning even with this first task to appreciate the isolated terrain within which small public works must be carefully built. But Maillart was restless and it was good news therefore when the company put him to work on some small bridges. From May until October 1895, Maillart worked on design and layout in the Bern office, kept up contacts with some former classmates,[28] and lived a quiet life in the circle of his family: his mother, his bachelor brothers Alfred and Max, and his married sister, Rosa Wicky.

Part of Maillart's work at this time was on one bridge for the last rail line completed by his firm: the Bière-Apples-Morges line running from Lake Geneva up into the Jura. Maillart's principal office work was on a branch line from Apples to L'Isle.[29]

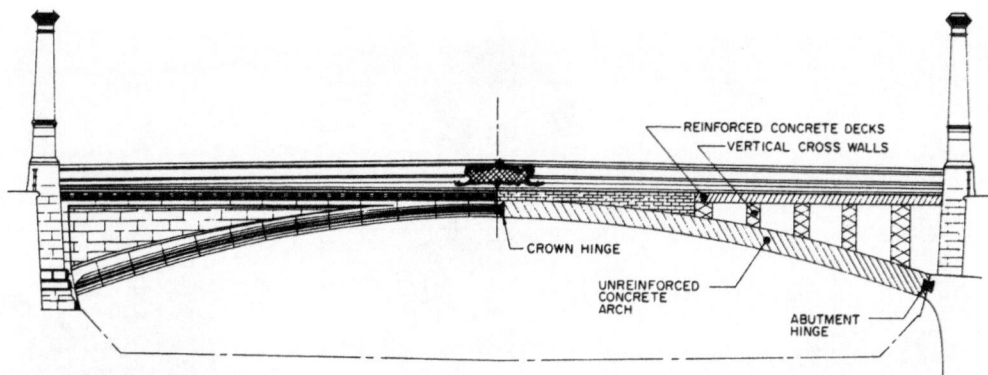

REINFORCED CONCRETE DECKS
VERTICAL CROSS WALLS

CROWN HINGE

UNREINFORCED
CONCRETE
ARCH

ABUTMENT
HINGE

Figure 9. Stauffacher Bridge showing deck, cross walls, arch, and hinges. (*Source:* Billington, *Robert Maillart's Bridges*)

Pümpin and Herzog exhibited confidence in their young engineer by sending him in October 1895 to Morges, where he would be more on his own, working directly under the chief engineer for the entire project, and supervising construction of the rail line to L'Isle and the little bridges that would carry that line over streams and roadways.[30] Now Maillart began to see built those designs on which he had worked in the office.

In Morges, he helped lay out the branch rail line to L'Isle, and then on March 27, 1896, he moved from Morges to Pampigny to supervise construction directly. Pampigny is a small village on a fertile plain; one of its features is the bordering brook Le Veyron across which Maillart's first bridge went up in the summer of 1896.[31] This 6-meter span arch bridge in unreinforced concrete had an arch measurement at the crown (midspan "keystone") of 60 centimeters, a considerable thickness for so small a span. Thus, Maillart began his bridge design career with a concrete form made in the image of massive stone masonry. At the time, concrete was generally treated merely as a substitute for stone.

The branch line over the new bridge was completed during the late summer ending with a 3-day celebration at L'Isle. On Saturday, September 12, the first train left Apples at 10 A.M., stopping just before Maillart's bridge at Pampigny, where there were speeches, toasts, and young ladies in green-and-white stoles who served refreshments.[32] An inscription noted that

Il est permis de s'égayer
mais interdit de dérailler

(it is permitted to make merry but not allowed to derail, i.e., to go astray). Rarely did 11 kilometers of narrow-gauge railway merit so much celebration. But for the local and somewhat isolated people, this new line was a connection to the more modern lakeshore life.

In late September, Pümpin and Herzog recalled Maillart to the Bern office. His first taste of real bridge construction was for Maillart what a first alcoholic beverage might be for most youths. The inebriation of it made office work seem so dull that he decided to look for another job. Encouraged by a classmate, Wilhelm Dick, then working for the Zurich City Engineering Office (*Tiefbauamt*), Maillart secured a position there.[33] The chief engineer of the *Tiefbauamt*, Viktor Wenner, had worked for Pümpin and Herzog and knew of Maillart already, beside which the firm gave him a fine recommendation.[34] So as 1897 began, Maillart left his home town and returned to Switzerland's largest city, the center of the Swiss engineering establishment.[35] There he was to design his first major

Figure 10. Stauffacher Bridge over the Sihl River in Zurich, 1899. (*Source:* Baugeschichtliches Archiv, Zurich)

bridge but the tradition of stone would cover the concrete.

RITTER AND THE BIG BRIDGE

The growth of Zurich in the late nineteenth century had led to relatively inexpensive bridges made of iron girders or trusses. These bridges were easily designed by analytical methods and quickly built – facts betrayed by their mundane appearance. They had a temporary look and they needed continual maintenance. Because of this, one of Switzerland's best-known designers, Robert Moser (1836–1918), had argued in 1895 for the wider use of conventional stone bridges; they were more permanent, needed almost no maintenance, and were prototypically Swiss – stone masonry in a country of mountainous rock. This was the attitude against which Maillart had to pit his advocacy of reinforced concrete.[36]

Modernization of city bridges had meant replacing wood with iron, a practice that went together with railroad building. Railroads, almost from the start, were made of iron: iron rails, iron locomotives, iron lines to exploit iron mines, iron terminal sheds, and iron bridges. The nineteenth century was, in the industrializing world of western Europe, a second iron age, even if many public buildings still tried to hide that fact behind stone façades.

For bridge designers, this situation presented a problem. Modern design then meant straight iron pieces that made up skeletal forms, but accepted architectural practice recognized only stone, with its mass and arches. To make a reasonably inexpensive bridge meant using iron (or by the 1880s, steel), whereas to make it aesthetically acceptable meant either curving the ironwork to give it a stonelike form or covering it with stonework. The debate over the Eiffel Tower characterized this tension. Traditional artists found the skeletal form an affront to the beauty of masonry-façaded Paris and even Eiffel compromised his design by permitting an architect to add purely decorative arches at the tower's base.[37]

This dilemma persisted throughout the early twentieth century. The only solutions that seemed possible were either cheap metal expensively decorated, like the Pont Alexandre III in Paris, or costly stone somehow simplified as in many of Moser's railway bridges. One of the few articulate voices with a different solution was Maillart's, who began in the late 1890s to argue that the only reasonable approach was to use a new material that was permanent and attractive like stone but light and inexpensive like metal. Maillart's design for the Stauffacher Bridge was a first tentative step in that direction.

In 1896, the city of Zurich had decided to build a bridge to carry Stauffacher Strasse over the Sihl River. The first Stauffacher design by the city – a three-span, steel-girder bridge – was rejected by the cantonal authorities because the two river piers would have obstructed the ice flow. The *Tiefbauamt* then made three more designs: a two-span, steel-arch bridge; a single-span, steel-arch bridge; and a two-span, stone-arch bridge. Maillart used a different approach, proposing a single-span concrete arch without reinforcement.[38] It was a sharp departure for the city authorities who, being unable to decide, called in Wilhelm Ritter. Ritter's ten-page report dated August 9, 1898, began with a general statement on the importance of aesthetics. He criticized the *Tiefbauamt's* plans for a two-span, steel-girder bridge and recommended a new design of only one span.[39] Ritter suggested "the study of a [single-span] concrete arch with three hinges," so that its relative economy could be assessed "after a comparable cost analysis." Ritter referred to the idea common in the late nineteenth century of building an arch in two halves, connected at the crown by a hinge, with each half joined to its support at the abutments also by a hinge. These hinges (literally made like door hinges) allowed the arches to rotate slightly up or down and thus prevented high stresses at those three locations (Fig. 9). Without hinges, the arches would still try to rotate and would often crack. Following Ritter's advice, Wenner agreed to let Maillart make the full design, which he did with a cost estimate of 220,000 francs, a substantial savings (more than 10 percent) over the steel alternatives.

Ritter's report and Wenner's willingness to follow its advised innovation allowed Maillart to design a major bridge in Zurich that received wide notice within the Swiss engineering community. Clearly, Ritter wrote his report to justify Maillart's proposal, and Maillart's more economical design in turn justified Ritter's confidence in his student.[40] But no one suggested that the concrete arch be expressed visually. And no one objected when the *Tiefbauamt,* following Moser's ideas, had the city architect, Gustave Gull, hide Maillart's concrete arch behind a façade of stone, thus conforming to the nineteenth-century urban ideal (Fig. 10).

Had Maillart's structure been left exposed, its visual effect would have been disappointing, because, knowing it would be hidden, Maillart thought little about its appearance. He created a work of high technical quality, but not one of any particular visual merit. Indeed, even its technical merit is restricted to a masonry view of concrete: The arch is not reinforced, it is relatively thick, and the arch must carry the entire bridge weight alone with no help from the horizontal deck. Once again, after the stimulus of a bridge, he found himself assigned to less exciting projects: the redesign of highways. So, as construction on the Stauffacher drew to a close in September 1899, Maillart resigned from the *Tiefbauamt* to take a job with the Zurich design and construction firm of Froté and Westermann. The city publically noted his resignation with regret.[41]

MAILLART AND HENNEBIQUE

The 2 years Maillart spent with Froté and Westermann completed his apprenticeship. Unlike Pümpin and Herzog, a firm near the end of its life in 1896, Froté and Westermann was founded in that year. Maillart went to work in 1899 for two contemporaries whose goal was to establish a business in the rapidly developing field of reinforced concrete. Both partners were promoters and developers, and between 1899 and 1901, they gave Maillart substantial design responsibility.[42] In addition to assuming the role of chief designer, Maillart also benefited because Froté and Westermann were licensees of the Paris-based Hennebique organization, which in the 1890s was the world's leading design firm in reinforced-concrete structures. Maillart thus got a firsthand view of the Hennebique system, both in the sense of structural design and of organization. Work with Hennebique's system exposed Robert Maillart to French design practice much as Ritter's critical writing had introduced him to German practice.

François Hennebique, born in 1843 at Neuville St. Vast in the northern part of France, apprenticed himself to a stone mason until age 18 when he began working on building sites, becoming a superintendent 5 years later. By 1867, he had established himself as an independent building contractor. In 1879, while building a villa in Belgium, a neighboring house burned down, and Hennebique's client asked for assurance against such an accident. Hennebique first considered enclosing the iron floor girders (exposed in the burned-down villa) with concrete. However, after some testing, he decided to replace the girders by iron rods embedded only in those parts of the concrete where cracks might occur, thereby making reinforced-concrete girders. Further tests proved the greatly reduced amount of metal to be sufficient and he therefore felt sure enough to guarantee the house as fireproof.[43]

For the next 12 years, Hennebique pursued reinforced-concrete research while continuing his regular construction business. On reading about American progress with the new material in 1892, he took out patents, retired from construction, and opened a consulting business in Paris. Between 1892 and 1902, business grew at an extraordinary rate from a total of 6 projects in 1892 to a total of 7,026 projects completed by the end of 1902.[44] These were executed through a system of concessions, whereby construction companies, usually already established in a locale, were given the right to build structures designed by the central office in Paris or sometimes by local offices in a few other cities. It was a highly centralized operation, directed personally by Hennebique from Paris.

Most of the projects were in Hennebique's native France, but at least as early as 1898, the foreign country with the most concessions was Switzerland,

where there were twelve in addition to the main Swiss office in Lausanne. One of the twelve was Froté and Westermann in Zurich. For such a small country, the intensity of Hennebique's Swiss activity was remarkable, accounting in 1898 for just over 10 percent of all Hennebique's work, considerably more than in any country outside of France.[45]

Lausanne, the earliest Hennebique center in Switzerland, was crucial to the entire international operation. When Hennebique's journal, Le Béton Armé, began to appear in June 1898, it concentrated largely on the Swiss works completed between 1894 and 1898.[46] Most significant about the works discussed were descriptions of the carefully controlled tests conducted under the critical eye of two of Switzerland's most distinguished engineers, Eduard Elskes from Lausanne (chief engineer for the Jura-Simplon line), and François Schüle (then a professor in Lausanne, and after 1901 professor at the ETH and Director of the State Materials Testing Laboratory). Such independent confirmation was crucial to Hennebique, and outside of a few examples in France, only the careful Swiss subjected his works to a rigorous testing from which the results were published.[47]

Maillart had seen Hennebique works during his time with Pümpin and Herzog at Morges and Pampigny near Lausanne. There the Hennebique organization built six small bridges between 1894 and 1897. These and a series of multistory buildings illustrated clearly to Swiss engineers how Hennebique, like most structural engineers of the day, saw concrete structures as similar in form to those of steel or wood: The system merely imitated those skeletal forms in reinforced concrete. Maillart, on the other hand, was beginning to see that concrete structures could have new forms that would be different from those in other materials. Hennebique's main idea was that concrete allowed beams, columns, and slabs to be built monolithically, cast together as a single unit. Without questioning this, between 1899 and 1901, Maillart went beyond Hennebique to develop thin curved forms not feasible for bridges in traditional materials.

Figure 11. Vienne River Bridge at Châtellerault by François Hennebique, 1899. (*Source:* National Museum of American History, Smithsonian Institution)

Since Wenner had credited him publicly with the Stauffacher design in 1899, the Zurich branch of the Swiss Society of Engineers and Architects had begun to recognize Maillart's talent. On January 30, 1901, he was invited to serve as the society's secretary. The invitation came on the occasion of a lecture on the Hennebique system (by Hennebique's chief Swiss agent), which Maillart wrote up and published in the May 25 issue of the *Bauzeitung.*[48] It was his first professional writing.

This weekly publication (the *Schweizerische Bauzeitung*) featured structural and mechanical engineering and architecture. It was privately owned and edited throughout Maillart's career by the Jegher family, who were independent both financially and in their editorial policy, often criticizing public officials. Already Maillart knew the second

generation, Carl Jegher, with whom he developed a strong friendship as a compatible critic of the authorities. Still the *Bauzeitung,* as the official journal of the Swiss Society of Engineers and Architects, was influential both within the profession and for the general public, so recognition in its pages could be invaluable to a struggling young engineer.

Maillart's 1901 article clearly shows a debt to Hennebique and it can be interpreted as publicity for a system of which Maillart's employer was a part. He hailed aesthetics, safety, and economy as the primary features of the Châtellerault Bridge over the Vienne River (1899), this "most important work to date" with the new material (Fig. 11).[49] But when Maillart wrote this article, his own design ideas had already begun to diverge radically from those of Hennebique.

The Shift of Vision
Zuoz, 1899–1901

ZUOZ: THE FIRST MAJOR
INNOVATION

Maillart's work at Froté and Westermann gave him the broad entrepreneurial experience lacking in public agencies. He designed a series of small bridges and supervised their construction.[50] He worked on the first reinforced-concrete bridge in Zurich, carrying a cog rail line over Hadlaub Street: Although called a Hennebique design, the bridge had been calculated according to procedures outlined by Wilhelm Ritter in a series of articles that had appeared in early 1899.[51]

At the same time, Maillart worked on construction of the immense reinforced-concrete frame building for the Schatzalp Sanatorium in Davos, for which the Froté and Westermann office had designed the structure, again following the Hennebique system.[52] Maillart also reviewed calculations for the Solis Bridge, then the largest stone-arch bridge on the narrow-gauge railroad, the Rhätische Bahn in the Graubünden.[53]

With this Hennebique experience behind him, in August 1900, Maillart began the design for a bridge over the Inn River in the Upper Inn Valley (Oberengadine) at the little town of Zuoz. The Canton of Graubünden had designed a steel-truss bridge to span the 30-meter-wide river, but Maillart was convinced that reinforced concrete would be superior. In September, while in Paris to negotiate a contract with a French engineer, Maillart received the news that Froté and Westermann had won the contract for his Zuoz Bridge.[54] It would provide his first experience with complete responsibility for the entire project: design, calculations, construction plan, and contract negotiations. Now he could relax and enjoy Paris's 1900 World's Fair, with its new works in reinforced concrete, among them Hennebique's, to whom the Fair had awarded the Grand Prix for his many exhibition structures.[55]

Being active in Graubünden, Froté and Wester-

mann knew by the summer of 1900 that the canton was planning a new steel-truss bridge at Zuoz. They had sent Maillart there on August 16 to inspect the proposed site and to make a presentation to the community council.[56] First, Maillart argued that his concrete bridge was competitive with steel with the added advantage that costs for concrete would be paid to local suppliers and not, as for steel, to manufacturers outside the canton. Second, he emphasized that "because concrete outwardly resembles stone, it will have a simple but elegant appearance and will unquestionably bring honor and embellishment to the community." He stressed that his design would have a fine finish achieved not by plastering over the concrete, but rather by a smooth forming and a special concrete mix.

Maillart was making a strong case for the aesthetics of his first major design along with his central idea: the principle of form rather than mass. The curved arch, the longitudinal walls, and the horizontal roadway slab "will together make the arch," he explained, where the bottom curved slab or arch is connected to the three vertical walls, which themselves connect to the top roadway slab. This connection forms two hollow boxes, the first such structures ever built in concrete (Fig. 12 shows

Figure 12. Zuoz Bridge design of 1900 (*Source:* Drawing by Mark Reed)

Maillart's initial design; in the final one he made the deep façade arch more shallow). Its three elements (arch, walls, and deck) together carry the load, as contrasted with the Stauffacher, where the load is carried by the arch alone.

In his quiet manner, Maillart stressed to the community council of Zuoz that the structure would have the virtues of stone but without its great weight for, like steel, the concrete sections are reduced to a minimum. In spite of this minimization, the strength of Maillart's hollow-box arch is far greater that his solid (95-centimeter-thick) Stauffacher arch. At the critical section midway between the crown hinge and the support hinge, the Zuoz arch strength is similar to that of a solid 130-centimeter-thick stone arch, with the approximate weight of a 40-centimeter-thick arch. Additionally, the stresses in the thinner Zuoz arch are less than those in the Stauffacher arch. Thus, Maillart's bridge was stronger, less expensive, and in his view better looking than comparable bridges being built at this time.[57]

If Maillart's new design was a radical departure so was his construction plan, for which he designed scaffolding to carry only the thin-arch weight. Once the arch concrete hardened, it would carry the walls and deck. Maillart got this idea from the ring method of building stone arches, such as he had seen in the Solis Bridge (1900), the first long-span arch in Switzerland to be constructed in three concentric rings of stone. In its construction, only the first layer was carried by the wooden scaffold, the second being carried by the layer already in place.[58] Besides his savings on materials, Maillart therefore saved also on scaffolding costs at Zuoz by carrying over this idea from Solis.

Unlike the Stauffacher Bridge, which could be justified theoretically, proof of the validity of the Zuoz design, as with most of the bridges Maillart designed after it, rested ultimately on full-scale load tests. Although common practice in Switzerland, such tests have never been accepted in the United States partly because the method was too time-consuming in view of the vast number of bridges built quickly as the country expanded and partly because of the belief, as in Germany, that such tests were too costly.

In October, Maillart made the arduous daylong trip by train and horse-drawn coach (*diligence*) to Zuoz for this all-important load test. Indicative of the importance attached to it, the canton had asked Professor Ritter to come from Zurich and direct the test with Maillart. In the clear cold afternoon, Maillart's workers lowered the scaffold for the arch by gradually emptying sand-filled cylinders supporting the vertical posts, transferring the bridge's weight from its temporary wooden support to the concrete structure.[59] This is a critical stage in concrete construction because it represents the first time that the structure must carry large loads on its own.

From the very beginning of his career, Maillart enjoyed the utmost self-confidence in his work. It was therefore no surprise to him that the bridge performed well under several different kinds of load tests.[60] Shortly after the town's elaborate celebration of the bridge's completion, he commented laconically that "the bridge held well and everyone is pleased now; naturally there are always a few little things to put in order, but all told everything has gone very well."[61] The "little things" consisted of some small cracks that Ritter did not regard as serious. In the conclusion of his report, Ritter stated, "the community of Zuoz has gained, with the new bridge, a fitting and trustworthy structure which will require a minimum of maintenance."[62]

Like Maillart, Ritter was a man of few words and this seemingly matter-of-fact statement represented substantial praise from Switzerland's leading structural consultant. The greatest teacher of bridge design thus handed over the future to his most promising student. Time has proven the result of this last close collaboration to be one of the most original concrete structures ever built: the first reinforced-concrete bridge in Graubünden, the first concrete hollow-box bridge anywhere, and the first concrete arch built by the ring method.

THE MEANING OF THE ZUOZ DESIGN

Maillart's innovation at Zuoz became one of the major concrete forms of the twentieth century, but it took a surprisingly long time before it was used outside Switzerland. This was so partly because Maillart's early designs received little publicity until he became better known in the 1930s, and partly because the idea, although relatively easy to understand now, broke so strongly from the masonry tradition of bridge design that engineers were unprepared to accept it. Maillart alone imagined this new idea. Why?

The answer lies first in his education, second in his early experience, and third in his personality (which I shall discuss later). Ritter had taught him about nineteenth-century bridges in metal in which the hollow box had been used since Robert Stephenson's 1850 Britannia Bridge. The Swiss professor also emphasized the interaction of elements in older wooden bridges, especially the arch-and-truss combinations used in the United States. Ritter's teaching encouraged procedures that used less material.

Building on images from Ritter, Maillart's early experience with the Stauffacher Bridge suggested to him the form for Zuoz. The flat arch had solid vertical walls, which in Zurich were decorative, whereas at Zuoz, they were essential structural elements. If Maillart found the Stauffacher form elegant enough to be repeated at Zuoz, he needed a new idea of monolithic structure to make Zuoz a true innovation. The vertical walls, connecting the deck and arch, turn the structure into a hollow box because the embedded reinforcing steel allows the entire cross-section to function as if it were one unbroken piece.

This idea is similar to the smooth-skinned airplanes of the 1930s that replaced the earlier biwing planes with wires and struts between the two wings. By making "streamlined" shapes, Donald Douglas created a stronger and lighter plane (the famous DC-3) in much the same way that Maillart had created a stronger and lighter bridge 30 years before. In part, Maillart's forms suggested such modernity to the artists who discovered his work in the 1930s.

In 1901, Hennebique was still designing arch bridges, as at Châtellarault, in which the deck beams, columns, and arches each functioned individually. This piecemeal system dominated arch bridge design up until World War II. In Austria, Josef Melan had patented a system in which the builder first erected a steel, I-section arch that then served as a scaffold to support wooden forms. Concrete cast into these forms covered the steel arch and made a concrete arch reinforced by the steel I-sections. This system avoided the expensive wooden scaffolding needed to support the concrete arch before it hardened; the Melan method was briefly popular in the United States at the turn of the century. It called for much more steel than otherwise needed, however, and was little used in the twentieth century.

The Hennebique firm designed the 1910 Risorgimento Bridge in Rome, which appears to be a hollow box, but it does not seem to have been designed as one. The Wayss and Freytag Company copied Maillart's idea for a highly publicized bridge at Liepheim in Germany in the mid-1930s. At that time, the concrete hollow box first appeared in the United States as well, but without reference to Maillart. It was not, however, until after World War II that the form became popular; in fact, it has dominated bridge design for medium- and long-span concrete beam bridges in reinforced and prestressed concrete.

Thus, the primary significance of the Zuoz Bridge was its hollow box. It was also a hollow-box, three-hinged arch; that was a secondary innovation on which Maillart would base a later, major innovation at Tavanasa. These new ideas illustrate how the design view of structures operates even where the means of calculation are not well developed. For the applied scientist, the hollow box in concrete had little appeal because it seemed to require a highly complex mathematical analysis. Maillart understood how the box would work in a general way even if he could not predict its behavior in detail. Indeed, that inability led later on to some minor cracking that, although not endangering the safety, would require a revision of Maillart's design ideas and lead to even more innovative designs. But to do that, Maillart would need more independence and the Zuoz Bridge provided indirectly a new route for a more independent career.

A PROPOSAL'

Robert Maillart was 29 when construction of the Zuoz Bridge began in the summer of 1901. His career was moving nicely toward greater business responsibility, the realization of new design ideas, and national recognition for his talent with concrete structures.[63] He enjoyed the role of leading engineer at Froté and Westermann. The firm found him serious; Alfred Koller, for whom he had worked in Pampigny, noted that "while being relatively young," he had "the sober character of an old man!"[64] He had become somewhat cynical, making fun of classmates who were falling in love and getting tied down with a wife and family.[65] For Maillart, life was first and foremost work, interrupted only occasionally by a handful of friends (Fig. 13) and visits with his immediate family. Such was the man who went in late July to Zuoz to follow the critical stage of casting the concrete arch for the new bridge. The trip was to change his life decisively.[66]

Going from Zurich to Zuoz in 1901 was not easy because the upper Inn Valley (Oberengadin) was then still accessible to the rest of Switzerland only through high mountain passes: Maloja to the south and Julier, Albula, and Flüela to the east. Maillart took the most direct route, which was by Swiss National railway to Chur, by the narrow-gauge Rhätische Bahn from Chur to Thusis, and then by horse-drawn bus over the Albula Pass.[67]

In late July, Maillart arrived in Zuoz, lodging at the Hotel Concordia (far left in Fig. 14), built in 1876 in the grand tradition of nineteenth-century resort hotels.[68] On Tuesday, July 30, after a day at the bridge site, he went into the hotel dining room for dinner and sat at the long common table known as the *table d'hôte*.[69] As usual, he was preoccupied with his work, but he did notice a family at the other end of the table and his eye was drawn to a striking young woman. She was about 5'4" tall with a handsome face, eyes set well apart, full lips,

Figure 13. Maillart and friends, circa 1900. (*Source:* Madame M.-C. Blumer-Maillart)

dark hair attached in back, and a slightly turned up nose in profile.

The next evening he sat closer to her and the following night at dinner he got up his courage to open a conversation. There was only 1 day left before he had to return to Zurich. That last day was August 1, the commemoration of the founding of the Swiss Confederation (August 1, 1291) for which Zuoz planned an evening of festivities. He had never known the experience of being completely absorbed by a woman; her lively responses made it easier to talk than he had feared.[70] He still felt awkward, but she seemed immediately to take an interest in him and they talked in French throughout dinner. He learned that she was Italian and that her name was Maria Ronconi (Figs. 15a and 15b).

He wanted to know all about her and she opened up to him quickly as if she had been waiting to talk about her life. Maria was an orphan and very much alone. She told him of her studies at convent schools near Florence and then at one in Namur, Belgium, where she had become fluent in French. She had returned to Italy and eventually to

Figure 14. View of Zuoz with the Inn River Bridge of Maillart in the foreground. (*Source:* Madame M.-C. Blumer-Maillart)

Turin to live with family friends named Wild whose children she instructed in the arts. The Wilds were among Zurich's most well-to-do and best-known citizens.[71] It was with the Wilds that Maria had come to Zuoz.

After dinner, they went out into the twilight to walk along the Inn River. The moon was out as they walked back toward Zuoz. They could hear the church bells ringing and the sky lit up with fireworks. It was then, after an acquaintance of barely 2 hours,[72] that he asked her to marry him.

She seemed to say yes and their talk turned to the future.

On their return to the hotel, Maria introduced him to Mrs. Wild, also an Italian. They discussed the lightning engagement and the elder woman quizzed Maillart about himself. This kind of precipitous commitment would be unusual under any circumstances, but in Switzerland at the beginning of the century, it was particularly startling. As a kind of personal reference, Maillart gave Mrs. Wild the name of a well-known colleague, Eduard Elskes. The engagement could hardly have come at a more symbolic moment, with the national celebration of an historic event that signaled a new political direction. It would open a radically new direction in Maillart's professional life as well as in his personal life.

AN INNER COMPANION

The day after Maillart's proposal, Maria arose before dawn to see him off on the carriage to Zurich. The weather was terrible, but for Maillart, as he rode through the Graubünden wilderness, the trip was the best of his life because, as he wrote Maria, "I did not go alone but with you, I spoke only to you and you answered me, and I assure you that this conversation was for me more lovely and more fitting than that at the *Table d'hôte* in Zuoz."[73]

To his surprise, his seemingly independent life alone, without distractions, now seemed like a kind of prison from which Maria had allowed him to escape. His first day back in Zurich, after everyone had left the office: secretary; draftsmen; the charming, wealthy, and easygoing engineer Max von Müller from Bern; and the military-minded Adolph Zarn, who kept the books, Maillart began his first letter to Maria. Like the structural calculations he could by then make so quickly and so accurately, his letter poured out in detail what his mind had already grasped in its entirety.

Maria's letters to Maillart on the other hand betrayed a worry that he would have difficulties because he had chosen an Italian and a Catholic.[74] In spite of these concerns, Maria felt that their brief encounter in Zuoz had bonded them irrevocably so that when Maillart wrote to remind her that she had not given him "a definite yes," she responded with a telegram from Bologna on Tuesday evening, August 6: "yes, yours for life – Ronconi!"[75]

His letter of the next morning is filled with joy as it expresses his initial feelings, but it becomes more serious as he thinks of her coming into a family of no independent wealth nor any particular social position.[76]

We are very simple people without pretensions and moreover people who have received nothing from their parents except their lives and their education and who now work for their living. I know very well that you would do the same and that you find that natural; but you have always lived in less bourgeois and less simple surroundings and I would wish that you would get accustomed already a little to this idea of change so that you will not perhaps be disappointed.

Maillart anticipated with considerable accuracy problems that might arise for a foreigner like Maria coming into the staid middle-class milieu of a Swiss engineer.

Robert responded to Maria's questions about a marriage date by admitting that "I am to such a small extent practical that I do not know how long it will take to get an apartment." Furthermore, he did not want to choose one or to furnish it without her advice: "I have so little taste for such things and if you could know just how cold and dull my present room is, you would understand that I am incapable of arranging all that." He concluded by stating his plan to come to Bologna sometime in August.[77]

Maillart's general self-evaluation of "to such a small extent practical" is not in conflict with his unique engineering gifts: As he had learned already at watchmaking school, he was a designer, not a craftsman. He represented the modern engineer, using careful calculations and closely observing the behavior of large-scale public structures. Unlike carpenter-builders of the eighteenth century, Maillart could not himself make things, rather he had been educated by Ritter to design and to direct construction using new, man-made materials. As the century began, the age of the field-trained engineer was coming to an end.

Even as Maillart wrote about domestic matters, his primary worry was about his career. He admitted to wishing for money of his own to provide for her, and he wrote, "I reproach myself for not having worked hard enough up to now. It is true that if I compare myself with all my classmates, I find myself among the most successful and I have

Figure 15. (a) Maria Ronconi, 1901; (b) Robert Maillart, 1901. (*Source:* Madame M.-C. Blumer-Maillart)

always been satisfied. But now, when I think about the two of us, it seems to me that I will have to do better."[78] On the morning of August 14, Maillart received a card from Maria in Florence, where she had gone, as she wrote, to "Make all of my religious visits."[79] The reference stimulated him to write a letter that elaborated his own ideas about religion and marriage and revealed his conviction that the natural was beautiful. Referring to an anticipated pregnancy, he wrote: "In general, the war against all that is natural begins already before birth. One tries to hide as long as possible that of which one should be proud; one is afraid of appearing ridiculous and yet only imbeciles can make fun of

that."[80] Maillart's preference for the natural world eventually came to dominate his ideas about structure, but for the moment his focus was on the forthcoming marriage.

TOGETHER IN ZURICH

August had begun with fireworks, bells, and a promise; it ended with a calmness that found Robert and Maria together in Zurich, she living with the Wilds, he in his "cold and dull" room on the Bodmerstrasse. Soon after her arrival, the couple went to Bern to meet his family, and in late August their engagement was formally announced.[81]

Shortly thereafter a notice appeared in the *Zürcher Wochen-Chronik* (*Weekly Chronicle of Zurich*):

ENGINEER ROBERT MAILLART FROM BERN, THE HIGHLY GIFTED BUILDER OF OUR STAUFFACHER BRIDGE, HAS JUST BECOME ENGAGED TO MISS MARIA RONCONI FROM BOLOGNA. OUR CONGRATULATIONS![82]

In September, they found an apartment at the Rigiplatz part way up the Zurich hill on the east bank of the Limmat River.[83] Maillart naturally observed conventional decorum and lived apart from Maria until after the wedding. But being together in the same city for a continuous period made September an idyllic month. Maillart's outward severity began to soften and he showed an unsuspected capacity for affection.[84] In early October, Maria moved to Rigiplatz and Maillart went back to his bachelor room. Before sunrise one morning, he went to stand under Maria's bedroom window, leaving her a note that he wrote there on the back of an envelope:[85]

Figure 16. Robert and Maria Maillart, 1901. (*Source:* Madame M.-C. Blumer-Maillart)

> Beneath your window
> 5 A.M. morning
> I will not have the pleasure of having you with me! But at least I have the satisfaction of coming to say good day under your window which alas does not want to be lighted up. It is still the rain which plays a trick on us.
> Good day, dear love, and have courage and this evening the rain will not keep us from seeing each other.
> A bit sad
> Your
> Robert

The engineer's early morning pilgrimage and love letter testified to the transformation of his personality this whirlwind courtship had produced. It was characteristic of Maillart to write Maria his thoughts even though they were now both in Zurich.[86] Basically shy in normal intercourse, he found it difficult to communicate emotions. When provoked, however, he could express his feelings strongly. For her part, Maria now started to write her future mother-in-law in Bern. Thus began a series of dialogues that would continue until Maillart's death in 1940: Robert and Maria, Maria and Bertha, and then beginning after the World War, Robert and his daughter Marie-Claire.

On November 11, Maria and Robert were married in Bern (Fig. 16) at the Church of the Holy Ghost, followed by a dinner at a local hotel for sixteen close family members. That evening the couple went directly to their Zurich apartment.[87] At the time Maillart met Maria, he had no experience with women. Not given to introspection, he was obviously unaware of the consequences that his own limitations might have on marriage, especially with someone of so different a background from his. Where Maillart's instinct was sound was in his uncanny sense of Maria's need to escape from her present situation. It was to that desire that he responded.

The courtship letters touched on those very aspects of the marriage that would shape it in the years to come. Although the couple's devotion to each other was undoubtedly sincere, Maria's loneliness and its effect on the relationship were to be profoundly troubling because it affected Maillart's concept of family life. His letters illustrate his tendency to speak his mind bluntly and often undiplomatically, a habit that would handicap his professional as well as his personal relationships.

His impulsive proposal to Maria illustrated another central feature of Maillart's personality: the quickness with which he made decisions and his tenacity in sticking to their consequences. The details of his structures were always to be fully rational, but the overall designs were surprising and cannot be explained by logical steps. Maillart himself sensed this mystery with respect to his courtship in a letter some 3 weeks after the Zuoz proposal:[88]

Ah how the laws of nature are just and beautiful; these laws have willed that we be one in spite of all and even in spite of us. When I first saw you at Zuoz, there was no voice saying to me "there is a pretty woman" but rather the voice of nature said over and over again to me "there is your woman, take her." It was not my eyes which dictated the feelings that time, because I swear to you that I hardly looked at you (I trust that this does not injure your pride in being a beautiful woman!) but it was something irresistible and unknown which pushed me toward you.

CHAPTER TWO

Designer and Builder
1902–1909

The New Firm,
1902–1904

INDEPENDENCE

Soon after his marriage, Maillart began talking to Maria about founding his own company.[1] With her encouragement, he tried out the idea on Max von Müller (Fig. 17), whose money could provide needed capital. Max was enthusiastic, but Maillart realized his carefree friend did not have the discipline necessary to run a business. He therefore approached his other colleague, Adolph Zarn (Fig. 18), proposing that he join him as business manager. Unlike von Müller, Zarn did not have much aptitude for personal relationships, but he was a good accountant. Unfortunately, he did not accept Maillart's offer at first. Nevertheless, in late December, Maillart informed Froté and Westermann of his intentions. The new firm came into being on Maillart's initiative alone; clearly, he would be its principal.

It was to provide a good life for Maria that Maillart had struck out on his own and she reacted joyfully, writing Bertha, "Probably he will leave [Froté and Westermann] on the first of January. Eureka!"[2] Following the excitement of that break, the couple took their delayed honeymoon, a 2-week trip to Italy in January, to visit Maria's relatives and friends. In Turin, Maillart met the Abeggs and the Wilds, both prominent Swiss families with large fortunes in textiles. Then followed visits to Genoa,

Florence, and Bologna.[3] By this time, Maria was beginning to refer to "Maillart and Company," Zarn signed on, and Robert began a vigorous correspondence to publicize his new firm.[4] Once back in Zurich in late January, Maillart completed the intense activity essential to founding the firm officially on February 1, 1902.[5]

A routine was soon established: Robert worked 12 hours a day at his office and Maria devoted herself to running the household.[6] Married life agreed with the young entrepreneur. As Maria wrote on March 10, "Robert begins to have a little double chin!"[7] Then came Maria's most important confidence to date to her mother-in-law: the discovery that she was pregnant. Maillart was frequently in Bern and thus could have carried such news as the pregnancy with him. In addition, there was always the telephone, by then readily accessible. But for Maria and Bertha, as well as for Maillart, the letters served as a kind of unbroken stream of consciousness that is possible neither in person nor by phone. For Maillart, this correspondence had a special importance because of what he took as his inability to articulate verbally his thoughts and feelings.

Outwardly, Maria and Robert appeared to many people to be of very different character: she outgoing, sociable, and charming; he quiet, aloof, and cool. Yet, the letters show Maria's serious side with a reflectiveness that could turn to melancholy, as well as Robert's keen sense of fun and capacity for strong emotional responses. She had already begun to refer to him as a grumbler and a "bear." *Ours* in French implies someone who is unsociable, and in fact Maillart's quick temper, when he would lash

Figure 17. Max von Müller. (*Source:* Madame M.-C. Blumer-Maillart)

out at those who opposed him, made him at times frightening. Maria soon learned to adjust to the discouragements felt by a man whose creativity could find an outlet only with great difficulty in a conservative society.

On March 19, Maria fell ill with the first of what was to be a lifelong struggle with gall bladder problems, in this case complicated by her pregnancy. Also despondent over her husband's hard-working Zurich routine that left her alone a great deal of the time, she decided to spend several weeks with Bertha in Bern. Responding to one of her letters in which she expressed concern that "events" might soon spoil their happiness, Maillart reacted strongly:

In most cases it is not events that create happiness or sadness in people, but rather the way in which they live and act. When one has no regret or remorse about past actions or omissions, one can be happy in spite of events which might be more or less upsetting. The moment that one can say that this "event" is not caused by ourselves then there is nothing to fear.[8]

Events were, however, commercially upsetting

that spring. From late April until mid-June, the new company continued without any apparent success. This bleak period was lightened only by Maria's return.[9] But once back in Zurich, she could see for herself what a toll the past month had taken on her husband.[10]

THE FIRST SUCCESS

This low period was a drastic change from the optimism with which the young couple had begun the year. In early February, Maillart had turned to his

Figure 18. Adolph Zarn. (*Source:* Madame M.-C. Blumer-Maillart)

first innovation and prepared a patent application for the ideas that had led to the Zuoz Bridge. This document is the first published record of Maillart's independent activity.[11] The five drawings with the patent show a hollow box whose top is a level horizontal slab for the roadway and whose bottom is a thin curved arch. Four longitudinal vertical walls connect the slab and arch and three transverse vertical walls connect the longitudinal walls together near each abutment (Fig. 19). Except for some minor details, these drawings are taken from the Zuoz design.

The patent clearly expressed Maillart's primary career objective: to build bridges in reinforced concrete according to his own ideas. Unfortunately, such an objective, particularly in a small country, must have other more profitable work to sustain it. Thus, as the months wore on, the young firm looked hard for any kind of project that would begin to establish its reputation and to produce a profit.

Maillart, filled with new ideas and self-confidence, had begun his company with no sure contract in hand. It was a risk made possible only by the financial support of his two associates. Four years Maillart's junior, Max had studied "Kultur" engineering[12] (a combination of agriculture and surveying) at the ETH in Zurich and upon graduation in 1898 had gone to work for Froté and Westermann. There he had done some engineering, but for the Maillart firm, his major activity was to travel in search of new work.[13] In spite of his charm, no business developed.[14] Rather, the first major works for the firm resulted directly from Maillart's own efforts, which required technical insight that von Mueller did not have. Meanwhile, Zarn, who had sensed Maillart's brilliance, but who had little interest in engineering other than as a business, bided his time until some contracts arrived to require his bookkeeping.

In late May, the municipality of St. Gallen had asked for bids on the water-retaining concrete cylinders for two large gas tanks. The city had made a design using heavy concrete walls [Fig. 20(a)], but it allowed contractors to submit bids based on their

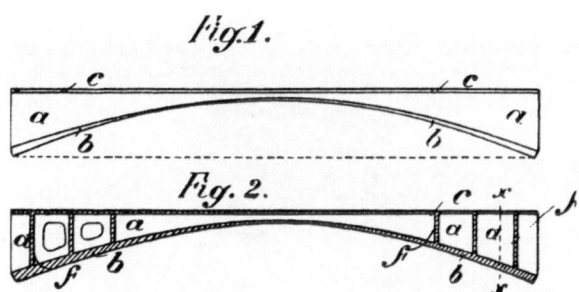

Figure 19. Maillart patent drawing: Zuoz-type bridge design, February 1902. (*Source:* Swiss Patent Office, Bern)

own designs. In overall dimensions, 9 meters high and 40 meters in diameter, these cylinders were far larger than any previous concrete water tanks. Maillart immediately perceived that the city design could be changed radically by using a new concept: Whereas the city had designed each tank as if it were made of heavy earth retaining walls, Maillart could drastically reduce the amount of concrete by using the idea that a light wall would behave as rings containing the water like the hoops of a barrel [Fig. 20(b)]. Because there was then no engineering analysis to justify his idea, he developed one based in part on the graphical analysis of Wilhelm Ritter.

Maillart solved the analysis problem through diagrams in which he guessed the size of the outward wall motion under internal water pressure. He next made a calculation of stresses based on the estimated motion and then plotted new motions in such a way that he could see how they deviated from the initial suppositions. By successive corrections, he approached the correct answer. Visualizing the behavior of the tank under water pressure [Fig. 21(a)] as the interaction of two systems [like the wooden vertical staves bending outward – see Fig. 21(b) – and the horizontal metal rings stretching circumferentally – see Fig. 21(c) – of a barrel], Maillart saw how they had to work together. More than merely the first correct analysis for a thin shell, it allowed the designer to see how the structure would perform while making the analysis, and it could be applied to any wall shape.[15]

On June 13, Maillart traveled to St. Gallen to

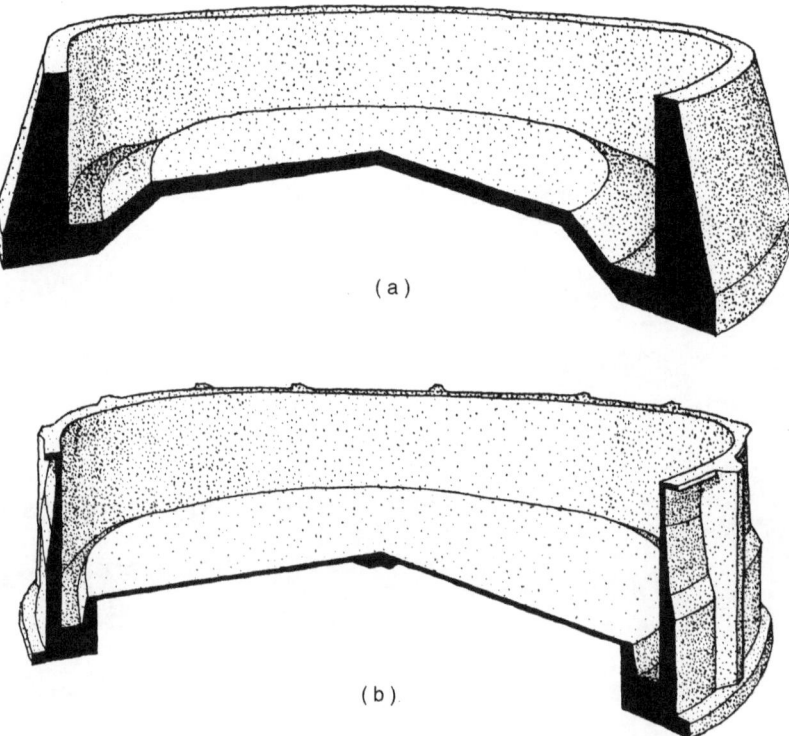

(a)

(b)

Figure 20. St. Gallen gas tanks: Sectional perspective of (a) city design and (b) Maillart design. (*Source:* Drawing by Mark Reed)

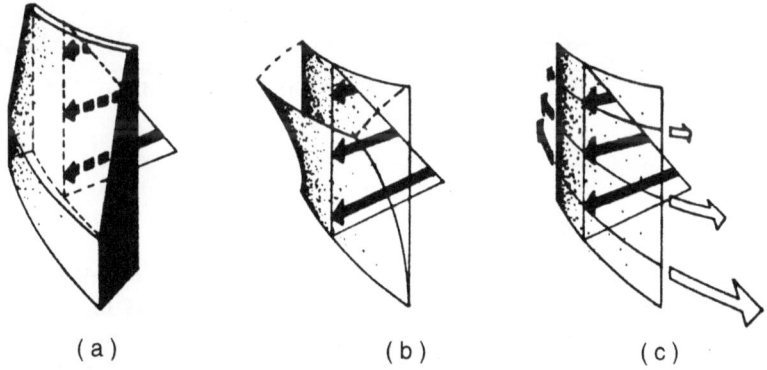

(a)　　　　　　(b)　　　　　　(c)

Figure 21. (a) Pressure on a wall-like surface; (b) resistance to pressure by wall bending; and (c) resistance to pressure by ring forces (*Source:* Drawing by Colin Ripley)

learn the outcome of the competition. Several other contractors had submitted bids based on alternative designs, and with Wilhelm Ritter sick, the city had hired Professor François Schüle (1860–1925) to make a recommendation.[16] When Maillart arrived, he found to his delight that Schüle had chosen his design. The new contract, with its promise of business success, revived Maillart's spirits.[17] He immediately began hiring people and ordered the reinforcing steel, which alone cost about three times the total construction cost for the Zuoz Bridge (Fig. 22)![18] Unwilling to let anyone else oversee the new

Figure 22. St. Gallen gas tanks under construction, 1902. (*Source:* Madame M.-C. Blumer-Maillart)

works, Maillart followed every detail at the site while negotiating continually with the city over prices and design modifications. His urge to control personally all aspects of a project first shows itself here.

In 1902, these were the largest reinforced-concrete tanks ever built and a survey 5 years later reveals that there had been none of such importance completed even at that later date.[19] Not only was their scale unique, but so was their design concept and the method of analysis invented by Maillart. Indeed, these tanks represent the first concrete structures to be correctly analyzed as thin shells – a major new form of the twentieth century. Water

tanks, roof vaults, and huge power-plant cooling towers are examples of the modern structures that arose from the ideas first put into correct mathematical formulations by Maillart in 1902.

We can appreciate Maillart's approach by contrasting his design to the city's. First, the tank walls required less than one-quarter the amount of concrete.[20] Second, the form is that of a true thin shell rather than a structure that would behave like a retaining wall or gravity dam. Before 1902, many concrete structures were built of thin walls or domes, roofs, or conical bases, but none had been of such a size combined with such thinness. These earlier works could be designed by simple methods of analysis already developed for stone or iron structures. Maillart's tanks were the first whose form and size required a new concept of structural behavior as the basis for design. Maillart had introduced this concept at Zuoz and his subsequent innovations in reinforced concrete would flow from it. He recognized that all elements in a monolithic reinforced-concrete structure will interact with each other in carrying loads. This interaction allows the development of new forms that could make concrete structures lighter and more elegant.

However, taking structural design advantage of *interaction* created an almost insuperable problem for the structural analyst; it was puzzling even to Wilhelm Ritter and the object of major research programs up until the 1970s.[21] Paradoxically, thousands of constructed hollow-box designs have performed satisfactorily over the past three-quarters of a century because the design idea is sound when in the hands of an experienced engineer. For the Zuoz Bridge, Maillart had developed a new form by recognizing that the deck, vertical side walls, and curved arch could be made to resist loads together as a hollow box. The same concept of interaction underlies Maillart's synthesis for the St. Gallen tanks. The St. Gallen thin shells were pioneering designs. In them, Maillart used a new form (the shape of the tank wall cross-section), invented a new method of analysis, and constructed the world's largest cylindrical tanks in reinforced concrete.[22]

THE ST. GALLEN BRIDGE: MORE RATIONAL AND MORE BEAUTIFUL

While Maillart was hiring staff for the gas tank construction, the city of St. Gallen issued an invitation (on June 28, 1902) for contractors to bid on a new bridge across the Steinach Brook. It was to overlook the famous rebuilt Benedictine Abbey founded in the seventh century by the Irish monk, St. Gall. As with the gas tanks, the city had made its own design, here based on an earlier competition; but in this case, there was no explicit request for alternate designs from contractors.[23]

Again, Maillart would not accept the official design. Between June 28 and July 7, he developed a new design, using two layers of unreinforced-concrete blocks.[24] The lower layer would form a complete arch of half its final thickness. The scaffold needed then only to be designed for half the total arch weight. Once built the lower layer could carry the upper layer. Maillart's ideas grew directly out of his recent experiences with the Solis and Zuoz bridges where the lower curved slab at Zuoz corresponded to the lower layer of blocks at Solis and Steinach. By implicitly criticizing the building department's bridge, he risked irritating the city engineer, a potential problem that most builders would have avoided by simply building to the city's plans.

This design, too, had a strong aesthetic basis that Maillart never separated from his ideas on construction techniques and on load-carrying actions. His proposal underlined the fact that it "offers *a more rational and a more beautiful* solution."[25] He argued that using prefabricated concrete blocks reduced the danger of cracking that would exist for the building department's design of a thick hingeless arch. By then, many hingeless concrete arches, including Hennebique's famous Châtellerault Bridge, had shown ugly cracks.[26] However, his main aesthetic argument was that exposed concrete is always better than a plastered surface. "With these construction advantages come also aesthetic ones for our proposal as opposed to one in [cast] concrete. When the concrete blocks are left completely unfinished you can see how much better they look than when they are covered even with the most beautiful concrete plaster."

Maillart also liked the look of stone bridges and that this façade would give the bridge a stonelike look did not at all upset him at this stage of his development (Fig. 23). Indeed, he argued that he preferred the stone look: "because of its less monumental appearance, it seems to us not the place here for reinforced concrete." Later on he would change his mind about what was an appropriate material, but in 1902, it was important to him to provide an alternative to the city design that would be technically superior and at the same time look better.

With Ritter still unwell, once again the city called in Schüle, who this time was more critical of Maillart's ideas. A long and acrimonious discussion forced him to reduce his price from just over 48,000 to 38,600 francs (approximately $390,000 in 1996) without even getting a clear design approval from Schüle. However, the city did give the contract to Maillart.[27] With this bridge, Maillart began to show openly his dislike of the authorities, especially when they attacked his design view by insisting on detailed stress calculations rather than by trying to understand the overall performance.

MAILLART AGAINST THE AUTHORITIES

In early August, Maillart had a serious disagreement with the city over the consultant's report. It was common practice in public works for officials to ask for academic advice on new design ideas. Ritter had been consulted for the Stauffacher and Zuoz bridges and would probably have been asked to consult on the Steinach Bridge had he not been ill. In his place, the city had engaged Schüle, with whom Maillart's relationship was quite different from that with Ritter. Unlike Ritter, Schüle could not easily visualize overall performance; he was more comfortable with the detailed stress calculations in which he could find fault with Maillart's work.

The nineteenth-century Ritter had a broad perspective; his teaching included metal, wooden, and stone bridges, and he saw structural theory not just as formulas and laboratory experiments, but also in terms of visually graphic analysis and full-scale load testing. Thirteen years younger than Ritter, Schüle was more of a twentieth-century academic for whom codes and code-related studies were central, and for whom detailed laboratory research took priority over practice.

Schüle's preoccupation with questions of stress analysis led him to a disagreement with Maillart when city building director L. Kilchmann (1852–

1926) sent him Maillart's design for review.[28] What worried Schüle was that Maillart's calculations assumed the arch to be built in one solid layer, whereas it would actually be built with blocks in two separate layers. From a theoretical viewpoint, Schüle was correct in noting that Maillart's calculations were not detailed enough to determine the correct stresses during construction. It was the same problem Ritter had with the Zuoz Bridge, but Ritter understood Maillart's ideas, knew that they were constructionally sound, and was content to let full-scale load testing confirm the structural behavior. In his questioning of Maillart's design, Schüle caused serious delays and disagreements between the builder and the city engineer. In addition to the construction stresses, Schüle objected to Maillart's plan to reduce the arch thicknesses from those shown by the building department. The contract required the approval of such changes by Schüle who left town in August and could not be reached for comment.

While on the one hand refusing to approve the new drawings without Schüle, on the other hand, Kilchmann criticized the construction delay in a series of acrimonious letters. The whole process infuriated Maillart, who felt the origin of the problem lay with Schüle and his reliance on analysis rather than design. It was the first time that Maillart found himself in open disagreement with a powerful public official. He was totally free of the intimidation most men in his place would have felt; on the contrary, his instinctive reaction was to fight Kilchmann as aggressively as possible.[29]

Maillart argued that he was obliged to make only those changes Professor Schüle required and that he would be responsible for delays only if he could follow his own plans.[30] He disagreed with the city's stress calculations. Moreover, he noted that he had already offered to build the heavier arch Kilchmann required if, on his return, Schüle agreed with the city's calculations and Maillart would get nothing extra. If Schüle agreed with Maillart, however, the city would pay for the added material. Maillart brazenly wrote Kilchmann: "If you are so sure that your calculations will be sanctioned by the

Figure 23. Steinach Bridge in St. Gallen, 1903. (*Source:* Madame M.-C. Blumer-Maillart)

consultant, it should be easy for you to agree to our proposal. Then the bridge will be built according to your wishes and it will not cost you a penny more."[31]

Things were coming to a head. Kilchmann softened his tone, even admitting some errors in his previous letters. In late August, he finally came to an agreement with Maillart:[32] Maillart would build to the increased dimensions required by the building department's stress calculations with the burden of extra cost to be decided by Schüle.[33] The bridge went ahead, but the delays in July and August meant that most of the construction could not take place until the following spring.[34]

Schüle and Kilchmann were not Maillart's only problems that summer; by July 25, Maillart and Company had 150 workers in the field at St. Gallen, all of whom had to be paid regularly. Financial worries also arose with the 6,000-franc required guarantee for the bridge contract – a difficult expenditure just then because of the large capital outlays needed for the gas tanks contract. All builders of public works face the problem of having to borrow funds because they are normally paid only for work completed, yet they must hire personnel, secure equipment, and order materials before there is any income. In July, the company had a number of outstanding loans and as yet no returns. It was not easy for Maillart to adjust to these responsibilities. He looked exhausted and was irritable.[35]

In the fall, these concerns were momentarily relieved by the birth, on Friday, October 3, 1902, of a son who was named Edmond Benedetto Maillart after Maillart and Maria's fathers.[36] Bertha came from Bern, a nurse moved in to live at Rigiplatz along with Maria, Robert, the baby, and the maid. Maillart was elated and his few postcards from St. Gallen trips reflect his good spirits. His company was by no means yet a lucrative business, but it was increasing its activity and gaining prominence. More importantly, Maillart himself was beginning to emerge in Switzerland as an engineer of exceptional talent willing to take on difficult structural problems. He was studying books about reinforced concrete and he was writing about it as well. All he needed now was increased opportunity to design and build.

With the works begun in 1902, Maillart experienced for the first time the central difficulty of his career – his concentration on design as opposed to business. Every technical problem required his personal attention and many of those problems arose because of his innovative ideas. He could not delegate the design or the negotiations on design to others as Hennebique had done. He was thus beginning to develop a business that depended on him alone.

By the end of 1902, Maillart had already confronted the barriers to creativity that would alternatively frustrate and stimulate him for the next 38 years: the authorities, traditional forms, and business management. Paradoxically, these obstacles were the foundations of a conservative society in which Maillart seemed so much at home. With no radical political or social tendencies, within his profession, he was already at 30 a radical designer. Unlike many architects of this period, he never connected radical building with radical politics. It never occurred to Maillart that in rejecting the ideas of a reactionary public works official, he might be rejecting the stolid Swiss tradition of conservative bourgeois democracy. Indeed, any connection between building and society would have seemed ridiculous to him at this time. With World War I, he would learn to his horror that there was a connection and in the late 1930s he would begin to write about it. But as 1903 began, there was for him only business and family; they were connected by the need for contracts and profits, which he hoped to realize through his radically new ideas about form.

The First Masterpiece Tavanasa, 1902–1905

BILLWIL AND ZUOZ

Up to 1905, nothing expressed Maillart's design view of engineering more clearly than the Tavanasa Bridge. His ideas and his confidence developed from the initial 1900 Zuoz design, to his commission in mid-1903 to build a bridge at Billwil, to his late 1903 study of defects in the Zuoz Bridge, to his tentative solution of those defects for a lost competition design in mid-1904, and finally to his mature solution of September 1904 for the Tavanasa competition.

With Zuoz clearly in mind, Maillart seized the opportunity to submit a bid for a new bridge at the village of Billwil over the Thur River in the canton of St. Gallen.[37] The canton wanted a two-span bridge to cost about 54,8000 francs, but the authorities permitted contractors to bid their own designs. Maillart realized that this bridge would be difficult, because it required crossing a length nearly twice that of the bridge at Zuoz.[38] But the Billwil Bridge would be a large structure for its time and Maillart was eager to do it in reinforced concrete using his new hollow-box form. In May 1903, Maillart had received a patent for his hollow-box form, enabling him to stamp "Bogenträger System Maillart" on the drawings that he submitted to St. Gallen on May 28 (Fig. 24). Five days later, he submitted a formal proposal with a price of 39,440 francs.[39]

Figure 24. Thur River Bridge at Billwil, 1904. (*Source:* Drawing by Mark Reed

On June 26, almost 1 year after he had received his first bridge contract, Maillart was told that he had won the contract for Billwil.[40] Of all designs submitted, the canton had chosen the two most economical, Maillart's and one by the firm of Jaeger and Company, and sent them to Wilhelm Ritter for review. Ritter recommended Maillart's design of two three-hinged arch spans over the Jaeger Company's design of two-hinged arches. Once again Ritter provided crucial support for his former student.[41] The Billwil commission was Maillart's first three-hinged arch as the head of his own company. Following the form of Zuoz, Maillart completed his design work and on August 6 sent in his calculations and drawings. The canton sent them back to Ritter for review, who questioned a few details and urged a second look at Zuoz, where some cracking at the crown had been observed. Maillart wrote back to point out how much more conservative he had made his Billwil design, but he noted the crack-

ing at Zuoz and agreed to add reinforcement to the crown at Billwil.[42] Ritter's report was his last; failing health caused him to enter a sanatorium in 1904; he died 2 years later. Ritter had helped Maillart secure approval for his early innovative designs. His absence would make it more difficult for Maillart to build his more mature ones.

By October, the concrete and stone facing for the foundation at Billwil was completed. Then began the delicate operation of building the arches, starting with construction of small temporary concrete piers in the waterway. On these piers, Maillart designed a light wooden scaffold to hold the curved platform that would support the arches. He had a field force of about forty men, laying the wood forms, placing the steel reinforcing bars, and casting the concrete slabs. Maillart operated with complete self-confidence, supervising the construction workers as well as laying out the entire building procedure. Construction resumed in March after suspension for the winter and in April 1904, the scaffold was lowered and the bridge stood on its own.[43]

For Maillart, Billwil was a crucial test. His design won approval in a competition that Ritter influenced. But Billwil was the first bridge that Maillart built as well as designed and for which he also assumed financial responsibility. He was at the site continuously, taking direct charge of all details. Then and later, he was unwilling to delegate authority, either in design or construction.

THE RETURN TO ZUOZ

With Billwil construction underway in September 1903, Maillart traveled to the Graubünden capital of Chur as a Zurich delegate to the biennial convention of the Swiss Society of Engineers and Architects. At Chur, he met Achilles Schucan, director of the local railway, the Rhätische Bahn, with whom he had negotiated the Zuoz contract in 1900. Schucan asked Maillart to inspect and officially report on some cracking that had appeared in the Zuoz Bridge. As soon as he could extricate himself from the meeting's concluding banquet,

Maillart boarded the bus to Zuoz. The next morning he studied the cracks that had appeared in the vertical walls near the abutments where the walls were highest and carried the least load. He concluded that the cracks were caused by temperature and humidity. The bridge was in no danger but required maintenance. These observations set him thinking again about the form that he had patented.[44] But before that the patent itself was challenged.

Froté and Westermann had taken Maillart to court on the grounds that he had developed the Zuoz Bridge while in their employ. On October 2, the judges ruled in favor of Maillart's former employer and voided his *Bogenträger* patent. When he arrived home that evening, he tried to toss the episode off with a laugh, but his disappointment was obvious.[45] The loss of his patent aggravated his constant concern about money. Already in late spring, money matters had begun to influence his life in small but annoying ways. In May, he met one of Maria's doctors at the Bern railroad station and rode first class with him until he got off at Burgdorf, whereupon Maillart immediately switched to second class. Later that month, he had to postpone a trip to Bern until Zarn returned with a railroad pass he could use. In June, he felt he could not afford to enter a shooting match attended by von Müller and Maria. After the court decision, Maillart reduced Maria's monthly household allowance from 280 to 250 francs and dismissed one of the servants. When Maria objected, he became furious, showing a touchiness about his reduced circumstances and a characteristic lack of tact in dealing with anyone who disagreed with him.[46]

His obstinance caused personal problems and in the short run it caused professional difficulties as well. He was determined to build bridges of the Zuoz type. In mid-1904, he made a design similar to Zuoz for the Uto Bridge competition in Zurich. There he thickened the exterior wall and all the walls near the abutment to strengthen the structure against the cracking observed at Zuoz (Fig. 25). He lost this competition but not his desire to build hollow-box arch bridges.[47] Maillart knew that his form needed improvement.[48] There are two types of solutions to any observed structural difficulty: one is to combat cracks with added material; the other is to avoid cracks by changing the form. Following the loss of the Uto Bridge, Maillart began to search for a solution of form over materials. Indeed, this is the central question in reinforced-concrete design, and what separates it in concept from both stone and steel.

To Swiss civic leaders at the time, steel characterized the unpleasant necessities of modern industry. Utilities, factories, railroads, and cranes were necessary for wealth in the modern world, but the industrial cities of the nineteenth and early twentieth centuries preferred stone for parliaments, bridges, and opera houses to embody civilization and culture. A central problem with urban building has been how to humanize the useful and at the same time make monumental works more accessible. Perceptive writers on architecture recognized this problem early in the twentieth century, and much of early modern architecture was an attempt to bring the practical and the beautiful together.[49] But in Zurich, city leaders rejected Maillart's bridge. To make his business successful, he realized that he would have to bide his time and find work other than public bridges.

To understand how Maillart saw design in 1904, it is essential to identify precisely those characteristics of both stone and steel that set these materials apart from each other and that set both apart from reinforced concrete. Scientifically, the major defect of stone is its low resistance to tension. The need to overcome this defect explains almost all of the forms used in stone construction up to the nineteenth century: the column, the heavy wall, the short lintel, the relatively long-span arch and the high dome, all are solutions to the problem of reducing tension. Socially, in the sense of cost to the public, stone was always expensive and used either where permanence was desired or where wood was unavailable. Cut stone involves substantial labor, and as the nineteenth century wore on, its use be-

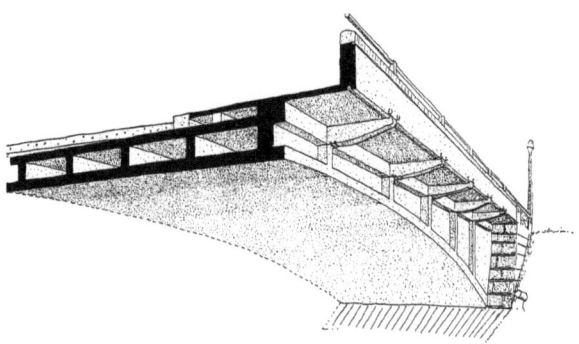

Figure 25. Uto Bridge over the Sihl River in Zurich, Maillart competition design (unbuilt), 1904. (*Source:* Drawing by Mark Reed

came more and more restricted to monumental urban buildings, adding to the sense of its appropriateness for important works.

Symbolically, stone recalled the past because of its long tradition and, most importantly, because its formal possibilities were largely untouched by the Industrial Revolution. No matter how much physics or mathematics one brings to bear on structural engineering, if one is restricted to using stone, then even the late twentieth century could hardly improve on Roman aqueducts and Gothic cathedrals. Thus, the permanence of stone symbolizes what in fact it materializes, a record of past glories that have survived. To most people in the last 200 years, stone buildings seem to symbolize cities that will endure.

This belief was so powerful that some of the most famous designers who tried to use new materials, in the end, succumbed to it. For example, Le Corbusier and Louis Kahn both designed solid concrete works that reflect, even if they do not copy, massive stone structures. Their ideal was in essence to create form by adding material because for many of these designers, sometimes engineers as well as architects, the discipline of structural behavior seemed either too demanding or simply not the controlling issue. Stone or a stonelike appearance came to be either the jacket that one put over the structure or the heavy contortions that gave a

more traditional form to the straightforward problem of supporting easily defined loads. Buildings, of course, must be closed in by walls and thus structural expression does not have the same priority there as it does for bridges.

Steel or wrought iron, on the other hand, provided a convenient foil against which to justify the use of stone. The primary scientific defect of steel is that it rusts and is not fireproof; hence, its uses have always been conditioned by its inherent lack of permanence. In buildings such as apartments, offices, and hotels, steel must be covered with some fireproofing material such as terra-cotta or concrete; where exposed, it must be continually painted. Steel is also an expensive material. It is used because of its exceptional strength, but to be competitive, very slender elements are necessary. Even then almost all steel designs must use standard straight pieces that are mass produced in centralized factories. Thus, steel construction involves the assembly of standardized pieces shipped to the field. At its most economical, a steel structure will be made of a number of straight pieces simply connected to make up a skeletal structure on which the "design" is then hung. Visually, steel is also skeletal, and its many thin pieces often create a cluttered image.[50]

Following Hennebeque and others, Maillart in his writings up to 1905 saw reinforced concrete as a combination of stone and steel that answered all objections. The embedding of steel bars in reinforced concrete overcame the low tension capacity of stone and the concrete cover gave endurance to the steel. In other words, the solving of one problem *automatically* solved the other.

Reinforced concrete overcame the high cost of stone cutting by substituting carpentry for masonry. Both had to be set on a wooden or metal falsework, which for reinforced concrete could be lighter than for stone because the structure could be lighter. Proper shaping of the concrete by wooden formwork led to a drastic reduction in steel. One reason for the high cost of steel structures is that when under compression, the light, straight pieces buckle

easily unless braced. Much of the cost of connections in steel frameworks goes toward preventing
this buckling. In reinforced concrete, the concrete
takes most of the compression and the steel takes
most of the tension, eliminating nearly all of these
costly connections. Finally and unlike many designers of the time, Maillart believed that the new material would lead to new forms not reminiscent of
the past. Its stonelike quality has a permanence and
its forms have a lightness that he felt could not be
explained by direct comparison either to stone or
to steel.

THE TAVANASA BREAKTHROUGH

With the bridge at Tavanasa, Maillart achieved
such a form, which he began to imagine during a
trip there in early August 1904. The primary reference point was, once again, the three-hinged,
hollow box of Zuoz. But for Tavanasa, Maillart
wanted to remove those parts of the longitudinal
vertical walls nearest the abutments (Fig. 26), eliminating those parts that had cracked at Zuoz and
leaving a lighter-looking form.[51] Moreover, the new
bridge exhibited a visual thinness, both at the two
support points and at the crown, which expressed
the structural fact of three hinges. By contrast, the
arch profile gradually deepens between the hinges
further to emphasize the thinness at the hinges (Fig.
27). This lenticular shape had the benefit of increasing resistance to loads at just the places (the quarter
span, i.e., the two points halfway between the
hinges) where the live load (the weight of motor
vehicles on the bridge) produces the severest effects
on a three-hinged arch. Finally, the visual impact
of the bridge was dramatically different from stone
or steel. It was clearly stone (artificial stone), but in
a slender, structural form that would be inconceiv-
able in pure masonry or unreinforced concrete (Fig.
28).

Maillart completed the elegant color drawings
for his Tavanasa competition design in time for
submission on October 1.[52] Shortly after the competition deadline, Maillart learned that his low bid
had been accepted, but once again the canton in-

Figure 26. Rhine River Bridge at Tavanasa, 1905.
(*Source:* Drawing by Mark Reed)

sisted on having the best-known consultant in Switzerland – now that Ritter had retired, Emil Mörsch
(1872–1950), a professor at the ETH – review the
unusual design.[53] The German-born Mörsch had established himself as one of the two or three leading
German students of reinforced concrete through his
work as technical director of Wayss and Freytag,
his 1902 book on reinforced concrete, and his design for the record-spanning, three-hinged arch over
the Isar river at Grünwald. For the Tavanasa, he
recommended that Maillart's design be built, with
only minor modifications. Maillart's ideas had
stood up to the scrutiny of a leading authority.[54]

At Tavanasa, Maillart broke the precedent of
deep spandrel walls that came from the Roman
arch bridge. He created that innovation by expressing the three hinges as places where the arch has
minimum thickness, and between hinges, the arch

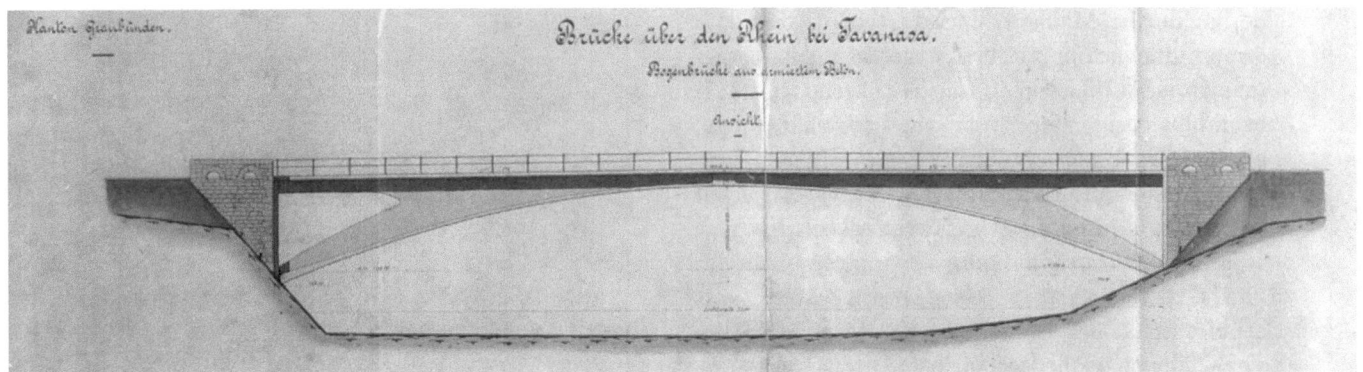

Figure 27. Bridge over the Rhine River at Tavanasa, drawing of elevation. (*Source:* Zurich Maillart Archive)

Figure 28. Bridge over the Rhine River at Tavanasa, drawing of sections. (*Source:* Zurich Maillart Archive)

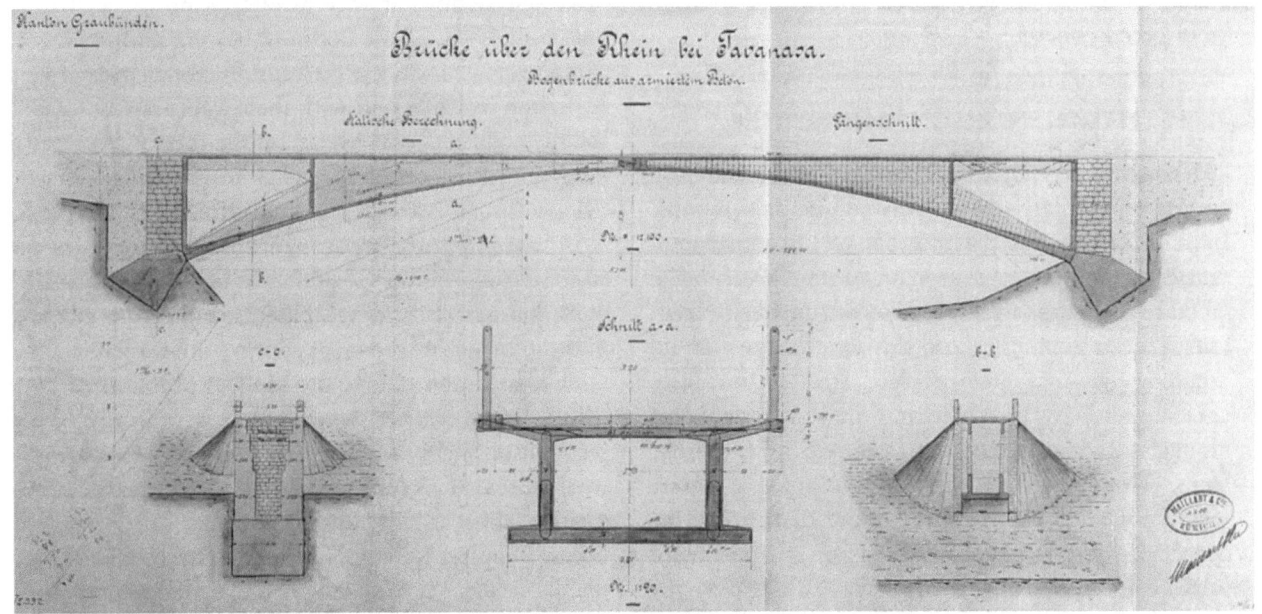

fuses with the horizontal deck to provide the necessary stiffening. Although the hinges make the arch insensitive to foundation settlements or temperature changes, they render it more flexible under truck loadings. It is the fusion of arch to deck that stiffens the bridge sufficiently to make the truck loads effect negligible. Of the many ways to achieve stiffening, Maillart chose the one that pleased him the most – an expression of great quarter-span depth with diminishing depth toward each hinge. Maillart kept a close watch on the bridge's construction throughout 1905, and on September 28, it was successfully load tested.[55]

The Tavanasa Bridge had little effect on bridge

design during Maillart's lifetime, but it exerted a strong influence on the best structural artists practicing after World War II. Christian Menn (b. 1927) began his design career (with the Letziwald Bridge, also in Graubünden) by following Maillart's Tavanasa design of which he said, "the principle of construction used for the Tavanasa bridge was still, 50 years later, the most advantageous one [in 1957]." Like artists in other media, Menn was deeply affected by a predecessor's [Maillart's] designs. Thus have his designs helped form a new vision of bridge design during the last third of this century.[56] Both Felix Candela (b. 1910) and Heinz Isler (b. 1926), masters of thin-shell-concrete roof design, have spoken of the stimulus of Maillart's new bridge form on their own vision of structural art.[57] Today's acclaim for the Tavanasa Bridge makes it one of the few seminal works ever built in reinforced concrete.

MELAN, MÖRSCH, AND RITTER

Although three-hinged arch bridges were common in 1905, Maillart's vision was unique; he saw possibilities for forms that the best of his contemporaries did not. The tradition of making three-hinged metal arches was widespread in the nineteenth century; the reasoning behind this was that one could thereby eliminate temperature stresses and also greatly simplify the structural analysis, which for hingeless arches at the time was a complex problem. Three-hinged arches of reinforced concrete were not uncommon either around the turn of the century – the largest (two spans of 230 feet with rises of 42 feet) being the highway bridge over the Isar River designed just before 1900 by Emil Mörsch. That bridge shows the beginning of the three-hinged form Maillart was to develop: The profile of the arch is deepest at the midpoints between hinges (47.2 inches) and tapers toward the hinges (35.4 inches at the springings and 29.5 inches at the crown).[58]

A yet earlier precursor of the form appeared in what was perhaps the first systematic set of arch tests, made in Berlin during February 1886. The

Figure 29. Three-hinged, arch-bridge forms: Emil Mörsch (Isar River Bridge at Grünwald), 1904. (*Source:* Photograph by D. P. Billington)

fifth of the six structures tested resembles in a primitive but significant way Maillart's bridge at Tavanasa of 1905. These Berlin tests were published in Emperger's handbook on reinforced-concrete construction in 1908, and with them appeared an essay by the engineer Josef Melan on the theory of arches and reinforced-concrete arches in particular. Here Melan discusses the best primary forms, or *günstigsten Gewölbeform* (most favorable arch form), and secondary forms, or *Gewölbestarke* (arch thickness) with mathematical developments; a diagram of the most appropriate secondary form for a three-hinged arch also shows clearly the kind of profile used by Mörsch and further developed in a radically new way visually by Maillart.[59] Melan's presentation was typical of the applied scientist who cared primarily about theory and mathematical formulations. Maillart's form is by contrast characteristic of his design view in which the mathematical developments are less critical than the physical performance of concrete.

Rather than mathematical complexity, the major problems of concrete bridge design are resistance to cycles of freezing and thawing, expansion and contraction due to temperature changes, proper roadway drainage, and details such as bearings and joints, and the pattern of reinforcing steel within the concrete. In 1990, a leading researcher in concrete structures singled out these same ideas, devel-

Figure 30. Bridge over the Rhine River at Tavanasa. (*Source:* Madame M.-C. Blumer-Maillart)

oped by Maillart (following Ritter), as being crucial to concrete design.[60]

Although there were these partial precedents for Maillart's forms when he began his own firm in 1902, it took his vision to bring out their potential coherently. Other designs, such as the Isar Bridge by Mörsch, however thin their arches, lack the visual power of Maillart's mature works; the profile variation is too slight, the springing piers are unnecessariy bulky, and there is no clear connection between the horizontal deck and the curved arch (Figs. 29 and 30). But the form used by Mörsch would become the common one for concrete arches over the next several decades. Even after Maillart's patent was invalidated by the court in late 1903, no one else would use his idea until the 1930s.

With Ritter gone, there would be no academic engineer to champion Maillart's ideas for another 20 years. Tavanasa was the last opportunity for Maillart to fulfill his vision until the 1920s. It was a technical and visual triumph, but it was only possible to build in the remote Graubünden. His hollow-box design for the Uto Bridge in Zurich, just before Tavanasa, was rejected largely on aesthetic grounds. The Uto was the first juried competition that Maillart entered; Billwil had been the decision of a cantonal engineer who acted after consulting Ritter. For the Uto Bridge, Maillart wrote his wife: "Our design does not at all please the general pub-

lic; what is necessary for them is the banal and the ordinary!"[61]

Maillart did, however, receive some early recognition from abroad. Among the few who grasped Maillart's vision by 1905 was the Austrian engineer, Fritz von Emperger (1862–1942). In 1902, von Emperger had founded the first technical journal devoted solely to reinforced-concrete structures, and in 1907, he began to publish his multivolume *Handbuch für Eisenbetonbau,* the most comprehensive compilation of structures in reinforced concrete ever produced. He sensed Maillart's innovations and published them immediately in a detailed critique of the Uto Bridge competition, including a full reproduction of Maillart's technical report, which would not have survived without the Austrian's publication. Von Emperger praised the Swiss system of design competitions, but he criticized the Uto jury for giving too much weight to "architecture," by which he meant stone-faced designs. For him, clearly Maillart's design was the best.[62] Several years later, von Emperger would publish the details of the Zuoz and Tavanasa bridges in his *Handbuch,* but these works were otherwise largely ignored by the profession however much they satisfied the design urge of Robert Maillart (Fig. 31).

A Farewell to Bridges
1904–1909

FROM BASEL TO DAVOS

Maillart's innovations went largely unnoticed in the established world. The Tavanasa Bridge gained Maillart no great profit and little favorable publicity in Switzerland. On the contrary, it aroused such strong aesthetic objections from public officials that his bridge design practice nearly came to a standstill between late 1905 and 1910. In 1909, he did build another such bridge at Wattwil, but there an architect put on a façade to cover the form. Maillart responded by seeking to develop a wider market for other types of concrete structures, a search that would lead to a major new form that would transform the design of buildings throughout the twentieth century.

In September 1904 at the invitation of the powerful Industrial Society of Cement Manufacturers, Maillart delivered a lecture in Basel, "Constructions in Reinforced Concrete." Maillart expressed his vision of concrete as *the* building material for the twentieth century. He spoke in a hotel made of reinforced concrete that had collapsed during its construction in 1901 – a fact he had been aware of at the time and that now prompted him to emphasize the importance of sound construction practice.[63] He demonstrated the variety of applications possible for reinforced concrete. This Basel lecture served as a program for the future of concrete and as a statement of his own willingness to take on almost any type of structure in order to demonstrate the new material's potential.[64] He was announcing a new direction for his energies and showing to the Swiss the examples of structures built in other countries.[65]

As a start on this way, Maillart obtained contracts to build two sanitoriums, one in and one near Davos. Inspired by the mid-nineteenth century observation that tuberculosis was almost never found in Davos,[66] a large new sanitarium was designed for that town by the Zurich architectural firm of Pfleghard and Haefeli. Called the Schatzalp, it was built by Froté and Westermann when Max von Müller and Robert Maillart were with the firm. In 1901, the same architects had designed the German sanatorium at Wolfgang at the northern end of Lake Davos. Three years later, the design and construction of structures for additions to these institutions were given to Maillart.[67] In addition, he completed there the Queen Alexandra Sanatorium in 1905 (Fig. 32). Like the sites for his most original work to date, Davos was in Graubünden, a region that fascinated Maillart. The largest and least densely populated canton in Switzerland, Graubünden claims the longest experience with participatory democracy. Its inhabitants revered independence and

Figure 31. Tavanasa with Maria and Edmond, circa 1906. (*Source:* Madame M.-C. Blumer-Maillart)

were more open than most Swiss to design innovation.[68]

The region had cast a spell over others, including Maillart's famous contemporary, Thomas Mann, who used it – and specifically Davos – as the setting for his novel, *The Magic Mountain.* Mann made the region a symbolic landscape for the cultural state of Europe before World War I. Its "sparse or scanty" setting, combined with a harsh but healthful climate, helped create the image of Switzerland as a sanctuary for the planning of new revolutions and the creation of great art.[69] Of all twentieth-century sculptors, Alberto Giacometti, who was born and brought up in Graubünden, best expressed the idea of wrestling beauty out of those "sparse or scanty" forms.

Another unique Graubünden form, completed also in 1904, illustrated Maillart's approach to work other than bridges: a gallery over the Rhätische Bahn near Tiefencastel. Building the Albula railroad line required many tunnels at the entrances to which steep rock walls often rose high above the openings, creating a continual danger of rock falls onto the tracks (Fig. 33). Galleries around these entrances protected them from such falls and Maillart designed and built one for the Rhätischen Bahn; it consisted of an arching cantilever over the tracks to support a buttressed wall behind which earth fill broke the rock fall and protected the tracks (Fig. 34).

Again von Emperger was alone in recognizing the quality of the design; he included it in his *Handbuch* with the comment "a totally original form . . . for the Rhätischen Bahn in Switzerland completed by the building firm of Maillart and Cie in Zurich."[70] This minor job was one of many innovative works that Maillart projected or completed in his early career. They were neither large enough nor were they repeated often enough to establish a successful building business. But the big sanatoriums did hold promise for great profits.

Maillart's shift to buildings and other functional projects helped establish his business and finally give him a measure of prosperity. In January 1906, the Maillarts gave up their apartment in Zurich and moved to Kilchberg, a few kilometers east of Zurich on the lake. Before settling there, however, they went together for a brief winter holiday in Arosa, where Maillart left Maria and Edmond and returned to business.[71] The fact that mother and son were able to remain in a resort for 2 months is an indication of the engineer's growing prosperity.[72]

On August 13, Maria gave birth to her second child, a girl, whom they named Marie-Claire. Maillart's jubilant mood was evident in the letters and cards he sent Maria in the months following the baby's birth.[73]

Figure 32. Queen Alexandra Sanatorium under construction, Davos. (*Source:* Madame M.-C. Blumer-Maillart)

BUILDING A REPUTATION

In recognition of his work in reinforced-concrete design and construction, Maillart was appointed in December 1905 to a new seven-man commission whose goal was the creation, by 1909, of a Swiss National Code for Reinforced Concrete.[74] Maillart had already served on an earlier Zurich Committee, formed in 1902,[75] a critical year for reinforced concrete because it saw the first major textbooks,[76] the beginnings of code writing, and a clarification of the 1901 hotel collapse in Basel that largely allayed misgivings about reinforced concrete.[77] If, however, the new material was being rapidly accepted as safe, enduring, and economical, it was far from being accepted as attractive or even appropriate as either an exposed surface or a determinant of visual form. The Zurich Committee's favorable 1903 report helped to guide design, but a National Commission would be far more influential.

Maillart's debates with the leading Swiss academics who made up this commission sharpened a conflict that was to continue throughout the century between practicing designers and professors. Maillart's ideas on form had already begun to challenge the urban bridge design of his day, now his ideas on formula – on the mathematical expression of analysis – conflicted with the growing tendency to treat engineering practice as the application of new academic research results. Maillart's ongoing debates with academics represent one of the earliest and clearest challenges to the idea that engineering is a rigid science devoid of artistry (Chapter 3 presents these debates). In spite of antagonizing some authorities, Maillart's building business prospered. By 1905, his reputation was such that he was able to win contracts even when confronted with lower bids. One such contract was for an addition to the Valsana Hotel in Arosa. Maillart was chosen for the quality of his work over the lower bidders, Froté and Westermann.[78]

Maillart continued to turn out a wide variety of work that gave him enormous breadth of experience with reinforced concrete. It also gained him a reputation for reliability in completing difficult pro-

Figure 33. Avalanche protection structure near Tiefencastel, 1904. (*Source:* Madame M.-C. Blumer-Maillart)

jects. One such project came to him when he began in the spring of 1907 to study a siphon (a tube in which water flows due to atmospheric pressure) to divert the Riehenteich stream. It was to be built underneath part of the area used for the new Badischen Bahnhof in Basel on the north side of the Rhine.[79]

In early 1906, the Basel Public Utilities Department invited bids for construction. In February 1907, when the department evaluated the bids, Maillart's proposal had two disadvantages: first, he was not the low bidder, and second, he was not from Basel. The building director, Paul Miescher, nevertheless recommended Maillart: "I have already had the opportunity to see large works by the [Maillart] firm under construction and have thereby been convinced that it works with great care. I should note that I would have preferred a Basler firm to have undertaken this work, only here is one case where one is justified in going outside the canton boundaries."[80] This kind of rivalry exists in

Figure 34. Avalanche protection structure. (*Source:* Drawing by Mark Reed)

most cities because a public work is usually the local object of political pride and economic benefits.

Like other major Swiss cities, Basel has a long
history and strong sense of its own distinction. The
city has been host to great religious figures from
Erasmus to Karl Barth, home of Switzerland's
greatest historian, Jakob Burkhardt, and is the
wealthiest city per capita in Switzerland because of
its industry. It was distasteful for Miescher to admit
that a Basler engineer could not do the work as well
as a Zurich engineer.

Miescher observed that the lowest price was for

Figure 35. Basel siphon, 1907. (*Source:* Madame M.-C. Blumer-Maillart)

a joint venture between a firm in Basel and the Zurich firm of Jaeger and Company, a frequent competitor of Maillart's. The director did not feel, however, that this joint venture could be relied on as Maillart and Company could be and he further noted that even Maillart's bid of 78,240 francs (about $800,000 in 1996) was well below the city estimate of 90,000 francs.

Before Maillart's 1907 design, siphons were circular. Maillart saw that for structures with large external pressures from covering earth, this form was inefficient. He therefore designed an elliptical tube that his tests proved to be sufficiently strong. To question the design of a utilitarian object such as a siphon required a certain degree of audacity. Clients for such objects were typically close-minded about any kind of innovation. Consequently, it was a rare individual who would take the risk of losing a commission for the sake of design innovation. On March 5, 1907, the contract was approved, and on March 14, Maillart carried out tests in his construction yard in the Giesshübel district of Zurich.[81]

The siphon was successfully built (Fig. 35) and put into operation in October.[82] The siphon contract demonstrated Maillart's reputation after only 5 years in business, but in fact its form had little influence on the history of concrete structures. It was one of his many clever but not fundamental innovations. Among these were prefabricated curbstones and concrete bases for wooden masts. Even in mundane problems, Maillart often saw novel possibilities.

Figure 36. Pfenninger factory, Wädenswil, 1905. (*Source:* Madame M.-C. Blumer-Maillart)

BUILDER AND ARCHITECT

Arguments about codes or studies of galleries and siphons do not lead to much business; and so in the spring of 1905, and through the remainder of that year, Maillart had focused his attention almost exclusively on buildings. A commission for a four-story cloth factory in Wädenswil, 22 kilometers south of Zurich, was his largest and most challenging contract of the year because it enabled him for the first time to act as architect, engineer, and builder of a prominent building (as he had for the Billwil Bridge).[83] Unlike bridges, buildings and even factories are usually designed by an architect with the engineer as a consultant. Maillart's design was intended to express visually the integration between skeletal structure and exterior wall (Fig. 36). The result was a façade of large windows showing a skeleton of unusual lightness for the time, with fewer elements than the typical Hennebique frame [Figs. 37(a) and 37(b)]. The large bays allowed natural light to fill the interior. Maillart was simplifying and eliminating elements.[84]

With his business growing in Switzerland, in November 1905, Maillart opened an office in St. Gallen.[85] The following March, the office received a major building contract – for the town Concert Hall. Its complicated reinforced-concrete frame was to be hidden completely, both inside and out, behind decorative façades. During the winter lull, Maillart took the contract design (Fig. 38) and began to revise the structure by making it thinner, yet stronger. His most striking change was in the support for the stage: Shown in the 1906 drawing by

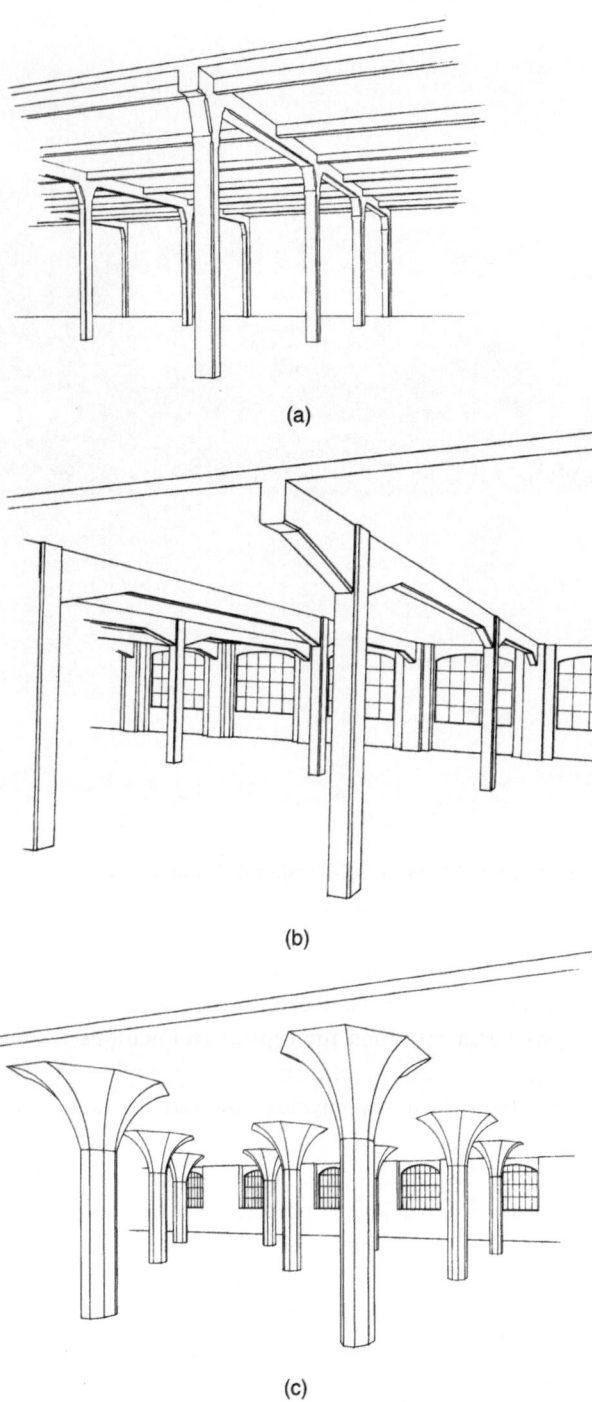

(a)

(b)

(c)

Figure 37. (a) Hennebique concrete frame: Perspective of columns, beams, and slab; (b) interior of the Pfenninger factory; (c) Zurich warehouse, 1910: perspective of Maillart's mushroom-column system. (*Source:* Drawing by Clark Fernon)

the architect, it appears below the floor slab of the stage as an arch that is reminiscent of the 1905 Tavanasa Bridge. The flat floor is fused with the curved arch so that it is thinner at the crown, thickest at the quarter-spans, and diminishes at the openings (Fig. 39). Even if only evident in the drawing, this structure represents a new integration of usually separate parts. It shows Maillart's relentless urge to lighten and to integrate structure in the interests of greater economy and better performance. If Maillart was not asked to build a bridge, he could at least slip one inside St. Gallen's largest monument to art in the early twentieth century.[86] Such were Maillart's little games; the following year, he began a more serious project, one of far reaching influence and of great business promise.

THE SAFETY OF REINFORCED CONCRETE

Five years after his decisive Basel lecture on the various uses of concrete, Maillart formulated his ideas on the nature of concrete in preparation for another lecture, "The Safety of Reinforced Concrete."[87] In this lecture, Maillart emphasized the originality of reinforced concrete in spite of its being composed of two much older materials, metal and concrete. He saw this new material arising from a synthesis of two traditional components almost the way oxygen and hydrogen together create something with properties radically different from either material taken separately. In the same way, he saw the synthesis of well-known structural elements – slabs, beams, and columns or slabs, walls, and arches providing a new efficiency of materials and a new aesthetic. Maillart also cautioned against having a national standard code that would be so detailed as to constrict the "experienced and conscientious" designer.

As a professional engineer, Maillart knew that it was essential to ensure the public safety of these new forms by some clearly articulated proof: he needed a new theory on which to base such designs. He could see that neither the theories developed for unreinforced concrete (essentially those used for stone) nor

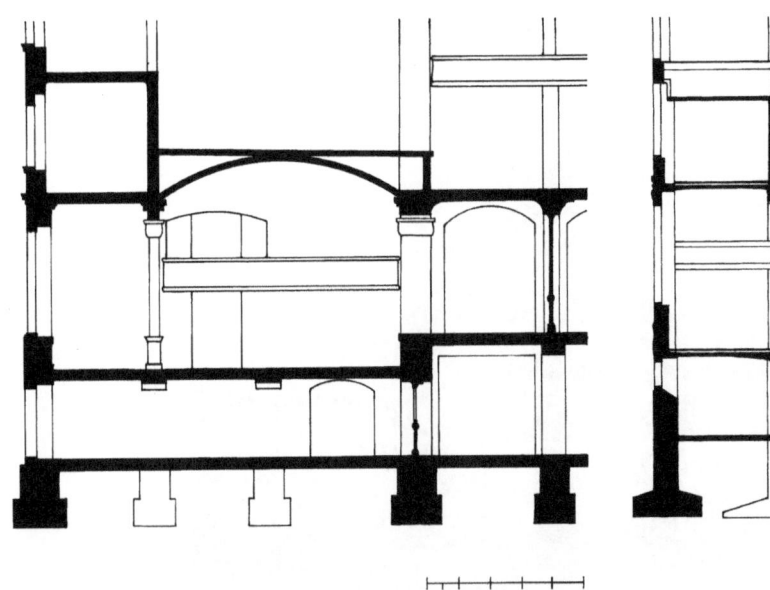

Figure 38. St. Gallen Concert Hall, original design, 1906. (*Source: Schweizerische Bauzeitung* SBZ 58 (17), October 21, 1911:229)

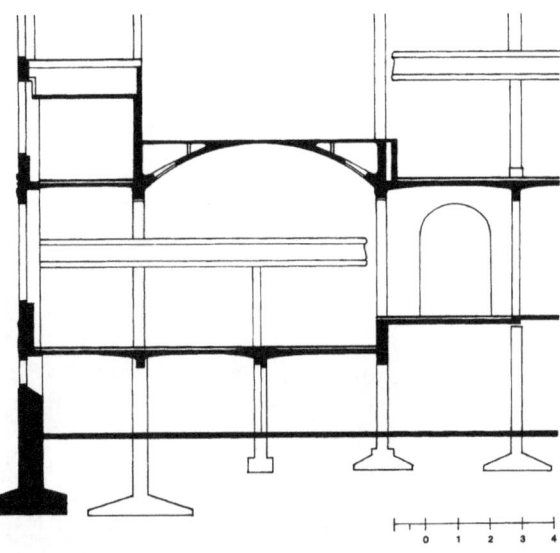

Figure 39. St. Gallen Concert Hall, Maillart design, 1907. (*Source: Schweizerische Bauzeitung* SBZ 58 (17), October 21, 1911:229)

theories devised for steel structures would suffice. But what would replace them? In his Zurich talk, he began to explore what that answer might be. He proposed a theory based on test results from large-scale models and full-scale structures rather than on detailed mathematical analysis based on unrealistic assumptions about reinforced-concrete behavior under load. As he frequently did, Maillart contrasted his view to that of the Germans, whose "requirements for calculation[s] . . . frequently lead to completely false results, as shown by . . . measurements from completed loading tests."

Maillart observed that concrete structures "cannot be calculated exactly and it is therefore better to pay attention to visible demonstrations." The water tank problem was one striking example of the contrast between Maillart's graphic and empirical approach to design and the algebraic and theoretical approach that was gaining favor by 1909. At this time, 6 years after his completion of the gas tanks, a competing method of thin-shell analysis appeared in Germany based on a general mathematical theory. Being complicated, it required substantial simplification to obtain practical solutions.

These solutions restricted the designer far more than Maillart's 1902 method, but twentieth-century engineers preferred them because of their supposed generality and because the solutions could be expressed in tables.[88]

Comparing the German method to Maillart's was not a question of rigor versus intuition, or precise and scientific predictions versus uncertain estimates. The primary uncertainties in reinforced-concrete behavior have never been removed by mathematical analysis however rigorously consistent. Rather, it was a conflict of emphasis. Would the designer's time and thought have to go into ever more complex attempts to eliminate analytic uncertainties, or could the focus be on physical behavior and on the choice of form? As Maillart increasingly adopted the latter course of action, the power of his designs grew. It was a choice that would engage him throughout his career.

MAILLART TRANSFORMS THE FLOOR

Three years after Tavanasa, Maillart proposed another major innovation. When in 1908 he began to

Figure 40. Flat-slab tests, 1908. (*Source:* Madame M.-C. Blumer-Maillart)

think about floor structures, the standard solution was Hennebique's monolithic design, using short spanning slabs supported by joists (small beams) that framed into girders (larger beams). These girders were monolithically connected to columns that carried all the load to foundations.

Already at the Pfenninger factory in Wädenswil in 1905, Maillart had lengthened the slab spans, eliminated the joists, and made the girders of larger span than in the usual Hennebique structures [Figs. 37(a) and 37(b)]. For 2 years, he had continued to design and build concrete floors in this manner, especially in St. Gallen, where the beams were mostly eliminated. Again load tests stimulated his imagination and he began to plan a series of research tests in his construction yard to develop a reliable design for a completely beamless floor. In the spring of 1908, he built several large models of such a floor in his construction yard in Zurich. During the summer, he tested these models (Fig. 40), satisfying himself that the idea was sound though even he had no mathematical formula for it.[89]

As the year ended, Maillart put together his thoughts on flat slabs, taking out a patent on January 20, 1909. The idea was to support floor slabs solely on columns, simplifying construction, and removing obstructions below ceilings. It promised to be his most successful business invention yet and one that shows clearly how he combined a technical and an aesthetic design to create a new and economic structural form. Maillart emphasized the great load-carrying capacity of monolithic concrete, the low cost, and the handsome appearance of the flat slabs with smoothly curved capitals over the columns.[90]

Maillart's patent of 1909 shows an uninspired triangular capital connecting the columns to the slab, a solution that was dramatically improved once he applied it to a real structure. His 1910 Zurich Warehouse is structurally a landmark, signaling a departure from all previous concrete floor designs. The beams are gone and the smooth, flat, concrete floors are supported solely by columns whose tops flare out in capitals that merge continuously into the slab above [Fig. 37(c)]. Maillart achieved this continuous curvature with a concrete

Figure 41. Zurich warehouse on Giesshübel Street, 1910. (*Source:* FBM Studio Ltd.)

form that uses only straight formwork. He was the building contractor as well as the designer, and he therefore had to be economically competitive. But he wished to make an elegant form that would express clearly the flow of forces between horizontal floor structure and vertical column (Fig. 41). It was again Maillart's design view that led him to this new form in spite of there being no mathematical theory to guide him. Whereas the applied scientists struggled for formulas, Maillart built numerous buildings with no difficulties and substantial profits.

Although Maillart appears to have developed his flat-slab invention independently, the idea had already been used in a different shape by 1905 in the United States. C. A. P. Turner, an inventive and

well-known American engineer, had built such structures, but with imitative Doric-type capitals and with a specification for a complicated arrangement of reinforcing bars in the slab.[91] In 1926, after his patent expired, Maillart wrote about his invention, comparing it favorably to the more awkward American system.[92] Maillart saw immediately that the bar arrangement could be simple and that the capitals should have a new shape. In 1910, this was a pioneering idea in Europe, and Maillart's invention quickly established a profitable business for him in the building of large warehouses, factories, and public structures.

In the summer of 1911, Maillart built his second flat-slab building in Zurich. And the following year, he proposed flat slabs supported on columns with mushroom capitals at top and bottom for a roof extending a water-treatment plant for St. Gallen (Fig. 42). Radically different from the city engineer's reinforced-concrete roof made of curved barrel vaults,[93] Maillart's design won him the building contract (Fig. 43). Unlike bridges, these buildings were easy to design once the basic idea had been tested; moreover, they were built mostly for private clients who did not require formal calculations or detailed engineering reviews. It was, in fact, the economy and attractiveness of his flat-slab structures that were to be the main reason for Russian interest in Maillart in 1912. At the same time, his success with factory structures won him a commission from the Pirelli Company in Milan to build a cable factory near Barcelona. With this contract as a beginning, he would feel justified in opening an office to seek work directly in that region, which, like Russia, was just beginning to industrialize. Maillart's simple reinforcing pattern has since become the standard for flat slabs in the United States and elsewhere.

THE HIDDEN WALL

Maillart's flat slabs were innovative, and widely applicable, hence profitable; but he could not resist invention even for a single structure. An example of 1909 for a simple wall design illustrates this tendency. In August, Maillart had secured the contract for a St. Gallen office building to be built for Otto Adler and Company.[94] He had already scaled down his office there in August 1908 because of very little building.[95] The Adler work consisted first of the office building and second of a 10-meter-high retaining wall to be built along one side of the building. The architect office of W. Heene had designed a solid retaining wall on a heavy foundation that connected directly to the foundation of the building wall. The solution is reasonable, conservative, and followed the state of the art in 1909.[96] The designer sloped the wall into the retained earth and used the concrete dead weight to resist the tendency for the mass of earth to overturn the wall into the building [Fig. 44(a)].

Once Maillart secured the contract, he reworked the retaining wall design completely. His guiding principle was, as usual, a reduction of materials by the integration of form. Just as in the gas tanks where he had replaced the heavy wall by a thin shell and in the concert hall where he had fused the stage deck with the barrel arch, in the Adler building, he built the retaining wall directly into the heavy outside wall of the office structure [Fig. 44(b)]. Maillart's retaining wall used only 23 percent of the concrete required by the Heene wall.[97] On the other hand, the Heene wall was easy to calculate, whereas Maillart's novel form required considerable new thought. The average engineer would have been satisfied with Heene's solution, but Maillart could not abide such waste even though he now had to make a design for which little precedent existed. In short, Maillart made more work for himself, but in doing so he created a new, hidden form that satisfied his urge to explore the limits of his profession by again synthesizing two structures into one.[98] Such synthesis leads to lighter structures, but because such structures can take more time to design and build, synthesis is not always more economical. In fact, it was not a commercial solution because it did not hold out the promise of wide application. In spite of its intrinsic importance, Maillart himself never publicized it and it was never published.[99]

ZURICH UNIVERSITY

In 1909, Maillart was also working on the main buildings for Zurich University that resulted from a 1908 architectural design competition. The winning architects made detailed drawings that then served as the basis for competitive bids by building contractors. The competition involved only architects, rather than builders or engineers. This essential difference characterized attitudes toward engineering and architecture in Switzerland and elsewhere in the early twentieth century. For monumental architecture, architects competed with-

Figure 42. Filter building under construction, Rorschach, 1912. (*Source:* Madame M.-C. Blumer-Maillart)

Figure 43. Filter building during maintenance, Rorschach, 1912. (*Source:* FBM Studio Ltd.)

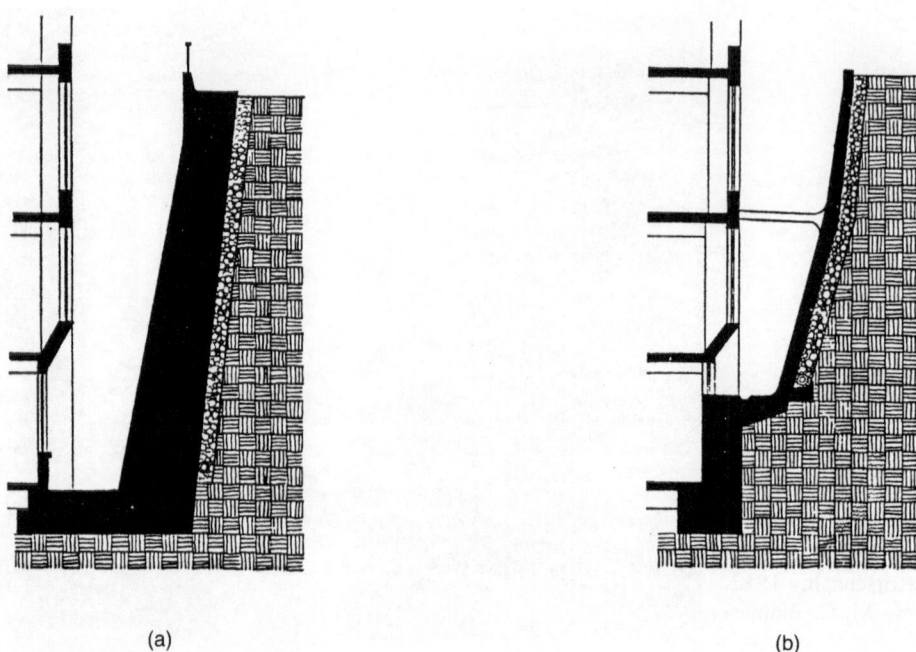

(a)   (b)

Figure 44. Adler office building, St. Gallen: (a) retaining wall design by Heene, 1909; (b) retaining wall design by Maillart. (*Source:* Drawings by Mark Reed)

out any binding cost computations, whereas for bridges, Maillart and his competitors frequently had to give binding bids with their designs. Thus, engineering played essentially no role in the design of prestigious architecture that in turn was .not judged with cost as an explicit constraint. Bridge engineering, on the other hand, had to bow to the fashions of architects, and builders were often forced to guarantee economy.

For the structure, Maillart had put in a bid of 367,000 francs. Because the twelve bids for the structure represented different engineering solutions, the canton turned once more to Schüle for an opinion.[100]

Again, Maillart was not the least expensive; this time the lowest bid came from a combination of the Favre and Jaeger companies. As at Basel, a careful review led to the recommendation that Maillart be given the contract: "Because of the quality of the firm of Maillart and Company good design and construction can be expected." The consultant had found defects in the designs of the three bidders that

were lower than Maillart's. But because Maillart was about 11 percent higher than the Jaeger group, the architects rejected Schüle's judgment.[101] In early December, Maillart restudied his bid for a second competition restricted to ten contractors. The architect's estimate, which had added a tower to the contract, was 420,000 francs, and Maillart bid 334,800 francs, with Jaeger-Favre still lower at 289,000 francs. The canton now made its own evaluation, dropping Jaeger because of inferior technical quality, splitting the contract in two, and giving the larger part to Maillart.[102]

Maillart's work at the university went so well throughout 1911 and 1912 that the canton asked only him to bid on the construction of the tower; the canton accepted his bid and he proceeded to build it in late 1912. Here is another example of Maillart's high-quality concrete work securing a contract. It was this experience in careful construction that served as a base for his spectacular later designs.[103]

The idea of separating design from construction

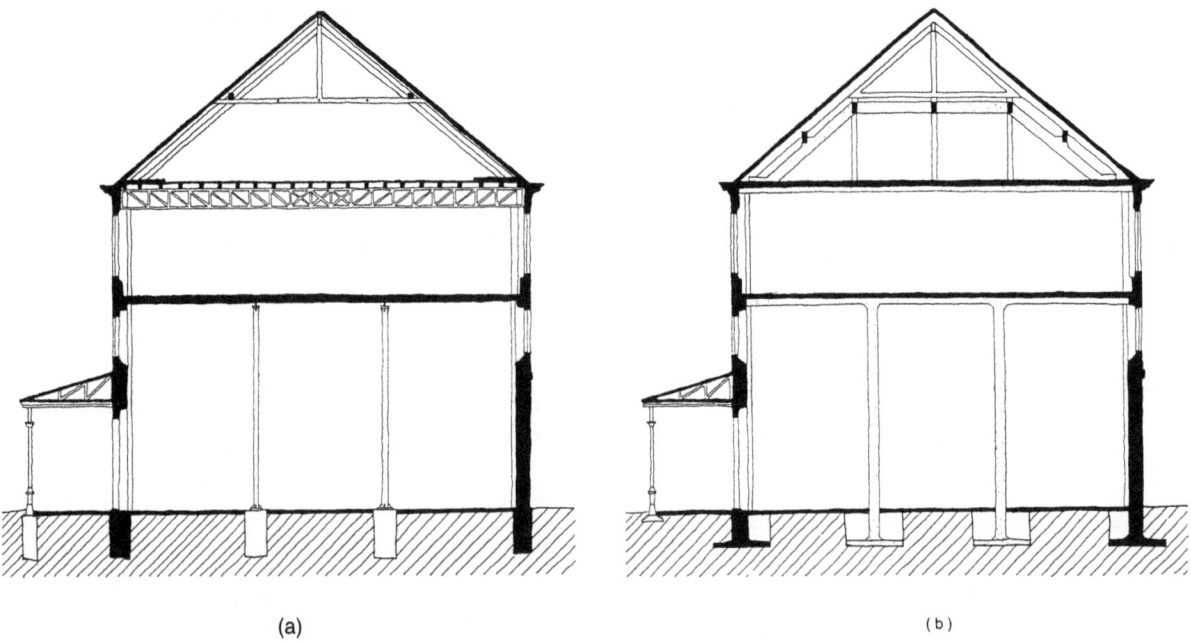

(a)   (b)

Figure 45. Moveable stageset warehouse, Zurich, 1909: (a) first design with floor truss; (b) Maillart design with floor supported by roof truss (*Source:* Drawings by Mark Reed)

was gaining acceptance as characterized by Jaeger's continued competition, always allied with a builder and, on bridges, with a local architect. Maillart worked in the opposite direction; he would have preferred to do the engineering, the construction, and for many types of buildings, the exterior design as well. Although he never proposed doing the architecture for anything so complex as the university buildings, Maillart did express the engineering ideal of doing all the work for buildings in which the structure and form could be unified.

He emphasized this view in writing late in 1910. With all the projects he had going on, he took the time to write an article for the *Bauzeitung* on the moveable stageset warehouse he was building in Zurich. He had made the structure of thin, straight beams and columns that expressed the efficiency of materials, and he was unhappy that the resulting lightness was to be hidden by heavy masonry walls:

"It is a shame that the whole building could not have been built out of reinforced concrete with an articulation of the lighter wall in order to improve the whole image, to lighten the weight, and to gain more [interior] space."[104] In short, he would have liked the same free hand with this warehouse as he had had in his 1905 Wädenswil factory (where he served as architect, engineer, and builder). With such freedom, he was convinced that he would have been able to improve the appearance, reduce materials and costs, and to increase the useable space.

For this warehouse, he did devise a new solution to the problem of a relatively long-span attic support to replace an earlier proposal using a truss [Fig. 45(a)]. Maillart instead made the roof structure into a truss [Fig. 45(b)]; it was an idea that he would pursue in the 1920s with visually striking results (Chapter 5) and in one case with a highly controversial structure (Chapter 8).

CHAPTER THREE

Builder and Millionaire
1909–1914

The Big Bridge Competitions
1909–1912

SURPRISE AT RHEINFELDEN

Maillart was at last able to return to bridge building in 1910. Beginning with Rheinfelden that year, and until 1912, he enjoyed a period of intense participation in a series of bridge competitions extraordinary for Switzerland at that time. Maillart entered five major bridge competitions held in Switzerland between 1908 and 1912; he did not win first prize in any, but he did succeed in getting the contract for three of them. Moreover, during this same period, there were two other nonjuried bridge competitions recorded in the *Bauzeitung*, both of which Maillart won. Finally, Maillart secured contracts for the dam bridge and a cable bridge connected with the project that was discussed most extensively during this period in the *Bauzeitung*: the Rhine River power plant at Augst-Wyhlen. Thus, by 1913, Maillart was Switzerland's leading bridge builder.

By 1909, juried design competitions had become something of a tradition in Switzerland, and they had been well publicized by the *Bauzeitung*. These competitions appealed to the decentralized Swiss because they allowed local officals to take part in the judging and they also permitted the local population to see the results and express its opinions either through newspapers or in town meetings. Such juried competitions for engineering works

were rare in other countries and practically non-existent in the United States. One of these juried bridge competitions, the international competition for a bridge to cross the Rhine at Rheinfelden, illustrates how they proceeded.

Late in 1907, the citizens of Rheinfelden voted to hold the competition according to strict guidelines[1] and on January 9, 1909, the *Bauzeitung* officially announced the competition, whose deadline for design submissions was set for April 30.[2] Maillart knew that a conventional approach to aesthetics would be crucial because of the attitudes of the two architects on the jury, especially Gustave Gull (who had covered the Stauffacher Bridge with a stone façade).[3] Already at work on the Wattwil Bridge with the architectural firm of Joss and Klauser, Maillart decided to enter the competition with them. He also recognized that, being a competition judged in part by German officials (the bridge would connect the German town of Rheinfelden to the Swiss town of the same name), detailed calculations would be essential, so in March, he set about developing complete arch computations based on a German rather than a Swiss source. Maillart's elegantly detailed computations prove that his growing reluctance to rely on such calculations were not due to any inability on his part to make them.

On June 5, the *Bauzeitung* published the results of the competition. The jury gave the first prize of 2,300 francs to Professor Josef Melan for a concrete arch design, a decision made largely on the basis of the design's architectural merits and in spite of its high cost (558,000 francs, or about $5,600,000 in

Figure 46. Rhine River Bridge at Rheinfelden, 1912. (*Source:* Madame M.-C. Blumer-Maillart)

1996 costs). The second prize of 2,000 francs went to Maillart (the cost of his design was 436,000 francs). The jury commended Maillart on his practical construction plan, and his detailed and well-conceived calculations, as well as for his "simple [*einfach*] overall form." Maillart designed the arch in precast concrete blocks, a system he was to patent in 1911; the jury criticized his use of these blocks for those submerged parts of the arch that could be exposed during low water. The jury wanted such parts to be clad in granite.[4] The drawings were displayed from June 1 to 14 in the town's sports hall. A decision as to who would receive the contract was postponed until the fall.

The city officials were in a dilemma. On the one hand, the first prize was costly and Maillart's design seemed clearly to be less so. On the other hand, the third prize had gone to Buss and Company, a firm with whom the community had previously worked and to whom the city officials were therefore favorably disposed. The officials decided to negotiate with Maillart in the fall, and in October, he gave them a lower price, 368,000 francs, for an alternate design using cast-in-place concrete instead of precast blocks. Negotiations continued until January 1910, when the Council revealed that it was also negotiating with Buss.[5] When asked to again lower his bid, Maillart raised it instead, defending that surprising move in a letter to the Council: "we consider that we have made *a very attractive offer which no one who is a serious competitor can easily underbid without assuming a great risk.*"[6] By this time, Maillart was convinced that he would not get the contract.

On Monday, April 25, he was astounded to read in the evening paper that, at its Sunday meeting, the community of Rheinfelden had voted to give him the bridge.[7] In fact, he had won the job thanks to the city's internal politics, which give an insight into Swiss society that is central to an understanding of its public works.[8] These politics began with the decision to hold a restricted, second competition in the fall of 1909 that resulted in a bid of 412,160 from the third-prize winner Buss and 367,969 from Maillart. The winter negotiations resulted in Buss reducing his bid to 391,780 and Maillart increasing his to 390,969. After failing to get Maillart to go back to his earlier bid, the Community Council decided to give the contract to Buss to build Maillart's design.

In an almost perfect example of Swiss democracy in action, on April 24, 1910, the entire community assembled to vote on the Council's proposal. Many did not like the idea of taking one man's design for which he had been awarded only 2,000 francs and giving it to another man to build, providing the latter with a contract worth nearly twenty times as much. Rather, "the citizens of the community, following the proposal of Herr Doser, vetoed by a majority the Council's plan to give the intellectual property (*das geistige Eigentum*) of the firm of Maillart and Company to another firm to build; the citizens decided to give the work to Maillart and Company"[9] (Fig. 46). In a fortuitous example of "justice being done," the report's central phrase, "*das geistige Eigentum*," is the very expression Maillart himself had used 2 years before in an article he wrote about a similar experience that had resulted in his loss of a large building contract for a bridge near Guggersbach.

On March 3, 1908, Maillart had been shocked to read in the *Bauzeitung* that the design of a new bridge near Bern was credited to Jaeger and Company.[10] He recognized the design as the one he had proposed 4 years before for the Sense River Bridge near Guggersbach.[11] He immediately wrote a letter to the editors (published on March 21) describing his design ideas, and the background for the canton's commission. Maillart then stated that "the constructed design of the firm Jaeger and Co. is in every part *an exact copy of our structural design*." He concluded with the observation that "Naturally the canton of Bern is allowed to let another company build a structure which they asked us to design and for which they have paid us. Our claim is only that the true origin of the design not be forgotten!"[12]

Jaeger's only defense was that he had received no information on Maillart's design from the district engineer who had represented the canton during construction. As the *Bauzeitung* noted, the

response was beside the point because the firm could easily have obtained the design through other channels.[13] Maillart took a risk in attacking Jaeger because in doing so he was implicitly criticizing the Bern canton building department. But that department made no repudiation and the incident contributed to a lessening of Jaeger's reputation.[14]

LAUFENBURG

In the same issue, the following Saturday (April 30, 1910), in which the *Bauzeitung* noted the Rheinfelden award to Maillart, it announced the details of a new competition for a bridge at Laufenburg.[15] The Rhine River narrows at Laufenburg and turns into wild rapids. The force of such fast water suggested hydroelectric power already in the nineteenth century. But the use of the Rhine at this location had consistently been a political issue both with neighboring countries and between adjacent communities in Switzerland. In 1889, several Swiss and German firms had formed a consortium that finally began construction of a dam in 1908.[16]

Already, in 1906, as part of the overall plan, this company had begun to discuss a new bridge. An old wooden bridge stood where the dam was to be built, so the power company was forced to build the new one in town. A German cable company had proposed a suspension bridge that the directors found aesthetically unacceptable, so they decided to hold an international competition.[17] As with Rheinfelden, the owner, in this case the power company, laid out the rules and designated the jury; it included Robert Moser as engineer, again Gustave Gull as architect, and Otto Zehnder, canton engineer of the Aargau.[18]

Maillart was under great pressure all through the month of May 1910. Not only did he now have the Rheinfelden Bridge, but he also was preparing competition drawings (deadline May 31) for a new bridge at Aarburg over the Aare River: a 60-meter-span concrete arch[19]; he was working on the Laufenburg design (due June 30), and even more important from a financial point of view, on competition entries for the new buildings at Zurich Uni-

Figure 47. Rhine River Bridge at Laufenburg, 1911. (*Source:* Madame M.-C. Blumer-Maillart)

versity (due June 6). At the same time, he was committed to his first flat-slab design and numerous lesser projects throughout Switzerland. He succeeded in submitting all these designs on time. For the results of both the Aarburg and the Zurich competitions, he would have to wait until late in the year, and in both cases, the final decision would hinge on the opinion of the expert who had been sympathetic to his gas tank design, François Schüle.

Late in July, Maillart learned that he had won second prize out of eighty-seven competitors for the Laufenburg Bridge. He did not then know who was first, nor did he yet know the jury's comments.[20] On August 6, he went to Laufenburg to see the exhibition of winning designs and to learn if he had any chance of securing the contract.[21] The first prize was given to a stone bridge designed by a joint venture of German architects and engineers who won

2,000 francs. The jury had awarded Maillart 1,500 francs for second prize; and 750 francs to each of two third-place designs.[22] Maillart had submitted two separate designs, one with two slender reinforced-concrete arches, the other with two relatively heavy concrete-block arches.[23] The jury only awarded prizes to bridges resembling stone, so they immediately eliminated Maillart's lighter design. In reading the jury's report, Maillart saw that for Laufenburg, it was more systematic than the report for Rheinfelden. Its clear distinction between heavy and light designs must, at least in part, have been due to the masonry-oriented Robert Moser.

The jury criticized Maillart's heavy arches for having too sharp a juncture with the abutments, but it applauded his idea of planting two trees on the central support. It found the calculations, the report and, the cost study complete and carefully done, coming from, as the jury stated, "the hand of an experienced engineer who is an expert at concrete construction. The design is so thoroughly worked out that, without any great modifications, it can immediately be detailed and built."

Maillart's report for the Laufenburg competition clearly describes his ideas and includes several general statements that appeared later in the jury's criticisms of other designers. For example, Maillart stated that a single-arch span, while technically possible, was aesthetically undesirable because part of it would be covered by high water once the power-plant dam was built. He also ruled out putting a structure above the roadway on aesthetic grounds and the jury later did the same. Again, Maillart's concept of design, even where the form is not one of his inventions, is a full integration of efficiency, cost, and appearance. He closed his report on the concrete-block entry as follows: "the form of the (central) pier is simple and strong and permits a resting place with several growing trees" – a nice touch in an otherwise straightforward structural design.[24]

This conservative jury found the first-place stone design to be aesthetically superior, but it asked rhetorically, "could the bridge in this form really be built for the cost given by the designer?" Because the power company was very much concerned with cost, Maillart's detailed plans appealed as a way of ensuring careful control of the price. He realized that he had a good chance of getting the contract because the first-prize winners were designers, not builders.[25] Once again, the client, recognizing that Maillart's second-place design was more practical than the one in first place, gave him the contract to build his block design.

Maillart's Laufenburg Bridge was not only priced below the owner's budget, but it was based on a fully detailed set of calculations.[26] Such detailed documents belie later accusations of his design by "intuition." A careful study of the hundreds of sheets of elegantly detailed and scientifically accurate calculations he made between 1902 and 1913 shows the firm numerical basis on which his design ideas rested. It cannot be stressed too much

that Maillart personally made all of his own calculations until mid-1904, when he hired Schneebeli, and even then he continued to do many of them until the office work became too extensive.

The completed work at Laufenburg (Fig. 47) does not give the observer any sense of the construction process, for which Maillart used an elegant temporary structure, consisting of three-hinged wooden arches (Fig. 48) on top of which he built the arch out of concrete blocks (Fig. 49). Visually, in this case, the construction was more intriguing to him than its product.[27]

Maillart had two other bridge commissions related to power plants at this time. One was a two-span structure over a canal at Wyhlen, which was perhaps the longest-spanning, hollow-box girder bridge built up to 1910.[28] The other was a dam bridge over the Rhine at Augst-Wyhlen. Construction of hydroelectric power plants was accelerating early in this century, especially in the densely populated and highly industrialized Basel area, which extended as far away as Laufenburg. Emblematic of Maillart's dominant position as Switzerland's leading bridge builder was his success in getting all four of the major bridges for these plants.

AARBURG

From mid-August to mid-October 1910, Maillart awaited word of his various projects, including the nonjuried Aarburg Bridge competition, which he had won.[29] The history of this last work provides further insight into the extent of Maillart's reputation in Switzerland after only 8 years of building activity.

The new bridge was to stand where the Aare River makes a sharp bend beneath an eleventh-century castle and where, since 1839, a suspension bridge had connected the little Aargau town of Aarburg (burg = castle) to the canton of Solothurn. As early as 1886, there was sufficient concern about this light bridge to justify the hiring of Wilhelm Ritter to study its safety.[30] There followed numerous proposals and a poorly defined 1908 competition, the results of which canton engineer, Otto Zehnder,

Figure 48. Scaffold for the Laufenburg Bridge, 1911: three-hinged wooden arches. (*Source:* Madame M.-C. Blumer-Maillart)

Figure 49. Concrete blocks for the arches at Laufenburg, 1911. (*Source:* Madame M.-C. Blumer-Maillart)

Figure 50. Aare River Bridge at Aarburg, 1911: solid reinforced-concrete arch. (*Source:* Madame M.-C. Blumer-Maillart)

found unsatisfactory. The canton, therefore, decided to hold a second competition restricted to those four firms that had previously submitted the most likely designs. The new program clearly specified that the structure should be an arch span of 68 meters (across the entire river width) with a rise at the crown of 6.8 meters, made of stone, concrete, or reinforced concrete.

The four firms submitted their drawings and bids in late May 1910, and Zehnder reviewed them in detail. His August 19 handwritten report reveals how local engineers saw Maillart at this time in relation to some of his stiffest competitors.[31] Zehnder rejected one design out of hand because of high cost and unsatisfactory appearance. Of the remaining three, all in concrete, two were of heavier and one of lighter construction. Both of the heavier designs had a number of questionable technical details even though one had the lowest cost.

"The light [Maillart] design," Zehnder observed, "is superior to all the others because of its lightness and elegance" and "technically [performance under load with a minimum of deflection and cracking] the Maillart design is the most outstanding of them

all." He reviewed all three together, noting their designers and builders and their prices:

I. Maillart and Company, Engineers in Zurich	84,000
II. Max Münch, Engineer-Architect in Bern	92,000
III. Jaeger in Zurich, Schäfer in Aargau, and Gribi in Burgdorf	76,000

Although Zehnder preferred Maillart's design, it was 8,000 francs more expensive than the Jaeger proposal, and for these small communities, such an extra cost was not to be taken lightly. Zehnder was therefore taking a certain risk in pushing his preference. He presented a careful argument about the quality of Maillart's design, noting that this was not a "World bridge" (Weltbrücke) and therefore did not require the massive monumentality that the other two designs gave it. In his view, Maillart's was more efficient and safer, and as a light structure it would become a "jewel" in the landscape. More important, however, was the quality of work to be expected, and here he was explicit, "I do not hesitate to give preference to the noted bridge building firm of Maillart and Company over the ad hoc firm,

Figure 51. Aare River Bridge at Aarsburg, 1911: scaffold. (*Source:* Madame M.-C. Blumer-Maillart)

put together for this project, of Jaeger, Schäfer and Gribi so that for this and the other detailed reasons design No. 3 must drop out even if it is the cheapest." Just as with the Basel siphon, Maillart's reputation overcame his higher cost. Maillart eventually got the contract in the fall of 1911 (Fig. 50).

It is somewhat ironic that this bridge, picked for its high-quality performance, would exhibit cracks in early life and stimulate Maillart to see a new form. Just as with Zuoz, he turned a performance defect into a major innovation. The Aarburg Bridge was rehabilitated after over 50 years of service, so that the defects did not prevent it from a useful life.

THE PERILS OF BRIDGE CONSTRUCTION

With construction of Aarburg held in abeyance pending the appropriation of funds,[32] in the fall of 1910, Maillart turned to the more immediate problems of his three major Rhine crossings: starting the

final design at Laufenburg, which he was awarded officially in mid-October; designing the temporary wooden footbridge for Rheinfelden, a major structure in its own right; and closely supervising a dam bridge at Augst-Wyhlen. This bridge was the furthest along, and on Wednesday, October 12, he carefully inspected the scaffolding over the fast-running Rhine.

The next day in Zurich, Maillart received word that part of that scaffold had collapsed and drowned one worker, the first fatal accident for his company. Upon inspection the next day, Maillart found that his superintendent had neglected to check the wooden scaffold construction, which had dropped five men into the Rhine, one of whom, an old mason, was lost.[33] Deeply upset by the accident, he lamented the fact that his activities made site visits increasingly difficult. With fewer visits, he had for some time felt that his personal control was dangerously diminished.

A similar event occurred in late June of the following year as Maillart closely followed construction of the Laufenburg scaffold, the series of light, three-hinged, lens-shaped wooden arches (Fig. 48). These forms avoided the need to put any scaffold into the river; Maillart took great pride in their appearance even though, as with the Rheinfelden footbridge, the structure was temporary. Unfortunately, workmen erroneously removed an essential bracing member, and the entire scaffold in one span gave way: four men fell, one of whom drowned. Coming one after the other, the two accidents were a cruel affirmation of Maillart's fears.[34] These were the only fatal accidents Maillart experienced, but during his early years of intense activity, there were other reminders of his need to supervise personally each site, including a costly settling of the supports for the scaffold (Fig. 51) on the newly cast Aarburg Bridge in April 1912.[35]

By 1912, Maillart had many more projects than he could follow carefully; he also had been unable, since 1905 at Tavanasa, to design a bridge according to his own ideas of visual form. In spite of such setbacks, he was making a good profit, primarily from his building contracts. But the internal conflict remained between the new designs he wanted to build and the conventional projects for which he received contracts, between a vision of bridge forms and an organization for big business. By 1912, this conflict had intensified and would soon come to a head.

THE STONE HAND OF MOSER

Maillart's successful activity between 1908 and 1912 belies the fact that he was not actually winning the bridge competitions that engaged so much of his energy. To understand the reasons for this failure, we must go back to the competition for one of the largest bridges in Switzerland, a viaduct over the Sarine River at Pérolles, outside Fribourg.[36] Determined to show that reinforced concrete could span greater distances than stone arches, Maillart submitted an arch design of unprecedented length.[37]

Even before the jury report for the Pérolles Competition in August 1908, Maillart could see by the exhibited drawings and the prizes awarded that the

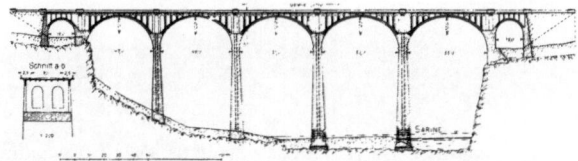

Figure 52. The Winning Pérolles Bridge Competition design by Jaeger et al.: longitudinal section, 1908. (*Source: Schweizerische Bauzeitung*)

hand of Moser had prevailed. The 70-year-old engineer believed deeply in stone and stonelike form, and now at the height of his reputation, he would be a leading jury member for a series of bridge design competitions, all of which Maillart would lose. Pérolles was the first of this series. The fact that Maillart's submission received any recognition at all (he won third prize) was probably thanks to Schüle, also a member of the same jury.[38] The first place for Pérolles went to Jaeger and Company, which as usual was in a joint venture with other builders and with architects; their design was a five-arch masonry viaduct, each arch about 50 meters in span and made of concrete blocks, giving the appearance of a stone bridge – Moser's idea of a proper concrete design (Fig. 52).[39]

Maillart, by contrast, had taken a completely new direction. His viaduct had a main span of 140 meters carried by a hollow-box arch made up of precast, reinforced-concrete elements. This immense structure, twice the span of any existing reinforced-concrete arch, represented a pioneering concept in prefabrication as well as in scale (Fig. 53). Unlike the small concrete blocks used at Steinach and Laufenburg, at Pérolles, Maillart designed huge precast pieces, each one of which included the entire hollow-box section. Precast hollow-box sections, now called segmental construction, would come into bridge practice only after 1945. Maillart's design was 35 years ahead of the profession.

His basic idea was that the design had to produce both economy of construction (simple straight pieces) and efficiency of materials (elimination of all unneeded parts). The jury's comments were at the same time aesthetically traditional and technically perceptive. They praised the well-thought-out cal-

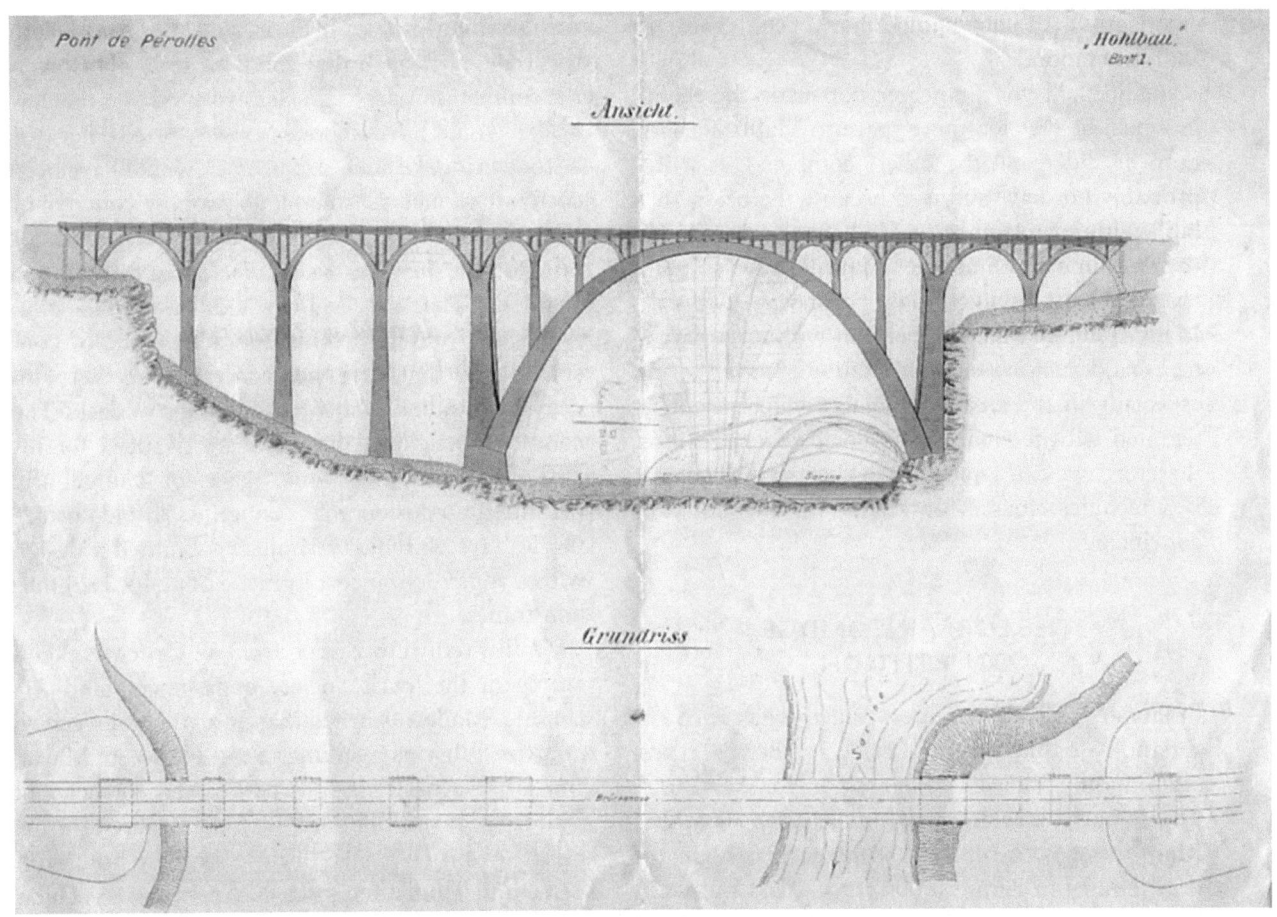

Figure 53. The Pérolles Bridge Competition design by Maillart: elevation, 1908. (*Source:* Madame M.-C. Blumer-Maillart)

culations and construction plans and commented favorably on the safety and low stresses in the design and on the detailed cost analysis. They concluded that "apart from the architecture which had been neglected, the design is interesting and shows a profound knowledge of reinforced concrete."[40] Because this was a concept competition held for the purpose of eliciting as many different designs as possible, no bridge contract resulted and the actual structure was not built until 1921, when Jaeger's design was constructed.

As with the 1904 Uto Bridge, the Austrian concrete pioneer, Fritz von Emperger, immediately recognized the significance of Maillart's Pérolles design: its originality and technical correctness. In October, von Emperger published a detailed analysis of the

competition with drawings and diagrams of the six winners.[41] He realized that the central problem facing the jury in such an "idea-competition" was the undefined relationship between architecture (implicitly hiding structure) and engineering that is apparent, with no decorative "architecture" to hide it. Just as with the Uto results, here, too, as von Emperger commented, "unfortunately architecture was put in the forefront." For von Emperger, this mistake discredited the classifications made by the jury. Each design given a prize for its appearance rather than its engineering, "we shall lay aside . . . since it has no lasting value – it is a pretty picture and nothing more. This is especially so for the first prize design (Jaeger's), where [additionally] the most important factor in the price, the cost of scaffold, was arrived at erroneously . . . [and should have been] three or four times more."

Again, had von Emperger not made the report, in which he devoted more space to Maillart's entry than he did to all the others combined, it would probably not have survived because the bridge was built without reference to Maillart's design. It was the first time von Emperger's journal gave so much space to an unbuilt bridge proposal. Although Maillart entered the competition without an architect, his design is visually imitative of stone forms, especially in the arcaded viaduct, which is not integrated with the main arch. For different reasons, therefore, we can agree with the jury that the design is a technical tour de force but visually not fully convincing.

THE LORRAINE BRIDGE COMPETITION

In January 1911, the greatest bridge project in Switzerland – for the Lorraine Bridge at Bern – represents another "failed" competition for Maillart.[42] The competition rules were similar to those for the Rhine bridges, except for the greater amount of prize money in Bern, 10,000 francs, and the much longer guarantee, of 1 year, for binding the bid prices.

For Maillart, this competition had particular significance, because it would cross the Aare River and connect the center of Bern to the street on which he had lived as a boy. Technically a far more difficult problem than any of the Rhine bridges, it required a much longer main span and a higher rise above the river. It was also more challenging aesthetically because of the deep valley and the difficulty of setting the new bridge alongside the existing old metal-truss railroad bridge.

As in all the jury competitions after Pérolles, Maillart felt compelled to enter with an architect and the natural choice was the Bernese firm of Joss and Klauser with whom he had already collaborated at Wattwil, Rheinfelden, and Laufenburg. Maillart realized that the bridge would have to "fit"

with existing bridges in Bern, and specifically with the 1844 Nydegg Bridge made of stone. However, the competition program clearly stated that the new bridge would have to be reasonably priced. Because a stone bridge of such size in 1911 would have been costly, it seemed reasonable to propose concrete or steel, if somehow one could satisfy the aesthetic taste of the jury (as usual, including Moser and Schüle).[43] That was Maillart's dilemma. His solution was to design a hingeless arch made of concrete blocks that were then concreted together with deep longitudinal walls and a roadway deck. The concrete-block idea, first used by Maillart for his 1903 Steinach Bridge and again for Laufenburg, permitted a considerable saving in scaffold costs.[44] On Friday, March 31, Maillart submitted a design with a construction cost estimate of about 1.24 million francs.

On his return to Zurich from an Easter weekend outing of the Swiss Society of Engineers and Architects, Maillart learned that his entry had received no prize.[45] It was a galling result. He wrote Maria, who after a brief stay in the hospital for her gall bladder problems, had left with the children for the resort town of Weggis: "I have seen the bridges of Bern, and I find that ours is the best one. Those which approached most closely the narrow ideas of Mr. Moser won the prizes! It was a competition to see who could please him the most."[46]

Maillart's perception of the Bern jury was shared by others. Indeed, that spring, a strong reaction built up, begun on April 18 by the Bern newspaper, *Der Bund*, which criticized the jury for exhibiting only the plans and not the costs. Even worse, the jury's report, accepted on April 8, was not released to the *Bauzeitung* until May 19, and to the contestants even later. Thus, the basis for the jury's judgment was not made public until 3 weeks after the exhibition of the drawings had closed, a delay that opened the way for public criticism of the jury.[47]

On Wednesday, May 31, Maillart went to Bern, where he met with architect Joss and the editor of the monthly *Die Schweizerischebaukunst* (*Swiss Ar-*

chitecture, referred to as the *Baukunst*), who decided to publish Maillart's design and to include a critique of the jury's report. As Maillart wrote to Maria the next day, "Our design for the Lorraine Bridge will appear toward the middle of June. Mr. Moser will not be happy but that is all the same to me. His criticism is too stupid."[48]

As promised, on June 16, the *Baukunst* published architectural renderings of the Maillart Lorraine Bridge design. The accompanying article noted that "very seldom has a jury's judgment been criticized as it has been in this present case . . . in the daily press." After a detailed discussion of Maillart's design from a visual and functional viewpoint, the article ended by focusing on the central problem:

What did the city of Bern achieve by this competition? It has, for the price of 10,000 francs, gotten four variations of a fifteen-year-old prize [a reference to the result of a previous competition in which Moser, working with Gustav Mantel, had won first prize[49]]: first prize design "*Ehre dem Stein*" [Honor the Stone] by chief engineer Dr. Moser . . . and one design after the Melan system; thus, nothing much new for 10,000 francs . . . this unhappy result is new proof that the correct choice of jury members is an essential requirement for the success of a competition, if it is to bring new ideas into modern technique rather than variations of older systems.[50]

For the next month, criticisms escalated, culminating on July 15 with the powerful *Bauzeitung's* article. It featured Maillart's design, with a detailed rebuttal of each of the jury's objections to it.[51] The unsigned article, probably written by Carl Jegher, son of the editor, begins by explaining the significance of a competition not only for the bridge at hand, but as "an image that reflects the level of structural design for its time." It repeated the criticisms made to date, including Moser's undue influence. Unlike earlier competitions, the engineering profession was ready for new ideas and sufficiently upset by Moser's obstinance to condemn openly his preconceived ideas. Public outcry about the competition results helped to postpone a decision on the actual design until after World War I.[52]

MOSER AND MAILLART

When Moser died in 1918, the respectful *Bauzeitung* obituary, without referring specifically to the Lorraine Bridge problem, nevertheless characterized him by his influence on that competition. The obituary summarized one type of Swiss personality with which Maillart had to contend all of his professional life. After reviewing the engineer's career, the *Bauzeitung* concluded with a long quotation from a close friend, Professor Albert Heim (1849–1937), to describe "Moser's inmost personality." Heim began by calling Moser

[i]n his being ever a Swiss of the good old type. He held fast to Swiss ways . . . cunning was far from his nature. Here and there he seemed rough and somewhat peremptorily harsh in his criticisms and opinions and he was not disinclined to conflict. . . . He belonged to no party, in the best sense; he created his own party. It must be admitted that whereas from his point of view he represented so much profound truth and experience, his personality pressed him more and more into oppositions and minority positions that made him feel misunderstood and forsaken . . . work, responsibility and trustworthiness were for him a kind of religion in which he had reached an inner, highly moralistic seclusion and contentment.

Heim ended by noting that for his personal friends and in his private life, Moser was totally different, "tolerant and good hearted and undisturbed by differences of opinion. To those whom he learned to esteem he remained truly devoted."[53]

It is a sympathetic portrait, one that could recapture – even with its reference to the Lorraine controversy – the major contributions to Switzerland made by such a "Swiss of the good old type." Looking now at a 1910 photograph of the stern but slightly troubled face, white-haired and white-bearded, of Dr. Honoris Causa Robert Moser, one feels a deep ambiguity about so talented yet so overbearing a personality (Fig. 54). A capable engineer, who produced works that were both elegant and sound, he was at the same time a man of deeply biased and questionable judgment.

What he symbolized, perhaps more than anything else, was the waning of another part of the nineteenth century, by honoring stone after its use had become obsolete. Moser could not see ideas outside the carefully built and deeply revered stone walls of his long and prolific career. He represented the traditional side of Swiss engineering that in later life could not give way to a changed profession. Because the city leaders in Fribourg, Luzern, Laufenburg, and Bern all picked him after he had turned seventy to judge their future works, Moser characterized a good part of Swiss society in the early twentieth century as it viewed building.[54]

In a country that works so carefully, that one might almost say makes a kind of religion of work, responsibility, and trustworthiness, it is not surprising that a man like Moser could gain great respect and thus be able to block the new ideas of a Maillart. In a way, Moser and Maillart ended their personal lives in somewhat similar ways, misunderstood, forsaken, and yet with great self-confidence. The difference was that one had built himself into the past, the other into the future. Their differences are not to be found in abstractions such as professional honesty or aesthetic sensitivity, but rather in the subtleties of mind that seek innovation within established disciplines.

Figure 54. Robert Moser (1838–1918), photograph of 1910. (*Source: Schweizerische Bauzeitung*)

THE OUTWARD TURN

If Maillart and Moser characterized the future and the past, they also stood for a basic shift from an inward Swiss perspective to an outward, international view. Moser was among the first generation (from the founding of the ETH in 1855 to the death of Culmann in 1881) of Swiss-educated engineers for whom the building of the great railways dominated his career. His mind was formed before reinforced concrete appeared while the Swiss struggled in their Alpine environment to modernize the passes that historically created their country and its economy. It took Wilhelm Ritter, of that first generation, to force Swiss engineers of the second generation (from the 1880s to the early 1900s) to look abroad and to develop the design of concrete structures.

Unlike other small countries in the early twentieth century – the Netherlands and Belgium – Switzerland had no colonies to absorb its surplus of highly trained engineers. Indeed, by the time Maillart entered the ETH in 1890, its quality and its potential were so great that it not only educated more Swiss engineers than the country could possibly use, but a substantial number of foreigners as well. Unlike Moser and many of his generation, younger engineers felt the constriction of a small, highly competitive nation that has little room to develop. Many engineers of Maillart's generation, like Othmar Ammann (1879–1965), left Switzerland permanently to attain international fame in the United States. The Lorraine Bridge illustrated the problem; seventeen designers competed for the only juried bridge project of 1911, and in the end, none of them got the commission. But after the Lorraine competition in mid-July 1911, there was less bridge building in Switzerland; and Maillart began to search for a new challenge.

The Tensions of Success

BUSINESS AND FAMILY

As Maillart's business became profitable, it was increasingly apparent that he was quite ill-suited to the life of a *grand patron* (big boss). He had founded the business as a result of his marriage and his wish to provide well for his wife, but he did not have the talent of a businessman. Diplomacy, salesmanship, and delegation of authority were foreign to Maillart, who preferred to devote his efforts to design and construction supervision. As early as 1903, Maria had urged Maillart to cultivate his influential acquaintances for business purposes, but he had adamantly refused.[55]

Nevertheless, because of his good reputation, business was comfortably profitable by 1907, but this absorption in work kept him on the road negotiating contracts or supervising construction; it represented a tension between family and business.[56] Maria was often sick. In addition, she was bored, restless, and alienated in Zurich and even more so in the well-to-do suburb of Kilchberg (where the family had moved to in 1906). Maillart could not provide the lighthearted company she craved and got from Max von Müller and a few others. An Italian fluent also in French, she was unable to master German, let alone Swiss-German, and found herself isolated from local society. Within a few years of her marriage, travel came to represent a means of escape from what she perceived as the repressive Swiss-German atmosphere. For almost 11 of the 17 months between April 1907 and September 1908, Maillart lived alone in Kilchberg while his wife traveled, mostly in Italy.[57]

Maillart's difficulty in understanding Maria's mood swings and need to get away strained the re-

lationship and created periods of personal unhappiness between them. He thus had two constant inner tensions: first, the tension between business success and frustration of his desire to build innovative designs; second, that between his business and his family life. In a letter he wrote to Maria just before Christmas 1907, he had given her a picture of his bachelor existence, typical of the long periods when she and the children were away. He closed the letter, somewhat pathetically, by asking her to "speak to the children a little about me so that they will not forget me."[58] These separations from his family were certainly not what the young fiancé had imagined for his married life, or more importantly, for his role as a father. In this respect, reality proved to be far from his ideals.[59] Nevertheless, he was proud of his growing family and never lost the pleasure of being with them (Fig. 55).

By 1908, it was obvious that his organization

Figure 55. Robert Maillart with his son Edmond in Arosa, circa 1906. (*Source:* Madame M.-C. Blumer-Maillart)

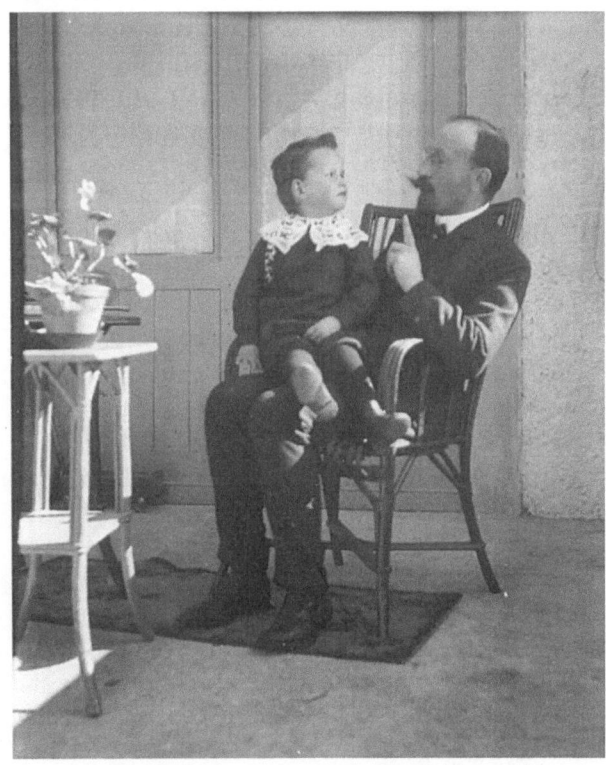

was unable to function properly without his continual presence both in the office (Fig. 56), his construction yard (Fig. 57), and – as we have seen – at the building sites. Unlike Hennebique, Maillart was unwilling to leave the direction of major works to others. Moreover, one of his key engineers announced his resignation in March to take another job. This loss combined with his loneliness and the annoyance of trying to find a place to live in Zurich. Both he and Maria had agreed on the move back to the city.[60]

As he struggled with his submission for the Pérolles competition – and in the absence of Maria, who was in Italy – Maillart's devoted sister Rosa helped him move.[61] Seven years older than Robert, Rosa had married a successful photographer, Arnold Wicky, and was raising three children, when she was suddenly widowed. From then on, she became a family fixture, particularly good to her long-widowed mother. She had a gaunt look even in her forties and a nervous temperament. Rosa was particularly fond of Robert, and in later years relied heavily on his judgment and advice. She, in turn, was efficient about household matters, which were difficult for her brother. By the fall of 1908, the family was together in a spacious apartment overlooking the lake on Hochstrasse in Zurich.[62] From there they would frequently travel by horse and carriage (Fig. 58).

One afternoon, on returning to the office from a trip, Maillart was astonished to learn that Maria had again departed, this time with the two children and without even informing him of her plans. She had the habit of traveling away until the family's move to Russia in 1914. The birth, on October 31, 1909, of a third child, René (1909–76) did little to alter it.[63] But Maillart as usual was also always in transit, as illustrated, in early September 1909, by his trip to Copenhagen as one of two Swiss representatives to the International Congress on Building Materials. He wanted Maria to go with him, but her third pregnancy ruled out such a long voyage.[64] Maillart accepted his wife's frequent absences and sometimes even took advantage of them to follow his own interests.

Figure 56. Maillart and Company, office in Zurich, circa 1906. (*Source:* Madame M.-C. Blumer-Maillart)

By the late spring of 1911, Laufenburg was under construction, the Rheinfelden footbridge was being built, and the works at Augst-Wyhlen were in full swing. Maillart had never been busier: he was constructing a number of buildings in Zurich and Chur, and in June alone, he was in Brig, Geneva, Chur, Tavanasa, and Neuchâtel. On a Saturday in June, traveling by train from Brig northwest along the Rhône, he decided to get off at Chexbres, high above Lake Geneva, to spend the night there. The next morning, he took the train to Estavayer, a picturesque old town on the eastern shore of the lake of Neuchâtel, where he stopped for the kind of hearty dinner he relished. (Maillart was also not indifferent to the fact that good rustic meals cost considerably less than they would have in town.) He proceeded to the small town of Avenches, the ancient capital of Helvetia, formerly an important Roman town. Here Maillart walked

Figure 57. Maillart and Company, construction yard in Zurich, circa 1906. (*Source:* Madame M.-C. Blumer-Maillart)

Figure 58. Carriage with Maillart, Rosa, and Maria in Zurich, circa 1908. (*Source:* Madame M.-C. Blumer-Maillart)

Figure 59. Paul, Alfred, Robert, and Max Maillart, circa 1908. (*Source:* Madame M.-C. Blumer-Maillart)

for about 7 kilometers to Morat. It was "a splendid tour with beautiful weather and views of the country, and interesting for the setting and architecture of these towns."[65] As before in Graubünden, Maillart enjoyed being alone in the country. Although he could now afford some luxury, he still relished the economy of simple living. He also kept in close contact with his mother, his sister, and his three brothers (Fig. 59).

VOLTASTRASSE: MAILLART AS PATRON

By September 1912, Maillart was in a position to consider the purchase of a private house. Before his departure for a business trip to Russia in October, he decided on one just around the corner from their apartment, at No. 30 Voltastrasse (Fig. 60).[66] Life there was full and comfortable (Fig. 61). For the first time, Maria felt at home and enjoyed the company of a small circle of good friends. Charles Lehr (1887–1978), one of Maillart's young engineers at the time, recalled years later that "I was invited to a number of parties at the Maillart home. I recall Madame Maillart as having been the perfect hostess, and I had the impression that Maillart was a wealthy man judging by the way he lived."[67]

With the move to Voltastrasse, the family's life was marked by an abundance of servants, rides in the new car, a convertible Hupmobile that Maillart bought in February 1913 and for which Armando the butler had become the chauffeur. But most striking are the glimpses of Maillart, who now gives the appearance of a prosperous *patron,* even though he remained well outside the conventions of such a life. Dinner was usually eaten with friends and followed by bridge. Maillart loved to play and often grew impatient with the small talk required by the length of the meal, protracted by the serving of dried fruits and nuts. Maria settled the issue by writing a menu that included "one-half hour for fruit and nuts." Her husband accepted the enforced socializing with good humor after that.[68]

Maria also liked to dance and especially to learn new steps such as the Tango. Once a week she

Figure 60. Maillart House, Voltastrasse 30, Zurich, 1913. (*Source:* Madame M.-C. Blumer-Maillart)

hosted an evening with a dancing master. The dining room table was pushed aside, and with four or five other couples, Maria would attempt to teach Maillart, who tried manfully to compute the movements, but it did not come naturally. As he frequently looked helplessly around in the middle of a new step, his young daughter was convinced that he was trying to calculate in space the positions to be taken.[69]

Maillart was not only remarkably tolerant of his wife's changeable behavior, he was willing to try just about anything to please her. His daughter recalled the following story:

At that time we had a little tortoise named Suzette; she was everyone's friend. At six each morning she would

Figure 61. Marie-Claire, Robert, Edmond, René, and Maria Maillart in Zurich, 1913. (*Source:* Madame M.-C. Blumer-Maillart)

pass by the door of my mother's bedroom, where she went towards [my mother's] bed. One day, she caught a cold. What was touching was that my mother, sad, evidently, and even horrified to see Suzette suffering – telephoned to my father who came immediately from his office. I can see him now climbing the stairs: he looked at Suzette, put her in a little basket and took her to the veterinarian, where she died. My father had these touching gestures . . . he did such things despite the discipline of his work.[70]

On weekends, Maria liked to ride in their car; sometimes they would drive along the lake, usually with Maillart and Maria in the back seat. On other occasions, they would organize a party with Max von Müller, who also had a car; they would go off into the country for the weekend. On one such excursion, there were three cars filled with elegantly dressed men and women, with the Maillart children, and their governess. They went to a luxury hotel in Gyrenbad, about an hour-and-a-half drive north of Zurich. Sunday evening on their way home, the caravan was stopped by police. An epidemic of hoof-and-mouth disease had been discovered among the local cattle, and everyone leaving the area was required to walk through a 10-meter-wide layer of sawdust saturated with disinfectant that was spread out before the party on the road.

The elegantly attired ladies balked at the idea of walking through such a mess, but Maillart took charge. He lined everyone up behind himself and began to march while beating time. The party followed. Maillart began to weave in a zig-zag path with everyone obediently tracing the same pattern. What looked to be an unattractive ordeal turned out to be a festive kind of conga line.[71] Maillart's sense of humor and almost boyish playfulness was in stark contrast with the severity with which he approached his personal and professional problems. His frequent lack of diplomacy with colleagues and even his occasional gruffness with Maria would not have been compatible in most people with the whimsy that was so much a part of his personality.

CAREER CONFLICTS

While Maillart began to turn away from Switzerland in 1913 and move toward lucrative Russian bidding contracts, he had already introduced ideas about design, research, and teaching that were in sharp conflict with those accepted by most European academics. Up until this time, Maillart's design talent remained relatively unexposed in Switzerland. What had become apparent, however, was his prodigious engineering skill; the profession recognized his talent for making light and inexpensive new forms, even if it did not appreciate their visual potential. As the Basel siphon demonstrated, when faced with difficult problems, public officials and fellow engineers often turned to Maillart. However, Maillart had too much of an independent mind ever to be widely popular. He openly questioned official judgments and publically criticized one of his stiffest competitors, Jaeger. Because his unusual engineering ideas went well beyond the experience of city and canton engineers, they had to rely on the advice of noted consultants to approve Maillart's designs – since Ritter's death, that meant François Schüle, who questioned many of his basic ideas. Therefore, throughout his career, Maillart was forced into a constant and often acrimonious dialogue with the academic establishment as well with his professional colleagues.

That dialogue had definite ethical overtones, as his public correspondence on the stolen Guggersbach Bridge showed. It also had a strong element of practical experience in reaction to the growing abstraction of many academics. Just as engineering faculties became more involved with analysis and less with design, so practicing engineers were becoming more focused on design and less on construction. Specialization, a hallmark of the twentieth century, was accelerating in the first decade of the new century; the result was a growing belief in the inherent necessity for collaboration within a compartmentalized field made up of academic researchers, architects, consulting engineers, and contractors. The idea behind this was that only by specializing could one keep up with fast-changing developments.

Maillart was averse to this trend. He did his own research and argued its correctness; he expressed his own aesthetic ideas at a time when other engineers were emphasizing only technical matters; he preferred to do his own calculations and detailing rather than build according to someone else's system; and, finally, he wanted to construct everything he designed because he found it foolish not to be able to realize his own designs.

If Maillart's career showed nothing else, it was that his greatest works were those he did without consulting advice. But for major Swiss bridge designs up to 1914, he was forced to collaborate with architects. Much in the profession was working against his conviction that calculations, construction, and aesthetics should be integrated and that only then would major innovations follow. With the formation of a National Code Committee in 1906, the Swiss began to formalize calculations, disconnected from building and design, an approach to which Maillart strongly objected.

PERSONALITIES AND THE CODE: PRACTICE OR RESEARCH?

On October 31, 1906, the Swiss Code Commission met to debate the recommendations of François Schüle (Fig. 62) for a new code to replace a provisional one established in 1903. Schüle's recommendations were based on a major research report on reinforced-concrete beams at the ETH.[72] Maillart and Schüle both served on the commission.

Schüle's report left the impression that much more research was needed before proper designs could be carried out. The clear implication was that more knowledge would lead to safer designs for as he said, "the usual calculation methods give incorrect and only approximate insights into the real values of the internal stresses of reinforced concrete." Schüle, along with Emil Mörsch, also a commission member, believed that structural design had to be based on methods of calculation developed from research in laboratories. The research results would correct errors and oversights in practice.

Maillart and his fellow committee member, Eduard Elskes, saw in Schüle's report more harm than help to the profession. The research, in beams, presented problems that Schüle never mentioned. First of all, in the laboratory, beams were supported on carefully made metal rollers and hinges that allowed the ends to rotate freely; such conditions almost never occur in practice. Second, Schüle was concerned about internal stresses; but Maillart believed (and pointed out publicly in 1909) such val-

Figure 62. François Schüle (1860-1925). (*Source:* ETH Archive)

ues have little to do with structural behavior. For him, it was critical to estimate the load at which failure occurs, and to predict deflection and cracking – the visual evidence of performance. By the late twentieth century, this is the accepted basis for codes. Third, Schüle designed test beams following some ideas of the Hennebique organization, which were soon found to be less safe than practices used regularly by Maillart well before the 1906 Schüle report, and eventually adopted universally.[73] Maillart, with solid scientific training, had the vision of a designer, delighting in the specific work; Schüle, despite sound practical knowledge, had the vision of an applied scientist, seeking to follow general rules.

Elskes and Maillart wanted full-scale tests on actual structures, a desire that ran counter to those in favor of scientific research in laboratories. Thus, when Schüle opened the Code Commission meeting, Maillart was primed with objections.[74] Even on the preliminary issue, the quality of materials used

for concrete, Maillart opposed any attempt to lay down fixed rules, feeling these might increase costs needlessly.[75] On the main issue, no one argued about the need for tests; the question was whether such tests would be exclusively laboratory studies (Mörsch and the other academics) or would include tests on full-scale actual structures (Maillart and Elskes).[76] The question continued the earlier debate between Ritter and Engesser on the same subject. Now Ritter's students were arguing his position against Ritter's German successor.[77]

Schüle had implied in his research report that the 1903 Provisional Code contained provisions that were unsafe. Maillart bluntly objected to Schüle's conclusions because they argued for changing the code on the basis of research rather than practice. Mörsch also opposed the changes, but because he wanted more research results (expected imminently from France and Germany). After some further discussion, and at Maillart's insistence, Schüle agreed to write up the present situation, leaving the 1903 code in force. (His article was published in the *Bauzeitung* in early 1907.[78])

The primary difference between Maillart and Schüle lay in the way in which they approached a project. Maillart saw a structure as a single complete form and strove to find an analysis procedure that would predict the performance of the finished object. For Maillart, the elements were studied after the design was set and their behavior had no meaning outside of the specific way in which they made up the entire structure. Thus, a curved slab not integrated to the deck (such as at Stauffacher) behaved radically different from a similar curved slab making up the bottom part of a hollow box (as at Zuoz). Maillart did not want to have rules for beams, for columns, for slabs, and for arches, which would require him to build from predesigned elements – like an erector set. Schüle, and other researchers, saw a structure as a collection of disparate elements and sought to define the behavior of each element in isolation.

Codes had begun to appear in other countries, and the process of standardizing calculations and systematizing design was well underway. Maillart took a leading part in this early activity, even though he had severe reservations about it. In addition to limiting a designer's freedom, codes and accompanying textbooks like Mörsch's allowed any inexperienced engineer to design in reinforced concrete and hence compete with Maillart and with other seasoned practitioners. Maillart was not merely attempting to restrict competition, but rather to emphasize the need for designers to have practical experience before making important designs on their own. The code was published in early 1909. Relatively simple, thanks in part to Maillart, it would remain in force for a quarter of a century.[79]

TEACHING AND THE CONFLICT OF IDEALS

Partly because of his prominence in the National Code Commission, the Federal Technological Institute invited Maillart in 1911 to give fifteen lectures on reinforced-concrete structures.[80] This honor gave him the opportunity to collect his ideas and to influence the next generation of designers in Switzerland. For the fall of 1912, he was asked to expand it from 1 to 5 hours per week. At this time, another practicing engineer, Max Ritter (1884–1946; no relation to Wilhelm Ritter), was giving a 2-hour course on the analysis of reinforced-concrete structures. His ideas contrasted starkly with Maillart's.

Ritter was beginning his career by following the same tradition of academic engineering as Schüle. Twenty-four years younger than Schüle, Ritter shared the latter's faith in theory and mathematics. Ritter was also a bridge designer and had worked for Maillart's former employer, Westermann, in competition with Maillart. Moreover, Ritter was related by marriage to the partner of Otto Pfleghard, an architect with whom Maillart had a falling out in 1908.[81] So, in their respective theories, designs, and business practices, Ritter and Maillart were rivals; they now found themselves teaching rival ideas.

Maillart taught in an unsystematic way, present-

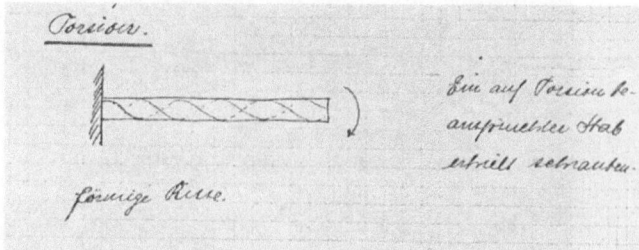

Figure 63. Lecture notes on torsion from Max Ritter. (*Source:* ETH Archive)

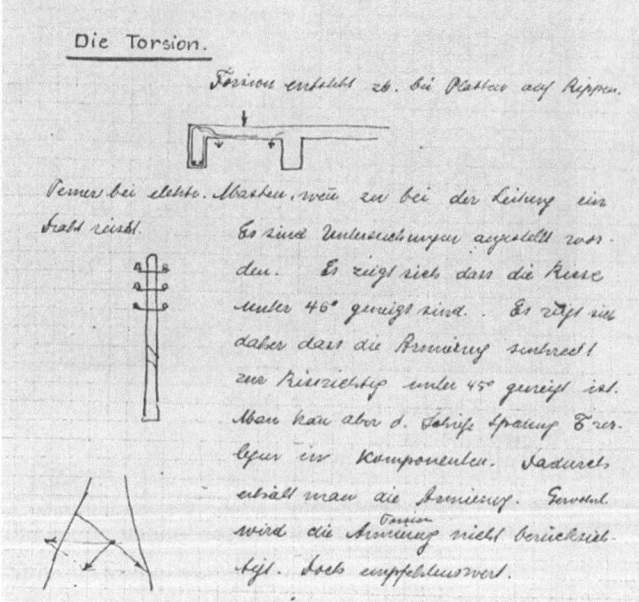

Figure 64. Lecture notes on torsion from Robert Maillart. (*Source:* ETH Archive)

ing a wide range of structures, none of which was discussed in detail.[82] He taught the elements of design organized within practice. Ritter taught the procedures of analysis in the context of theory. Maillart and Ritter presented diagrams and mathematics in entirely different ways.[83] Ritter's diagrams were predominately a set of abstractions, each intended to illustrate a single principle such as torsion (Fig. 63), whereas Maillart's were mainly illustrations of actual structural behavior (Fig. 64).[84] For silos, for example, Ritter and Maillart both gave formulas, but Ritter's sketches are line abstractions, whereas Maillart's show the actual silos (one of Maillart's diagrams even shows the conveyor systems used to fill a set of six silos).

Ritter began his lectures by giving a bibliography and he continually referred to the most up-to-date analytic methods. He discussed material from the *Journal for Mathematics;* Maillart discussed the ventilation of silos by air in hollow walls (part of a patent Maillart was drawing up). Maillart's lectures were those of an experienced practitioner, whereas Ritter's were those of a mathematically inclined analyst. Students were fortunate to have both approaches. Unfortunately, as time went on, it became harder and harder to sustain the more design-oriented approach.

The curious impression that one gets in looking at these two sets of 80-year-old notes is that the unacademic Maillart material is hardly out of date; his details and forms look like buildings today, whereas Ritter's analyses are almost all obsolete. The basic mathematics are timeless, but the analytic procedures were soon replaced by newer developments, from the simplified methods of the 1930s to the more complex numerical procedures of the 1970s, which rely on electronic computation. Because Maillart worked from full-scale performance, his ideas are as timeless as temperature changes, gravity, and the properties of steel and concrete.

Of course, as the practical calculations for his designs showed, Maillart knew modern analytic procedures, but he wanted his students to think about practical problems during their academic studies, just as had Wilhelm Ritter. Unlike his own teacher, Maillart lacked the time and the temperament to systematize his lectures even though he felt strongly that analysis had displaced practice in teaching at the ETH. But education in analysis cannot be neglected because even by 1912, there was a good deal of questionable design based on faulty analysis and an incomplete understanding of reinforced concrete. Also, as uses for the new material spread, it was no longer the sole province of pioneers with long experience, but became part of the average engineer's necessary competence. Finally, reinforced concrete is a more complex material than steel, and its behavior depends more on design. All

of these questions argued strongly for a systematization of the subject in a way that Culmann, Wilhelm Ritter, and others had systematized structural analysis in the late nineteenth century.

Thus, by 1912, there were two lines of teaching: one coming from practice, the other from research; one coming from the experience with completed structures, the other from the development of calculation methods. These two lines appeared as early as 1911 in Zurich in the persons of Robert Maillart and Max Ritter and are typical of what was happening in most technical universities at this time.[85] A conflict of ideals can become personal, and with Maillart and Ritter it did; but the story of their deep and bitter rivalry did not become public until the 1920s when it would seriously influence Maillart's career.

By 1914, Maillart had been teaching for 4 years, he had served on major Swiss commissions, and had pursued his own research and designs, not to mention the struggle – successful at last – to establish his own construction business. That business was growing abroad thanks to the rapid industrialization of countries like Spain and Russia, and it was to those foreign sites that Maillart began to travel in the second decade of the twentieth century.

The Move Toward Russia

ST. PETERSBURG, 1912–1913

Large building contracts were rare in little Switzerland. So Maillart in 1912 began to look elsewhere in earnest while still responding to a few bridge opportunities. The tension between design innovation and construction profits waned; he had little time left for bridges. The last Swiss bridge of Maillart's prewar building career as a designer and builder was for the Muota River in Schwyz, one of the three oldest cantons in Switzerland. His design for a reinforced-concrete, three-hinged hollow box, following the model of Tavanasa, won the nonjuried competition against canton engineer Herman Gubelmann's three-hinged, reinforced-concrete arch. Finally, he might get a challenging design to build a bridge according to his own ideas.[86] Just as he was given this project, the prospect of larger works appeared abroad. One of Maillart's classmates was a Russian named Benjamin Person (1867–1937), who by 1912 had established himself as a successful builder in St. Petersburg. It was through Person that Maillart would enter the booming economy of Russia on the eve of World War I.

Swiss banks were seeking ways to invest in the rapid industrialization of relatively underdeveloped countries like Russia and Spain; and advanced technical ideas originating in countries like France, Germany, and Switzerland were ripe for use in these countries.[87] Maillart's mastery of reinforced concrete in general and his flat-slab idea in particular were ready for such export. Thus, as he finished a series of major works in Switzerland, he began to explore vast potentials in the partially westernized cities of Russia from Riga to St. Petersburg. In early April 1912, Maillart departed for his first trip north into the Russian Empire.

Founded in 1702 by Peter the Great, St. Petersburg was a major civil works project in its own right. Built in the marshy delta of the Neva River, it was Peter's attempt to turn Russia toward Europe: From Peter's time, the city continually looked toward Western Europe for its building skills and for its cultural life.[88] In the later nineteenth century, St. Petersburg experienced a period of rapid industrialization and urbanization, marked by a huge jump in population and by an extraordinary acceleration in building. The city's population doubled between 1850 and 1890 to 1 million, and by 1912, and it had more than doubled again to 2.2 million. This "tumultuous pace" was fueled by large foreign investments and the widespread use of foreign technical personnel, resulting in a 60 percent increase in the value of production between 1908 and 1913.[89]

Maillart arrived in St. Petersburg on Monday morning, April 15. Person met him at the station and took him straight to his office, where he introduced him to his associate, a Mr. Bernstein, director of the firm that was to build a cold storage warehouse for which Maillart was to be a subcontractor (Fig. 65). Maillart spent the next 6 days in St. Petersburg settling the contract, inspecting the site, and making calculations. He was warmly received by Person and his wife, who in the evenings took him out to fine restaurants and the ballet. He arrived back in Switzerland early the next week, convinced that the warehouse was "only a modest beginning."[90] Maillart returned to Russia in early July to supervise the start of construction and to taste Russian night life, as he wrote Maria:

My friend Person comes to get me and we go each evening to a different "jardin-variéte" where we have dinner around 10 or 11 o'clock and where one could have the opportunity of finding more than enough pretty Russian girls to aid one's digestion. Person is a grass widow like me and also like me he is very proper.[91]

While amused by this mild night life, Maillart remained, as he reminded his wife, a serious Swiss. As such, he worried because work at the warehouse was moving slowly; Maillart found the local workmen lazy. He stayed on long enough to get one of his Swiss engineers started on the job. The following week he returned to Zurich via Stockholm, Bucharest, Budapest, and Vienna.[92]

The Aarburg Bridge had been repaired, and Rheinfelden neared completion, but the weak foundations found at the Muota River Bridge site forced Maillart to abandon his Tavanasa-like form and to rethink that design completely. Thus, at a time of large-scale business negotiation and high-profit construction, he had to spend considerable time on a new design that would not bring him any profit.[93]

With these and other works in progress, he was able to buy out his partner Zarn, giving Maillart more financial control to go with his technical supervision. His new letterhead showed the variety of work he could now do and it emphasized the fact that he was completely independent:[94] But at the same time, he was forced to delegate authority.

Figure 65. Warehouse, basement, St. Petersburg, 1912. (*Source:* Madame M.-C. Blumer-Maillart)

With Zarn away in St. Gallen and von Müller in military service, in the summer of 1912, Maillart had to forego a trip to Wangen to negotiate a contract for a new project in order to take care of the office. Instead, he sent his chief engineer Arnold Moser, who concluded the affair but only "after having been bargained down somewhat!" It clearly annoyed Maillart that he could not go himself.[95]

In spite of Maillart's apparent concession to the amenities of upper middle-class life at Voltastrasse, the family spent little time together there during the 19 months they lived in their first house – a total of only 10 months not including Maillart's own travels within Switzerland.[96] Between December 1912 and July 1914, Maillart made ten separate trips to Russia, each one from 10 days to 3 weeks, while Maria continued her prolonged visits to It-

aly.[97] During a trip in April 1913 to St. Petersburg, he wrote that there was not much new work but that there was "much to discuss for the bridges that they will build." Four days later, he wrote that he was to "visit an architect of the 'crown' in St. Petersburg and in several days I will visit two generals at Peterhof, the residence of the Czar since the 'crown' must say if our bridge design is handsome enough. Signor!"[98] There are, however, no remains of any Maillart bridges in Russia.[99]

The social side of his Russian adventure amused Maillart as he reported to Maria that "we went to see a Russian Operetta and the 'Variéte' [show] which followed it at midnight. Madame [Bernstein] did not come with us and so Mr. B. invited the primadonna of the operetta for a little [late] supper. I gave the impression of being a bit stupid (or very

Figure 66. Pirelli factory, Villanueva y Geltru, Spain, 1914. (*Source:* Madame M.-C. Blumer-Maillart)

much so) because she spoke only Russian and thus I could only act as the acolyte. After that I decided to study Russian energetically". But he reassured her, "You need not fear that I am going to take on these 'primadonnas' right away, rather more likely the little ballerinas. At all events that is more economical, because business does not appear to be brilliant. However I hope to be able to profit from my presence here to still catch one. They are going to come get me to go and see the thing." The running together of ideas about ballerinas and business expresses Maillart's quiet sense of humor even in the face of persistent practical problems. The idea that the socially awkward Maillart would perfect his Russian to seduce a show girl must have been very funny to anyone who knew him well – especially to Maria and to Maillart himself.[100]

Maillart's international work expanded in September 1913 to include a new office in Barcelona, a center of the rapid industrialization that Spain, like Russia, was experiencing in the early twentieth century. There he was building a large cable factory (Fig. 66), which encouraged him to send one of his best young engineers there to expand the business.[101]

RIGA, 1914

Like many others at the time, Maillart did not take much note of political events in 1914. As he traveled freely through Italy, France, and Spain on one trip, and through Germany, Russia, and Austria on the next, he seems to have had no more premonition of any major disruption than Europe's leaders and diplomats. At the beginning of 1914, Europe was in high gear economically, the scene of intense international activity. Much of this activity focused on Russia. France was extending large loans and Germany was pouring in exports.[102] It was for one of these French investments that Maillart began building a large rubber factory in Riga, where he traveled at the end of 1913.[103] On Tuesday morning, January 6, 1914, he was back in Zurich to prepare for the visit of Benjamin Person and to make another large bid.[104]

Demanding as the travels and the office work were, they had begun to pay off handsomely. By 1914, he had built two warehouses in St. Petersburg and was well along with the factory in Riga (Fig. 67). In mid-February, Maria, his sister Rosa Wicky, and friends Max von Müller and Paul Nissen were having coffee in the Odéon Café by the lake in Zurich. On one side of a piece of paper Maria wrote, "Memories of the month of February," under which she wrote the dates of major family events in that month including for February 14 "1st Russian Million." Maillart's efforts had produced by early 1914 1,000,000 Swiss Francs (about $10,000,000 in 1996), confirming his status as a wealthy man.[105]

The city of Maillart's major activity in 1914 had long been the dominant Baltic capital. Founded in the thirteenth century, Riga had become an industrial center well before the cities of Russia proper. As the capital of Latvia, it benefited from a more accessible port and a closer proximity to the more

developed Western European nations. By the late nineteenth century, it had important industries such as metalworking, railroad car manufacturing, and shipbuilding yards. In the early twentieth century, its electrical and chemical industries grew rapidly, as well as the production of rubber goods.[106] Its population rose dramatically from 282,230 in 1897 to 517,522 in 1913.[107] Rapid expansion not only brought in numerous Latvian workers, but also greatly increased the foreign population. Like St. Petersburg, Riga was a place of feverish building activity in 1914.

In such a climate, speed of construction was essential; the economics of a boom period pressed for the rapid establishment of factory or warehouse operations and consequently made difficult the careful control of design and construction on which Maillart's Swiss reputation was based. This pressure for speed led Maillart to work in very different ways from his Swiss practice. At Riga, the Russo-French rubber company itself made the basic design and Maillart built only part of it; a German firm built the rest.[108] This divided effort required close coordination and detailed planning of construction at the site – Maillart's forte. He had to take someone else's design, drawings, and specifications and imagine how the structure would be put together in minimum time. He was usually given a construction contract and had to plan, hire local workers, make calculations for scaffold structure, and order materials. He had to associate himself with a local engineer, as he did with Person in St. Petersburg. In Riga, he worked with an engineer named Schneider in whose offices he could have drawings made and where he could carry on the business of negotiation, planning, and buying materials. As soon as construction would actually begin, Maillart saw to it that one of his trusted Swiss engineers was on the site to oversee construction and ensure economical procedures.

Maillart had to create a new organization for

Figure 67. Factory under construction in Riga built by Maillart, 1914. (*Source:* Madame M.-C. Blumer-Maillart)

each project, and by 1914, he was a master at this. It was this talent for specific, single-site organization of major construction that allowed him, almost alone, to carry out highly profitable ventures. Because he had low overhead in Russia, the profit margin was high. Maillart's major effort at Riga in 1913–14 was to complete a huge rubber factory in little over a year. The structure was completed on June 14, 2 months ahead of schedule.[109]

By 1914, Maillart still had about a half dozen talented engineers working in his Zurich office along with another half dozen draftsmen, but the Swiss business was dwindling and he had already sent one engineer to Russia. He also had a hundred or more foremen and workers in the field during the building season in Switzerland, but these people were not permanent staff – they were hired only as building volume required. Maillart still shuttled back and forth between Zurich and different cities in Russia, keeping close track of construction. He was tired and lonely and he urged Maria to come with him to Russia.[110]

In Russia, people had no measure by which to judge the quality of Maillart's work, and thus he could not count on his technical reputation to win him contracts. He had to operate more as a salesman. In April, he traveled again to St. Petersburg to secure a large contract for yet another rubber factory.[111] He did not get the contract, and the next day he wrote to Maria: "The big contract for the rubber factory . . . escaped us because the competitor had already worked for them and they did not want to cut them off! And that in spite of our bid being less expensive!" To make matters worse, while driving, he knocked over a pedestrian. Maillart gave him 10 rubles on the spot so as not to be further harassed, prompting the man's miraculous recovery. Maillart remarked that he "looked as if he wished to repeat the same test!"

RIGA COAST IN THE SUMMER OF 1914

Because nothing was going right in St. Petersburg, Maillart decided to go to Riga the following day.

With celebrations and music forbidden during Lent, Riga was a somber city, so that after a few days at the construction site, he left for home.[112] Then for a little over 1 week, the family was together again in Zurich. Maria finally agreed to take a family trip to Riga for the summer and leave in June.

But it was not to be: 4-year-old René contracted scarlet fever, making the June trip impossible. With the immediate travel date in doubt, Maillart could not wait. On May 29, he took the train alone from Zurich north.[113]

As early as 1825, people had begun to discover the summer pleasures of the Riga coast. By 1844, a regular ferry service connected it to Riga proper, and in 1877, the rail line made access simple and rapid. But the real beginning of the coast's international reputation came with the European boom economy early in the twentieth century. In 1895, there were 20,000 summer visitors; by 1914, the number had grown to 70,000 with only 11,000 permanent residents.[114]

With work slow one afternoon in early June 1914, Maillart left Riga at 4:30 P.M. for the coast. Once at the resort of Edinburg, he went to inspect the rooms he had engaged for his family and himself.[115] The season would not begin officially until mid-June, but preparations were already in evidence: bath houses setting up, little cafés getting refurbished, places for skating – almost no people, but much anticipation. He was eager to have the family join him. He wrote Maria that "The season will begin in two weeks and it is loveliest at the beginning. Do not leave later than the 26th [of June]."[116] Maillart's preoccupation with work and family left little time to follow European news, of which one seemingly minor event on June 4 – the day he returned to St. Petersburg[117] – was the announcement of a planned trip by the Archduke Franz Ferdinand, heir to the Austrian throne, to Sarajevo, capital of the Austro-Hungarian province of Bosnia.[118] Maillart returned to Zurich on June 23 to help prepare for the family trip north.

Then on June 28, at 11:30 A.M., the Archduke Franz Ferdinand of Austria and his wife were assassinated. Although diplomats recognized the dan-

THE MOVE TOWARD RUSSIA

Figure 68. Trip to Riga showing Maillart on deck, 1914.

not enough work there to justify a full summer away from Switzerland. Moreover, there were new projects in both France and Spain and the fading Swiss business needed attention. But, Maillart decided to overlook these professional considerations in order to spend the summer with his family.[120]

On July 11, the chauffeur-butler, Armando, drove the Maillarts to the Zurich-Enge station in the Hupmobile. It was a festive departure with many friends – the women elegantly dressed in summer white – at the station to wish them a bon voyage. Off they went in sleeping cars: north to Hamburg and Lübeck, where they took the boat (Fig. 68). Four days on the sea brought them to Riga. There, much to Maria's delight, they were met by another chauffeur in the same Hupmobile: Maillart had the same model shipped several months before to Russia so that Maria would feel more at home.

The family settled in at the Pension in the village of Edinburg near Riga and enjoyed an idyllic summer.[121] Maillart commuted regularly to Riga to oversee a series of undemanding, small works there: two warehouses and a canteen building for the rubber factory. The events of early August were not expected to last long. On July 23, Austria sent ultimatums to Serbia about the assassination; unsatisfied with Serbia's reply, Austria declared war on July 28. Strikes in St. Petersburg were immediately called to support fellow Slavs in Serbia, and Russia mobilized on July 30, followed shortly by France and Britain. Within days, Germany had declared war on France and invaded Belgium, thereby bringing Britain into the conflict. By August 4, World War I had begun.[122]

gerous implications of the assassination of the heir to the Austria-Hungarian throne, most of Europe believed that 1914 would be no different from any other year. As one historian stated, "If there were a fatal flaw threaded through this era [1900–14] of confidence – and this was the great illusion – it was the humanitarian belief that a general war among Europeans was really unthinkable."[119] It was in this context that the Maillarts busied themselves in late June to make the summer trip to Russia. There was

CHAPTER FOUR

War and Revolution
1914–1919

Russia
1914–1916

THE DUNES OF AUGUST

What Maillart considered a haven appeared quite otherwise to his friends in Switzerland in July 1914. "Do you not feel the rumblings of war?" one wrote, and another lamented that the family should have come back in mid-July as the threat presented by the war became clear to many.[1] During this time, Russian troops advanced into East Prussia, and by August 20 had defeated the Germans at Gumbinnen, prompting the German Chief of Staff von Moltke, to send three corps and a cavalry division from Belgium to the east, together with Erich von Ludendorff, his best general from the Belgian campaign. By the end of August, the Russians had been stopped.[2] But the German offensive on the western front failed and the war settled into a stalemate.

So the Russian Baltic was at peace that summer. Behind the smooth beaches rose sand dunes and behind them cool woods fed by freshwater streams. There the Maillart children joyfully ran and enjoyed weekend picnics, often with Maillart's engineer Hans Bircher. Maillart commuted into Riga to oversee some small works there.[3] It was an idyllic time in a peaceful spot that characterized the tranquil by-product of Europe's frenzied industrial prosperity up to mid-1914.[4] But the nature of the

war had changed, and Maillart failed to recognize this change.

At the end of September, Bircher was called back to Switzerland for military service. The Maillarts saw him off at the main railway station, where the scene was chaotic: the station overflowed with people and the trains were jammed beyond capacity. Bircher had to go first to St. Petersburg (now renamed the non-German Petrograd), then by way of Finland and Sweden to Berlin and Zurich.[5] It was a long, uncomfortable trip and one Maria was loath to make. At Maillart's suggestion, they rented an apartment in Riga in which they were settled by early October.

RIGA WINTER

Maillart continued to search for new work while his Swiss business nearly stopped and his Russian projects were expected to be completed by the end of the year.[6] Cut off from Switzerland and his office force, Maillart was a general without troops. Russia was confronting the central problem that all underdeveloped countries would face in the twentieth century – the problem of making its military, political, or industrial operations successful at individual sites. Grand strategies were fine, but without specific local organizations to carry them out, they were bound to fail. Maillart possessed just what Russian society as a whole lacked dramatically, the ability to organize a field operation successfully. Like the Swiss mercenaries of old who were prized for their battlefield talents, Maillart at the building site was always in command and successful. But in 1914, Russia had difficulty matching Maillart's talents to the realities of its industrial needs.

Just after the New Year of 1915, Maria and Robert traveled to Moscow, leaving the children with a young Swiss lady and the governess in Edinburg.[7] Edmond was getting lessons from this *Demoiselle* to make up for his absence from school; neither Marie-Claire nor René had any formal schooling in Russia. While on the winter beach, the children were captivated by the unearthly sights of the coastal north. The ocean waves had frozen in the form of ice breakers, which the children called "Russian Mountains"; and they delighted in sledding over the snow-covered dunes, down the beach, and out onto the motionless waves in the Riga Gulf.[8] Nature stood still, and children played on the silent sea while 300 miles to the south in February 1915, Generals Ludendorff and Hindenburg crushed another Russian force. It was below Wirballen, through which Maillart had passed so many times since April 1912, that the winter battle in Masuria consumed the Russian Tenth Army.[9]

At winter's end, the whole family took a trip to Petrograd, which left one indelible impression on 8-year-old Marie-Claire. While walking along the Nievski Prospekt, they noted activity in the Winter Palace. Just as the Maillarts passed the main palace gates, they flew open and a procession of troikas emerged at a great gallop. It was the Czar.[10]

Once again in Riga, Maillart and Maria stayed in their apartment and the children went back to the Pension Kevitch at Edinburg. They could now see that it would be a while before they could get back to Switzerland.[11] In early May, the Germans broke through the Russian lines near Krakow, causing yet another massive Russian retreat and weakening their entire position from Romania to East Prussia. By June, it was clear that even the north front from Dvinsk to Riga was endangered. Nevertheless, as the warm weather returned, the Maillarts went again to the sea. Within a month, the sound of German guns outside Riga convinced the engineer to retreat to the Russian capital.[12]

RETREAT: PETROGRAD TO THE UKRAINE

Maillart packed off Maria and the children with their young teacher-governess aboard the wide-gauged train to Petrograd while with his chauffeur, Ernest, he drove the Hupmobile north over the deteriorated roads to the Russian capital. At Petrograd, the family was met by Person, who took them over 25 miles west of the city to a group of wooden summer houses at the edge of a seemingly impenetrable forest. Within a clearing was the Persons' summer dacha.[13] Person gave the Maillarts one of these wooden houses, where they set up housekeeping for the summer of 1915. The lack of running water and electricity was amply made up for by an array of servants. Much of the time from late July to mid-September, Maillart was either in Petrograd or farther off in search of commissions. His only work was the design and construction of a series of twelve small bridges for the new electric rail line between Petrograd and Oranienbaum.[14]

Throughout the summer, the Russian war effort continued to deteriorate. By late September, the eastern front stabilized well inside Russia and millions of the Czar's subjects fled from the west to the supposed safety of the interior. The misery and death of soldiers and civilians were so horrible that even seasoned generals recoiled at the sights.[15] On September 5, the Czar himself assumed command of the army, and plans were laid to move back the industrial war support system far beyond the German advance. A major example of the preparations for this move was an immense factory complex that had been in operation at Riga. Intended for the production of electrical equipment, it was to be replaced by a new factory at Kharkov; this huge work was entrusted to Maillart.[16]

Although the Russians had suffered major losses, Europe generally thought the eastern empire capable of holding together; few sensed its internal weakness. Maillart also remained optimistic, and once the Kharkov contract came, he was determined to move his family to that more remote city, which he felt would be safe from fighting. By September 21, the Maillarts were temporarily settled in a Kharkov hotel by the railroad station. Only 13-year-old Edmond had been sent back to Switzerland with his young teacher. The engineer plunged into the challenge of organizing a colossal building site within a limited time.[17]

The allied strategy was to arm and equip Russia's abundant manpower. But speed was now crucial. The eastern front had to be kept intact to divert German troops from the French and British in the west. Britain controlled the seas, but even so, getting equipment to the Russian lines was difficult, especially after the 1915 allied failure to open up the Dardenelles. Therefore, local industry had to be pressed hard to produce. Maillart's Kharkov factories for the General Electric Company of Russia were an essential element in this plan.

KHARKOV, 1915–1916

Maillart was undoubtedly helped in obtaining the contract by the Swiss director of the electric company (the director's brother ran the Swiss boarding school for boys in Glarisegg, which Edmond now

Figure 70. René and Marie-Claire in front of apartment building, Kharkov, February 2, 1916. (*Source:* Madame M.-C. Blumer-Maillart)

Figure 69. Maillart apartment (second floor) on the Hospitalnaja, 22, Kharkov, 1916. (*Source:* Madame M.-C. Blumer-Maillart)

attended). Indeed, the director eventually gave Maillart additional work, and once in Kharkov, the Maillarts found a thriving Swiss colony. One Swiss family, named Bender, rented them a large, elegant apartment at Hospitalnaya (Fig. 69), bordering a city park, where they stayed from early October until the summer of 1916.[18]

Maillart quickly began assembling a trusted staff of Swiss engineers. Richard Wyss came from the now closed Riga office, Hans Bircher, released from military service, returned from Switzerland, and Maillart could offer Victor Tchiffely a job. Also Ernst Eigenheer, one of his most talented engineers, came from Zurich. He wrote to other Swiss engineers in the hope of luring them to Kharkov, where work was suddenly booming in contrast to the war-induced depression in Switzerland. Von Müller

Figure 71. Factory in Kharkov built by Maillart, 1916: interior. (*Source:* Madame M.-C. Blumer-Maillart)

promised to come in several weeks when he, too, would be released from military service. Maillart designed the entire project, consisting of five major buildings, so that most parts were standardized to simplify construction.

In early 1916, the war resumed with new ferocity. Verdun was the German objective in the West, where the famous battle began on February 21. In response to the French and British call for Russian help, the Czar obligingly began a March offensive just south of Dunaberg.[19] The way home from Russia to Zurich through Sweden was now closed; there was even great difficulty in getting mail through.[20]

Between the end of February and mid-March, the Maillarts made the fateful decision not to return to Zurich for Easter (April 23), but to remain in

Kharkov at least into the summer. Maillart had just obtained more work and the pressure to complete it quickly increased as the war accelerated at Verdun in the west and near Lake Narotch in the east. The Russians replaced silver coinage with paper money; Maillart began to amass a small fortune.[21]

Frequently Maillart would take 9-year-old Marie-Claire (Fig. 70) with him to the site where she remembered the great length and height of the 1,000-foot-long building (Figs. 71 and 72).[22] This plant was famous throughout Russia. In the British commercial publication, *The Russian Year Book of 1916,* the project was singled out: "the electrical works of Riga have been removed to Kharkov . . . [which] after the war [will produce] on a large scale . . . small electrotechnical equipment, such as before the war came from abroad, chiefly Germany."[23]

Von Müller arrived in May, and Maillart began to think ahead to when the plant would be completed in midsummer. He planned to seek more work then as he was loath to have assembled such a large and efficient force of engineers, workers, and machines only to see it dispersed upon completion of the one project.[24]

In June, General Brusilov began a great Russian offensive on the southwest front between the Pripet Marshes and the Romanian frontier, about 500 miles west of Kharkov.[25] Throughout June and July, while the Germans and French fought at Verdun, the Russians moved steadily west. These advances improved Maillart's chances of work for the government, whose stability depended on military success.

In late June, another engineer, Karl Lehr, arrived from Switzerland to help with construction supervision; he moved into a cottage on the outskirts of Kharkov near the Pastroyka (building site) with Ernst Eigenheer, who had arrived earlier. These two together with Richard Wyss and Hans Bircher were Maillart's key staff along with Viktor Tschiffely, who acted more as a business manager than as an engineer. They had worked for Maillart in Zurich before the war and could be relied on for high quality. Thus, along with von Müller, Maillart had the majority of his principals with him during the feverish summer months when the Pastroyka reached its climax.[26]

Concrete construction for the factory buildings had ended in mid-August. By October, the plant began operations, with an official inauguration on November 1. Some incidental work did continue for Maillart throughout the next winter, but 1916 was to be his last year of major building activity in Russia. During that year, he also built a steel mill in Kamemskaya, but he was unsuccessful in finding new projects to use the 1,000 concrete workers he had assembled, the machines he had imported, and his Swiss staff of engineers.[27] Reluctantly, he dispersed his efficient organization. He had earned a small fortune from Kharkov, but by October 1916, he was left with few new prospects.

Maria in Russia
1914–1916

THE INWARD TURN

Maillart's large building contracts – first in Switzerland, then in Spain and Russia – focused his attention on field organization, contract negotiations, and construction quality. His big Russian projects, although unremarkable as forms, were completed quickly, and unlike his work in Switzerland, required little travel, being localized first in Riga, then briefly in Petrograd, and finally in Kharkov. So despite the greater distances in Russia, the move, and virtual entrapment, they allowed him more time with the family.[28]

The war prevented mobility, but even without this deterrent, Maillart felt a need to spend more time with Maria, as her health deteriorated (Fig. 73). He was a born family man: Although his work satisfied his intellectual needs, his wife and children brought out the playfulness in his personality. Even his work was wholly meaningful to him only when he could share it with them. For example, the Billwil Bridge was joyfully inaugurated with Maria, as compared with Tavanasa, a greater accomplishment, but robbed of complete satisfaction for Maillart because of his wife's absence at its completion.

Those who knew the opinionated and demanding engineer professionally might have doubted the degree to which he could relax with his family. But relax he certainly did, as demonstrated by an incident at the hotel on the Riga coast.

Hearing hysterical screams from the fashionable ladies who, on their way to dinner, had encountered a bat, the bald Maillart played on the French word for the offending creature (bald mouse). Swathing his head in a towel to protect his nonexistent hair, he told the women: "Look, do as I do, put a towel around your head if you are afraid that the bat will alight in your curls. . . ." As with

Figure 72. Factory in Kharkov built by Maillart, 1916: façade. (*Source:* Madame M.-C. Blumer-Maillart)

the conga line he had led on the road in Switzerland, Maillart managed to get the well-dressed company to follow him downstairs, safely away from the supposed danger.[29]

An example of the engineer's devotion to family was his all-night search in Kharkov for the family's lost dog, Bobby (who found his way home alone). Marie-Claire never forgot those many times when Maillart would without hesitation go to great lengths to please her mother, whether in rushing a turtle to the doctor or searching for her dog, not to mention the frequent instances of trying to comfort her through illness.[30]

An adventurous trip by a three-horse sleigh to a snowbound monastery, and the pleasures of Christmas 1915, did not, however, conceal from everyone that Maria's abdominal pains were worse.[31] On the evening of January 20, 1916, she left for a cure some 500 miles south of Kharkov at Kislovodsk, the most fashionable of the north Caucasus thermal spas. There she spent 3 weeks in baths she judged to be "as good as Karlsbad."[32] She returned home

in mid-February in time for a large costume ball for Shrove Tuesday.[33]

Maria felt well enough to organize the party. It was to be a gala event. She and her husband invited their close friends, hired a small orchestra, and engaged a singer who played the balalaika. Everyone dressed in costume. Maillart adopted a dashing Mephistopheles disguise: a red outfit including a red hood with black horns, that went well with his pointed beard (Fig. 74; Maillart is at the back). But at the last minute, Maria was not well enough to attend. The party continued listlessly until Maillart organized a cotillion and marched everyone slowly into Maria's bedroom so that she could greet each guest. He stood throughout at Maria's bedside, happy to see her smiling at the solicitous procession.[34]

When Maria decided she would not return to Switzerland in the spring of 1916, Maillart rented a small vacation house (*datcha*) five miles outside Kharkov where the family could spend the summer.[35] At the time, in spite of her growing weakness, Maria still expected to be in Switzerland before summer's end, especially because local doctors did not recommend an operation. She missed Edmond, consoled only by their *rendezvous du soir,* when she would go out under the evening sky to gaze at the stars she knew he was also looking at.[36]

At Easter, the Maillarts had to give up their sumptuous Kharkov apartment as the Benders were coming back. They collected everything they had, and moved completely into the wooden *datcha.* From there, Maillart commuted daily in his Hupmobile to the Pastroyka on the other side of the city. In addition to von Müller's arrival in Kharkov in the spring of 1916, the Maillart's English friend, George Kitts, was now there also. So, in spite of being far from home and more seriously ill, Maria had the company of some old friends.

Maria passed an agreeable time at the *datcha* that spring, mostly in the house but also in the pavilion outside, where she could greet visitors and enjoy the parklike atmosphere of the birch forest, cleared of underbrush, in which one could walk in good weather. There were stables, small houses for the servants, and nearby, other *datchas* with foreigners.[37]

MARIA

Maria played the gracious hostess as usual, but her friends realized that she was not herself. There were large weekend gatherings of the Swiss colony with Maria always at the center. Then came the biggest event of the summer, the Swiss national holiday of August 1, for which a midday feast was offered at the Maillart's home. In the photo (probably taken by George Kitts) recording the celebration, nearly everyone looks somber and Maria, smiling wanly, appears to be ill (Fig. 75).[38] She was, in fact, in great pain. The following week she was taken to the hospital in Kharkov where she underwent a gall bladder operation. Over the next 2 weeks Maria lay in the hospital, where children and friends visited her. Maillart was there continually.

Then very early on the morning of August 21, Maillart suddenly appeared in the children's room

Figure 73. Robert and Maria Maillart in Kharkov, 1916. (*Source:* Madame M.-C. Blumer-Maillart)

Figure 74. Party in Kharkov, February 1916. (*Source:* Madame M.-C. Blumer-Maillart)

at the *datcha.* He gently awoke Marie-Claire and René; he sat down on the edge of the bed, put one arm around each child, and told them that their mother had died.[39]

The next day was the funeral, and to the surprise of some, it was filled with people, including a large number of workers from the Pastroyka.[40] Maria was interred in Kharkov with a prominent gravestone and the local Russian-language newspaper carried a front-page announcement in French of her death and burial in the Catholic church at 10:00

Figure 75. Celebration of Swiss Independence Day, August 1, 1916, at the summer Datcha outside of Kharkov.

A.M. on Tuesday, August 22, following a Holy Mass.[41] There she came to rest in the church she had loved far from her sunny Italy and her adopted Switzerland.

MAILLART AND EDMOND

Barely 2 weeks after his wife's death, Maillart wrote (on September 3) a long letter to his 13-year-old son. He had telegraphed the family in Switzerland right after Maria's death, but now he had to console the one person, beside himself, who would be the most lost without her:

When this letter arrives you will already have received the terribly sad news of the death of your poor Maman, that Maman who loved you infinitely and for whom your departure was an enormous sacrifice to which she con-

sented in thinking only of your well being. She longed to see you again and her thoughts were always with you.

Maillart knew how painful Maria's separation from her son had been, but he had difficulty conveying any understanding of Edmond's loss:

It is infinitely sad to think that we shall no longer see her on this earth. But you must console yourself with the thought that her spirit surrounds you always and that she will accompany you in all that you do. Thus often when you are undecided on any action or resolution, you should ask yourself, "What would Maman think of it?" And then your heart will tell you what you must do and in so doing what is good, you will honor the memory of your mother and you will thereby correct the faults that you would have committed during her lifetime and recompense her for all the pain that she underwent in order to make of you a fine boy and a fine man.

These are lofty expectations to lay before a lonely boy. Maillart, who had fulfilled such ideals in his own life, who had indeed kept his own mother's sacrifices in mind during his youth, was here using a time of sorrow to admonish his son about the future. For only a brief moment was he able to cast off the role of counsellor in responding to Edmond's August 3 letter from Arosa: "What pleasure Maman would have had in reading it! Especially since it is so well written. For me it recalls the time when I used to go frequently to Arosa and to Davos where I was building some hotels. Then I also passed a dozen times over the Furka and twice over the Flüela!"

Maillart reveals in this rare burst of emotion a deep homesickness for Switzerland and in particular the wild Graubünden. But once more he laid a burden on his high-spirited son in referring to Maria:

You must know that your teachers were not very pleased with your recent performance and Maman was worried about you. But I reassured her by saying that it was merely momentary, careless acts, that you would think them over, and that we would have better news from you. Try to make it so because Grandmaman also worries about your last report card.[42]

Edmond had written his father from Geneva after hearing of his mother's death but before receiving his September 3 letter. On October 3, Maillart replied to his son that his letter did me a great deal of good. I am sure that what you write me are not just words but that truly you will act as you have promised to. For my part, knowing full well that I can never replace all that you have lost, I shall try to do my best to fill the great void that now surrounds you. And you must be assured that, even if my affection for you cannot be as apparent and as beneficial [*bein-faisante*] as that of your mother, it is no less profound.

Maillart was aware – as he was in courting Maria – that he often appeared to be cold and indifferent. He simply could not express his emotions in the same open and effusive way as his Italian wife could. Unlike his father, Edmond had within him both the heritage of Italy and of German Switzerland. He had much of his father's professional talent – at school he had already distinguished himself in technology – but he had also a deep affinity to his mother and to her love of the unpredictable and frivolous.

Later on with René, Maillart would come to accept such behavior, as indeed he had learned to with Maria; but for his eldest son, he held a higher standard and expected a more serious return. Perhaps he was thinking of the disappointment that his oldest brother Paul had caused his own mother. In any case, his attitude did not make life easy for young Edmond. Maillart wrote this second letter on Edmond's fourteenth birthday:

Voila ta fête!, but a sad *fête.* It is fourteen years since we received you and that you have been for us a source of satisfaction and hope. As a baby, as a little one, and as a big boy, she regarded you as the apple of her eye and to see you once a man was for us a thought that we cherished together. There you are now, young man, no longer a boy and I do not doubt that you do honor to our hope.

Again, instead of loving reassurance, Maillart sent Edmond a reminder of his obligations. The letter concluded with the consoling thought that they would be together at the end of the war, and that Marie-Claire and René longed to see their big brother.[43]

With these two letters, the correspondence from Kharkov ceased. In October, Brusilov's advance was stopped, Hindenburg was placed in command of all German forces, and there followed a rapid

recovery for Germany. This meant that the war would continue.[44] The Maillarts faced another Russian winter and renewed uncertainty.

Revolution and Ruin

MAILLART'S MILLIONS

As winter approached, Maillart rented a large apartment in a big building on the Torgovaya Square in the center of Kharkov. There on an upper floor, the family settled in with Eva, an aristocratic refugee friend from Riga, with their Russian cook Masha, and with Maillart's niece Marguerite Wicky, recently arrived from Switzerland. Being a nurse, she had joined the Red Cross and somehow got to Kharkov to take care of the children after Maria's death.[45] Maillart set up an office in the apartment, where he and his remaining engineers sometimes worked. These colleagues, Wyss, Bircher, Tschiffely, and Lehr, would often take meals with the family, and without any particular discussion of the matter, they began to treat 10-year-old Marie-Claire as the hostess. They would thank her after the meal and kiss her hand as was usual in Russia; Maillart, too, began to treat her not only as his child, but also as a replacement for Maria.[46] Although it was not uncommon then for a daughter to replace a mother who had died, it was unusual for one so young to be treated with such respect.

Maillart could still have left Russia as late as the winter of 1916–17, and in spite of the eastward retreat of the Russian front that summer, the war seemed to be distant and without much danger for a Swiss family. Maillart had never been interested in politics, just as he had never become deeply involved with finance. Beside his family and a few close friends his primary interest was in his work. This lack of awareness, combined with the incredibility of a Russian defeat, explains what he now did with the considerable fortune he had amassed from the Pastroyka: Except for a small amount of easily transportable jewels and gold, he invested most of this money in coal mines in the Donetz.

Even the fateful political events of the fall and winter of 1916–17 did not bring home to Maillart the precariousness of his situation. As the Russian war effort deteriorated, the army rebelled, its officers openly supporting the overthrow of the Czarist government. By March 20, 1917, Czar Nicholas II had abdicated; the first of the two-part Russian Revolution had taken place.[47] A provisional government led by constitutional democrats took power.

Like most people at the time, Maillart viewed these events as the inevitable transition from monarchy to constitutional democracy experienced already by a number of European countries. There was no doubt in their minds that the new government would soon restore stability. In fact when, in early March, Karl Lehr told Maillart of his need to return to duty in the Swiss army, his employer urged him to leave his rubles in Russia pending a more favorable rate of exchange with the Swiss franc. As Lehr recalled, "Maillart could have converted his own money as well, but he decided to wait for a better rate of exchange. Maillart had borrowed money from Swiss banks in order to finance his Russian work." Fortunately for him, Lehr did not follow Maillart's advice and took his money with him to Switzerland that spring.[48] Meanwhile, coming from Switzerland in a sealed railway car, Vladimir Ilyich Lenin arrived in Petrograd on April 19 to call for a radical change.

THE CONTINUING REVOLUTION

Coming from the carefully crafted democracy of Switzerland, Lenin brought to Russia the diametrically opposite idea of a total dictatorship resting on an all-embracing ideology in which there were no objective moral laws for human behavior. For Lenin, the real enemy was not the Russian aristocratic monarchy against which the first uprising had been aimed, but the parliamentary liberal democrats who had replaced the Czar. For the bourgeois

Maillarts, Lenin represented by far the more serious threat, the consequences of which were not yet fully apparent in the spring of 1917 as the democrats were in control.[49] But with the Russian army exhausted, it was a precarious control, and by July 15, it had failed. Kerensky resigned as Minister of War only to return to the government a few days later as Prime Minister. But now the army ceased to exist as a vital force and when in September, Kornilov, as Commander-in-Chief, tried to march on Petrograd to restore order, his army dissolved before he reached the capital. The culmination of chaos came on November 7, when Lenin and the Bolsheviks overthrew the provisional government.[50]

Once again during a time of deep troubles, the Maillarts had found an idyllic place to spend the summer. With his work in Kharkov completed, Maillart had decided to go south to the resort town of Feodosiya in the Crimea. There, overlooking the Black Sea, he had found a summer house surrounded by an orchard that extended to the sandy beach.[51] Having returned at summer's end (1917) to a relatively tranquil Kharkov, it was only after the November coup that the family began to experience the full thrust of the revolution. They often watched from their windows while the Bolsheviks lined up townspeople against their courtyard wall and shot them. Only their Swiss nationality saved them from a similar fate. It was not easy to get enough to eat, and without their loyal servant Masha, the Maillarts would have suffered even more. She was always able to obtain food from the local peasants, being on good terms with the Bolsheviks.[52] René Maillart recalls that she was the mistress of their local chief.[53]

The famous conference at Brest-Litovsk began a series of complex negotiations between Germans and Russians. In his anxiety to make peace in order to gain control of the vast Russian empire, Lenin sacrificed huge territories to the Germans, who in the spring of 1918 were allowed to occupy Kharkov.[54] Now it was the Bolsheviks who were lined up in the courtyard of the Maillarts' apartment house and shot by the new conquerors.[55]

During most of 1918, Maillart kept up some incidental work at the Pastroyka, where his manager, Tschiffely, still lived and worked. The other Swiss engineers left by the end of the summer. Maillart had his mines and even his faithful Hupmobile; to protect it from theft, he had a secret space built into the barracks at the factory site. This long, wide building also housed Tschiffely and his staff. As late as September 1918, Maillart was still able to live in reasonable safety under German rule. However, there had been no communication with Switzerland for almost a year and little opportunity to build.

On November 11 came the armistice. A relief to most of Europe, it terrorized members of the middle class like Maillart, who had remained in Russia. He and his peers represented the hated values of capitalism, democracy, and the religious traditions of Western Europe, all of which Lenin was committed to destroy. And with the German withdrawal, the Bolsheviks returned to the Ukraine.

ESCAPE

Kharkov was now buffeted between violently opposed Russian factions and the Germans – still nominally in control but desperate to hand over power to anyone who could assure them safe passage back to Germany. On December 20, the Bolshevik Red Army seized Belgorod, 40 miles north of Kharkov, and prepared to move south. Every middle-class person left in Kharkov frantically looked for ways to escape from the city. They paid exorbitant prices for railroad tickets south.[56]

Within a week, Kharkov lacked electric power, its water system was threatened, and continual gunfire in the streets signaled total anarchy. Fear was ever present and every ring of the apartment bell gave the Maillarts anxious moments. Early in the morning of Monday, December 30, such a ring came. Marguerite Wicky answered and in rushed one of Maillart's foremen from the Pastroyka. He was wildly agitated and implored Marguerite to get Maillart, who was already at work in his apartment office. When Maillart appeared, the foreman exclaimed, "I know that you are on the list, perhaps even tomorrow morning they will come to get

you." It was clear that the Bolsheviks, now in control of the city, intended to kill Maillart, and it is very likely that they would have succeeded had it not been for the loyal foreman, who Maillart had helped when the foreman's wife had become critically ill (Maillart had paid for a specialist who had saved the woman). Now this man was risking his own life to save his employer. Maillart at last made the necessary decision: "*il faut partir.*"[57]

They packed everything they could carry, leaving only the jewelry lest it be stolen en route. Maillart still believed he would soon be able to return. He gave some money to Masha, who cried bitterly as they left. Tschiffely would stay on at the Pastroyka; apparently as an employee, he was not yet on the list. He remained at least until late June 1919.[58] The family – Maillart, Marie-Claire, René, and Marguerite Wicky – left that evening for Odessa with only three small bags. Maillart fortunately had obtained some train tickets through the French Consulate at Kharkov, but when they arrived at the station, there was only one train with room in something that looked like a cattle car with wooden benches. There was no choice but to get in; no one could even guess when the next train would be available.

The 371 miles to Nikolajev, a major seaport on the Bug River, normally took about 14 hours by train followed by a 7-hour steamer trip to Odessa.[59] But nothing was normal in December 1918. The Maillart's car rolled out of Kharkov and through the fertile Ukrainian countryside blanketed in snow. The region was in a state of civil war.[60] At one stop, a young Bolshevik in command of the station asked some of the people in the train to come in and eat with him. It was a frightening invitation because, although they ate well, the man was drunk and amused himself by going up to people while precariously balancing a grenade in one hand.[61]

Finally, after nearly a week on the wretched train, they arrived at Nikolayev. Maillart immediately tried to get tickets on a boat to Odessa, but too many others had the same idea, so he found an office building where they could sleep – all hotels being full. Fortunately, after only a few days, the family was able to embark for Odessa. They arrived on January 10, 1919, weary, but relieved to be finally in a place that would provide an escape from revolutionary Russia. That escape was not to be as easy to make as they had hoped.

At the time of the Maillart's arrival in Odessa, one historian has described it as " 'containing' in a nutshell all the problems of southern Russia."[62] The French were more or less in charge, but there was an ever-present three-fold threat: from the volunteers (White Russians), the Ukrainian Nationalists, and above all the Bolsheviks. The allies had just captured the German Black Sea Navy (the former Imperial Russian Navy), so that, with the defeat of Germany and Turkey, the way was clear from Odessa to the Mediterranean. But the few weeks that separated Maillart from Kharkov had reduced the wealthy, socially prominent engineer to a penniless refugee. As such, obtaining passage was nearly impossible.

The Maillart troop found a single room on the third floor of a fleabag hotel run by a woman, as Marie-Claire recalled, who was "horrible and dirty. She must have been a heavy drinker. . . ." Marie-Claire and René slept on the floor while Maillart took the only bed. Then, after so much strain, Maillart's health gave out. Marguerite immediately recognized pneumonia. For several days, Maillart's life seemed endangered. A doctor was found who told them the invalid must have some warm milk every day to build up his strength; Marie-Claire was charged with the task of descending early each morning to heat milk in the kitchen, black with cockroaches. She persevered and, finally, Maillart could get up haltingly and eventually go out.[63] There followed weeks of dickering with the Swiss Consulate, which periodically provided pieces of paper, valid neither as authorization to leave Russia nor as a passage home. One document attested to Maillart's work in Kharkov as a "builder of important works and the owner of considerable construction yards." It noted that he was forced to leave Kharkov "because of the Bolshevik invasion" and requested that the military and civilian authorities help him return as soon as possible.[64] Maillart

fought hard for this paper in the belief that the Bolshevik threat would be temporary. Familiar with Russia's desperate need for technical assistance, he was convinced that he would soon be invited back from Switzerland to continue work and thereby reclaim his construction property, his capital, and his personal possessions.

Finally, in early February, Maillart's persistence paid off: permission came to leave Odessa on February 3 aboard a passenger steamer commandeered by the French Navy and sailing for Constantinople. (They actually had tickets as far as Saloniki – Thessalonika – in Greece.) The day arrived and the debilitated little band gathered their three small suitcases and walked toward the docks, Maillart leaning on Marguerite and Marie-Claire. Before them appeared the famous 193 granite "Potemkim" steps descending from the city to the harbor. Supporting the sick engineer, they slowly made their way down through chaotic crowds of all nationalities. Suddenly Maillart realized that someone had grabbed the briefcase he had under his arm. It contained the documents of his major works in Russia.[65] Too weak to give chase, he could only watch helplessly as the thief disappeared into the throng.

CONSTANTINOPLE, THESSALONIKA, AND HOME

With all the losses, what impressed 9-year-old René was his father's concern about losing the stiff collars that were in the stolen briefcase! In fact, Maillart spent considerable time at their first port of call, Constantsa in Romania, looking for more collars. He may have been poor and sick, but he hated the idea of appearing shabby. He also hated being inactive; faced with the unusual situation of being without work, he proceeded to instruct René in the elements of bridge.[66]

They stopped at Constantinople long enough for Maillart to take them through the old city and to visit the great dome of the Hagia Sophia. The group ate at a little restaurant in a narrow street of the old quarter where Maillart insisted that they try cups of thick Turkish coffee. The boat trip had revived him and his escape gave him new energy. Up until now, he had been forced into an almost constant preoccupation with money matters – worrying either about making ends meet or about preserving what he had finally earned. With no money left, he seemed to enjoy a euphoric liberation.

Finally, eleven days after leaving Odessa, the family landed at Thessalonika, the ancient northern Greek port on the Aegean Sea. The ship went no farther and again Maillart had to find lodging and yet another ship passage. Fortunately, he knew a Swiss industrialist living in Thessalonika who helped the bedraggled family. He found them a house, lent them his horse and carriage, and invited them to several meals. After exactly 1 month, they left on March 14, aboard a small Italian steamer bound for Naples, where they arrived on March 21.[67]

From Naples, they boarded the train to Rome, where Maillart took the children to see the dome of St. Peter's, which he compared with that of Hagia Sophia. In spite of their forlorn condition, Maillart took advantage of a brief layover to instruct the children and to enjoy for himself the great structures of past eras. Then they set off again to the north, reaching Bologna on March 24 and the Swiss border above Domodosola the next day.

The Maillarts' entrance into Switzerland could hardly have been less impressive. Of course they had to take third-class tickets, and they felt like third-class refugees as Marie-Claire wrote:

We were badly dressed, without suitcases, only a few sacks. It is not very pretty to tell it, but we were polluting the air. The conductor made us sit all the way at the back of the car because I believe he was embarrassed to have such "clients."[68]

That evening (March 25), 4 years, 8 months, and 22 days after their elegant first-class northbound departure from Zurich, the Maillarts disembarked from the foul-smelling train in Geneva and fell into the arms of welcoming family: Bertha Maillart at the center with Edmond, now 16 years old, Rosa Wicky, Paul, and Marie Maillart among others. Despite personal and material losses, it was a happy reunion. Coming from a sophisticated international life, Maillart nevertheless felt much more at home in provincial Switzerland.

CHAPTER FIVE

Scholar and Designer
1920–1927

Burning the Sacred Fire

DEEP IN DEBT

Maillart's return to Geneva in March 1919 was a mixed blessing. He was, of course, happy to find his native land, his mother, his siblings, and other relatives after nearly 5 years' exile; but, having left Switzerland at the height of his career, he returned recently widowed and practically destitute. Not only did the engineer have no money, but with no business organization, he had no means of making money to repay his debts to the Swiss banks that had financed his Russian work. He did not even have a place of his own in which to house his children and himself.

It was a bitter comedown from the luxurious life in the Voltastrasse house (which he had sold after Maria's death, when he had also closed the Zurich office). The four Maillarts moved in with Bertha, Rosa, and Rosa's two daughters, Marguerite and Edith, Maillart's favorite niece, in a two-apartment complex. In the same building, Paul and Marie, with their two teen-aged children, had an apartment as did the Zarn family. Here in generous and attractive quarters, the family lived decently, but as invited guests. Only gradually over the next months did Maillart begin to sense the changes that had come over the next generation.

Once he had settled his children, Maillart left for Zurich to try and reestablish some contacts. Concentrating on pulling together remnants of his old organization, he visited Paul Nissen and saw Madame Tschiffely, whose son still lived tenuously in Kharkov. He stayed in Zurich for several weeks suffering, as he wrote his children, "great disappointment not to find a home and the many people that one loves. Thus we must try to find happiness where we are now [Geneva], and where everyone surrounds us with affection."[1]

Too deeply in debt to revive his construction company, Maillart decided to open an engineering design office in Geneva and to try and resuscitate his branch office in Barcelona. To do this, he had to borrow 237,000 francs (about $1,320,000 in 1996) from family and friends.[2] During spring and summer, while the children vacationed with Paul Maillart, the engineer shuttled back and forth between Geneva and Zurich, but found no work of consequence.[3]

In the fall, Maillart prepared his first trip abroad since his return. In 1914, he had sent Viktor Hässig, one of his best engineers, to Barcelona, and during the war, Hässig had completed two large utilities there – a thermal power plant and a gas and electricity station – but if Maillart hoped to make the Spanish business profitable, new work was essential. On October 8, Maillart left Geneva by train, arriving – after several stopovers in northern Italy – in Barcelona a week later. Here he began intensive discussions with Hässig and potential local clients.[4] Maillart spent a productive two weeks in Barcelona, eventually securing four major building contracts. Still in search of work, he left the capital of Catalonia in early November and traveled without business success to Oviedo, Madrid, Cordoba, and

Seville.[5] Early in December, he was back in Geneva, where he began serious work in his office at 18 Rue du Marché.

PARIS AND GENEVA

As the year 1919 ended, there was still little work in Switzerland. By contrast, northern France was in the process of rebuilding after the devastation of the western front. Over 800,000 buildings in the ten northern departments had been leveled and thousands of miles of canals, highways, and rail lines had been rendered impassable. Countless bridges needed to be reconstructed.[6] Maillart felt sure that France's enormous reconstruction effort must offer him some work, even though it would be difficult for a foreigner to compete with French engineers who also needed work to reestablish themselves after the war.[7]

He left for Paris in mid-January. To reduce his expenses, he stayed near the Place d'Etoile in a hotel managed by the brother of Marie Maillart. Responding to the promise of initial explorations, he returned to Geneva with the idea of putting together a detailed brochure in French similar to the one he had done in 1913 for Russia. For this new brochure, he collected photographs, wrote descriptions of projects, and solicited testimonial letters from former clients. Then, for several weeks, he commuted back and forth between Paris and Geneva in pursuit of job possibilities.[8] Maillart disliked Paris, vastly preferring to work in "a small place with pure air rather than this ant hill where there is always a light fog which blocks out the sun's rays and makes the air heavy." He missed Marie-Claire, but urged her to forgive the now financially pressed and dispirited von Müller for being unpleasant to her, advising that "It is not bad to be kinder than other people are."[9]

Maillart continually encouraged his children to apply themselves to their schooling, which had been so drastically curtailed during the Russian episode. He knew that after the excitement of the revolution and their escape from it, they were having difficulty in adjusting to the orderly, severe life that was so typically Swiss and so much a part of his own experience. He failed to see that Edmond's apparent frivolity masked a deep depression brought on in part by the loss of his mother, but even more by his sense of inferiority to his father.

Maillart's impatience with underachievement was apparent in his correspondence at this time. There was a marked difference in tone between his letters to Marie-Claire and those he wrote to Edmond. He tried to counsel them both, but the playfulness and irony that softened his advice to Marie-Claire were largely missing in his written conversations with Edmond.[10] For his part, Edmond had so little time with his father that he felt he hardly knew him. Now they grew even more estranged as Maillart devoted his energies to reestablishing his business. Like his mother, Edmond was subject to mood swings and, moreover, had little sense of the value of his own accomplishments.

Although not insensitive to his children, the engineer was incapable of providing them with much comfort. His inner resources were foreign to most of the younger generation. Indeed, his dogged pursuit of new ideas by age 42 had given him a more successful career than any of his own generation of engineering classmates would ever have. He had always been emotionally isolated; only Maria had been able to liberate him. After her death, he looked to education as a substitute for the home life the children were now deprived of. For Edmond, Maillart saw a technical career, and he compulsively urged him to improve his performance at school.

Without at first realizing it, Maillart saw in Marie-Claire a surrogate for his wife, and for her part, Marie-Claire began to feel a special bond with her father. Unlike Edmond, she did not instinctively compare herself to her parent. In part, this reflected the difference between a son, expected then to measure up professionally to his father, and a daughter whose life would be to run a household. In part, the difference was one of personality, Marie-Claire being more stable emotionally than her mother or her older brother.

Again Maillart established a pattern of life: weekdays in the Geneva office and evenings almost always playing cards with either the Zarns or the family. Late in the week, he would often go to Zurich, where he would stay in a hotel and take his meals with Paul Nissen; on Saturday afternoon, he would return to Geneva.[11] He almost never worked in the evenings or on Sunday, which he reserved for family outings.

THE BURNED-OUT EUROPEAN

Three years of hard work in Switzerland did little to relieve Maillart's debt. Desperately seeking means of restricting his budget still further, in August 1922, the engineer made the drastic decision to abandon the family apartment and move in with the Zarns.[12] This allowed his mother and sister to take a smaller place; Marie-Claire was sent to live with her uncle, Hector Maillart, a well-known doctor and collector of portraits of Geneva personalities.[13] It was galling to Maillart to live on Zarn's charity. From the beginning, he had considered the partnership with Zarn and von Müller as a compromise of his desire for complete professional independence made in order to expand his business and better provide for Maria. He liked Mrs. Zarn, but he was never close to her husband. Nominally, Zarn worked for Maillart and Company, but he accomplished almost nothing for the firm except to provide some of the financial help needed to sustain it. Von Müller had not returned to the firm after the war and now, in addition to Zarn, Maillart had only two engineers and a draftsman in Geneva.[14] Small as this group was, it was still too large to support with the work available in late 1922 and 1923.

Much of Maillart's own time was devoted to activities that brought no immediate return: looking for work, making new design proposals, and attending committee meetings. Having exhausted all possibilities in and near Switzerland, in the summer of 1923 as a last resort, he decided to investigate possibilities in Riga. Debt, dislocation, and rejected designs weighed heavily on his mind as he prepared to travel again to the Baltic city.[15]

Maillart went to Berlin first (August 9 to 11). He could hardly have picked a worse moment to be in the German capital. In January, the French had occupied the Ruhr Valley in an effort to force the Germans to pay reparations for the past war; Germany countered by ordering a passive resistance. Meanwhile inflation had grown with dizzying speed throughout the winter and spring, in part stimulated by the government's need to support the non-working resisters in the Ruhr.[16]

All of this turbulence focused on Berlin and came to a head at just the time of Maillart's visit. Huge crowds collected around government buildings on the Wilhelmstrasse; pressure on Wilhelm Cuno's government provoked its fall on August 12. Maillart contemplated this mounting tension with deep foreboding; it reminded him of Kharkov in 1918, and he had no regret in leaving Germany for the north port city of Stettin. From there he traveled by boat to Riga. The dining car prices quadrupled within days and stamps of huge denomination were required even for a postcard to Switzerland.[17] The engineer stayed 2 weeks in the capital of the newly independent Latvian nation. He saw some friends from the war years and visited his structures. He made two preliminary designs: a bridge in Riga and a building in Kaunas (Lithuania), but neither became a commission.[18] It was an unproductive trip, saddened by memories of former times and the unlikelihood either of recuperating his Russian holdings or of resuming work there.

For a while after his return from Latvia, he still hoped to be reimbursed for the 3,300,000 francs that he could document as his Russian losses; but with the Communists in charge, those hopes faded rapidly. By the time the Fifteenth Party Congress would meet in December 1927, Stalin had begun to take full command. The Soviet Union began its first socialist 5-year plan and all hope ended for payments to capitalist entrepreneurs from Switzerland.[19]

In the fall of 1923, Maillart designed numerous new works, but only two projects were actually completed that year. Financially, it was a poor year. On New Year's Day in 1924, he wrote a brief message to Marie-Claire. In it he expressed his feelings after four years back in Geneva: "I am a burned-out European."

Maillart repeatedly urged Marie-Claire not to forget to answer letters.[20] This need to keep in touch expressed his growing loneliness and his increasing dependence on those few friends and family who were left, such as Nissen, Jegher, his mother, his sister, and above all his children.

Maillart lived the simplest of lives, he took no vacations, he coveted no possessions, he refused to ingratiate himself with those who might benefit his business, and he sought no companionship outside of family and a few old engineering friends. Such a person could easily judge others harshly, especially those who gained fame, money, and position without the fullest moral integrity. We will see how he did so judge his antagonists in the engineering establishment. For different reasons, his judgments of Edmond seemed at times also to be harsh.[21]

FATHER AND DAUGHTER

Maillart worried about his daughter Marie-Claire, especially in early 1924, when she engaged in a romance with a well-known older man, causing a minor scandal in Geneva. Harsh criticism from her family and from most of her acquaintances offended Marie-Claire so deeply that she decided to leave the city. She had friends in Geneva, but she did not feel a part of that old, patrician city. Her extraordinary experiences since 1914 made her feel older and more independent than her schoolmates, who were growing up in a secure traditional society into which they would marry and settle down. Marie-Claire had a strongly developed will of her own; she had a wanderlust, a love of adventure like her mother, and the sharp judgmental tendencies of her father. Maillart could give her no home, hardly any allowance, and no hope for a secure future. When Marie-Claire left Geneva, her Italian aunt in Bern (Alfred's wife) took her in, and by late May, Maillart enrolled her in a domestic science school at nearby Worb. Marie-Claire hated this type of school and resented her rejection in Geneva.

Once established at Worb, she began a correspondence with Maillart similar to the one he had with Maria. Gradually, she began to confide in her father her loneliness and her loss of faith. Her confidence gave him the deepest pleasure: "I am so happy that you wrote me so openly and I will give you my thoughts," he wrote on August 1. Maillart had found the companion he most needed for the remainder of his life: not the continuous physical presence of someone to fuss over him, but rather someone who would occupy his thoughts and who would become part of his inner life.[22] Father and daughter had found a bond that could mature and strengthen. In early fall, she entered the Moravian Girls School of Montmirail, where to her great relief, she could study landscape gardening and design instead of housekeeping.[23]

After she had only one semester at Montmirail, Maillart had to admit to his daughter that his financial straits were such that he would have to delay payment of her tuition.[24] Responding to her concern, he explained how poor finances, while restrictive, had little to do with happiness, which he attributed to how one "takes life."[25] He observed that her loss of religion had made her "find that life is ugly." Maillart did not attend church, but he attributed his strong moral sense to Christianity. Rather than counsel a return to conventional religion, he advised her to develop her character to the point where it gives the "greatest possible return (as one speaks about a machine)."

Maillart tried to express to his daughter his own satisfaction with a life that was outside conventional society and free of the vagaries associated with worldly success. The solutions he was beginning to find in his work helped the profession, effected economies for clients, and benefited the public without producing great wealth for himself.

He did not wish her to think of work as a necessary evil but as something that could "fully fill your being. Do you have the 'sacred fire' [for work] or not, that is the question!" Then he expressed his faith that "if yes, the just reward will come by itself; if no, you risk losing satisfaction even if you achieve a good report." His advice was a confession of the

Figure 76. Arve River Bridge at Marignier, 1920: 1911 design by French authorities, modified in 1920 by Maillart. (*Source:* Slide by David P. Billington)

inner struggle he had fought since his return to Geneva. His burned-out feeling had not extinguished his own "sacred fire."

Attacking the Establishment Education and Theory, 1920–1924

THE GENEVA BROCHURE

The promotional brochure that Maillart prepared for the French market in 1920 provides a summary of a most remarkable one-man career in reinforced concrete up to World War I. It includes three major aspects of Maillart's experience: a broad spectrum of special structures (eleven of them), ranging from the 1902 gas tanks to the concrete-sheet piling (a wall of thin strips driven into the ground) for the canal at Wangen; the ten most significant bridges; and his major flat-slab buildings plus the Riga factory and the 1920 Barcelona power plant. Throughout the brochure, Maillart emphasized aesthetics as well as efficiency.[26]

Considering that the brochure summarizes 18 years of work by one designer, it is a record that had no equal at that time. Much larger operations such as the Wayss and Freytag, Dyckerhoff and Widmann, and Hennebique companies had completed many more works, but none with such an array of unusual forms, and certainly no individual designer exceeded the breadth of Maillart's total productivity. And yet if Maillart's career had ended here, he would have become now only a footnote in the history of engineering. Some historians of architecture might have stumbled on the Tavanasa Bridge and some historian of building might have recognized his flat-slab innovation. But his technical mastery of concrete structures and his stunning visual sensibility would only become evident during the last two decades of his life. The building preparation, illustrated in the Geneva brochure of 1920, was indispensable to the later works, but in itself it would not merit a biographical study.

In the early 1920s, events led Maillart in an unanticipated direction that would demonstrate publically his technical mastery of structural engineering. But first he needed some ordinary contracts to get an office organized.

Armed with his new promotional material, Maillart began serious talks in the French capital and finally succeeded in landing one project, for a bridge near Bonneville in France, 35 kilometers southeast of Geneva. This three-span arch bridge over the Arve River at Marignier had been designed in 1911 by the public works department of the Haute Savoie; apparently because of the war, it had not been built. The detailed drawings and calculations for which he was hired would at last occupy his office. The bridge form (Fig. 76) was given to him; he could therefore turn over the work of calculation, drawing, and specifications to his small staff in Geneva. The Marignier Bridge marked the start of Maillart's postwar consulting practice, but because it does not represent a Maillart form, he never spoke of it as his design.[27]

Other prospects eluded him.[28] He was, if not depressed, simply bored; the vain search for commissions left him intellectually unsatisfied. And so for the first time in some years, he turned to the technical literature and began to study once again the state of structural engineering that had ceased to engage him by 1905, when the activity of his own company dominated his thoughts. The events that

stirred his creativity were his own inactivity and the vigorous actions of a man soon to become Switzerland's most politically powerful engineer.

THE CRITIQUE OF EDUCATION

Maillart's reflections came about primarily as a result of his reaction to the technical research and aesthetic ideas of Zurich's fastest rising academic star, Arthur Rohn (1878–1956). Because of Maillart's reputation as a designer and builder in 1920, the alumni association of the ETH had appointed him a member of its six-man committee to review the institute's curriculum.[29] With his appointment, there began a series of debates between Rohn and Maillart that would make public all the ideas that Max Ritter, François Schüle, and Robert Moser had brought to bear against Maillart before the war. By 1919, Rohn had begun to speak publicly in various Swiss cities on bridge aesthetics[30] while intensifying the research program in structural engineering at the ETH, where he began to supervise a series of advanced students.[31] In 1921, he was elected president of the Swiss Society of Engineers and Architects, and he quickly became the principal Swiss engineer to judge major bridge competitions and to provide structural consulting (Fig. 77).

As far as published documents can tell, although Rohn often consulted on bridges and sat on juries, he designed few (if any) bridges in Switzerland. Even his 8-year practice in Germany is referred to in obituaries and testimonials not as evidence of design but of administration.[32] He was a natural director, organizer, and administrator, but his technical understanding was not in the same class with that of his predecessors in the bridge structures chair: Culmann, Wilhelm Ritter, and Mörsch. He made no lasting contribution to structural theory as they had done and he created no structures that can be identified easily as his designs. Rohn felt the pressure of comparison, but in Switzerland, few challenged him. In such a small country, the influence of one man can be great, particularly when he is put at the young age of 30 into the most honored Swiss professorship in civil engineering. Rohn is re-

Figure 77. Arthur Rohn (1878–1956). (*Source: Schweizerische Bauzeitung*)

membered as "the lion" with a regal, some would say, arrogant personality. Maillart was annoyed not only that Rohn dominated bridge design competitions and pressed for more academic research, but also that he seemed to identify himself with German technical traditions.

In an extraordinary action perhaps unparalleled in the history of structures, between 1920 and 1923, Maillart criticized publicly each of Rohn's research projects in a series of published papers: on channel beams, on rivets, and on dams. Maillart's articles illustrate his contentious nature as well as his insight into structural performance and especially reinforced concrete.

Unlike designers in steel, those who designed in concrete also had to determine how it would be fabricated, thus imposing a severe discipline on form. The new material, literally an unformed semi-fluid mass, provided the means for structural art only when its nature was understood.

There were at least two impediments to this understanding: first, the danger of scientific intimidation wherein the designer dwells too much on the uncertainties in mathematical formulas; this was what Maillart criticized in Rohn's research program. The second impediment was the belief that the creation of the most poetic forms demanded an aesthetic superimposed on the new material. Le-Corbusier wrote that the engineer created harmony through calculations, but that the architect "realizes an order which is a pure creation of his spirit." The engineer was bound by calculation and rational thinking, whereas the architect added fantasy and hence beauty.[33] Of course, LeCorbusier wrote for architects and thus uses engineering aesthetics as a means to prod his own profession to escape the same twin pitfalls that Maillart felt in the early 1920s, that is, either a thoughtless rationalism, sometimes called functionalism, or a sterile imitation of past styles. With Maillart, these two pitfalls seemed to be obstacles erected by the work of Rohn.

Moreover, these two obstacles to the new art of structural engineering reinforced one another. Rohn tried to see engineering as mere calculation, and, consequently, when he turned to aesthetics, he believed in the need for an architect. LeCorbusier saw beauty in architecture, which is certainly true, but he used engineering only as a reflection of calculation. Why do applied aesthetics (building as art) and applied science (building as formula) seem to reinforce each other and to have elicited such a strong reaction from Maillart?

The reason lies in the intimate connection between calculation and aesthetics. As any mathematician knows, there is a strong aesthetic inherent in calculation; the word elegant is as much a part of the mathematician's vocabulary as it is of the philosopher's. Mathematical elegance refers to the sparse beauty in a proof with the least amount of calculation – the fewest steps in the argument. The simplest, most direct solution is always preferred and is felt by the mathematician to have a rare beauty.

This type of elegance is part of the beauty inherent in the structural engineer's aesthetic. For the engineer, calculation represents not only a mathematical argument, but also a physical model. It is a model of something made by people and hence subjected to the patterns of society (economics of construction) as well as to the laws of nature (gravity and wind). These natural laws do not explain those social patterns; the natural and the social are independent variables. Any attempt to create great accuracy in the physical model is rendered useless by the imprecision inherent in the social model – represented mostly by costs and politics.

One way out of this complex interaction between material and social forces is by specialization, whereby the elements of each project are addressed by a team – architecture (aesthetics), engineering (calculations), building (costs), and ownership (functions). This scheme assumes that somehow all the elements will be connected in a final product of quality; the leader becomes a manager.

Another way out of the problem is for one person to integrate all the elements into a single design; the leader becomes a designer. The only way one person can encompass diverse aspects of the design of a large structure is to simplify radically each element while retaining its essential features. Here the aesthetic of elegance becomes essential because no mind can hold together all the subtleties of natural science, mathematics, social science, and appearance. Integration is the key and it can unlock the designer's imagination only if the elements are clear and simple.

Therefore, the designer who seeks to integrate all aspects of a structure will be forced to simplify each element. Many of the best artists of the early 1920s suspected that this sacrifice of refinement was the essential basis for an aesthetic compatible with modern technology, but they could not visualize how anything beautiful could arise from such an austere art. In the new material of the twentieth century, reinforced concrete, Maillart was pioneer-

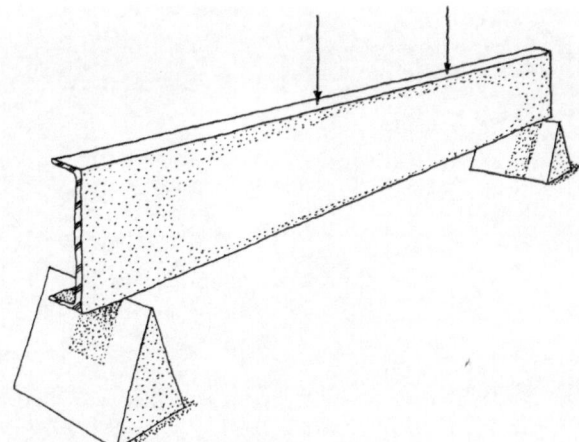

Figure 78. Channel beam. (*Source:* Drawing by Mark Reed)

ing a new art form that allowed for a personal aesthetic expression – a sense of creative freedom – out of the search for elegant forms that could unite efficiency and economy but transcend them.

Taken together with his earlier inventions of the hollow box and the flat slab, Maillart's new studies have in common his search for new forms that gain strength and formal refinement from an integrated shape rather than through mass. He sought to free the designer from arbitrary rules by describing structure through simple mathematical formulas and through the performance of works in service. He delighted in seeing larger issues within the purely technical debates of this time and that delight included challenging the highest authorities. Maillart respected the authority of nature – the ravine, the rock, the laws of gravity and wind – and the properties of concrete. But he was suspicious of the authority imposed by society – the textbook, the codes, the teachings of professors, and the professional consensus.

Maillart began to attack Rohn's approach indirectly late in 1920 by criticizing the work of Carl Bach, Germany's leading experimental research professor for structures.[34] Maillart's later review of Bach's major publication, *Elastizität and Festigkeit* (*Elasticity and Strength*), expressed ideas that he had developed earlier in arguments with Schüle (in 1906)

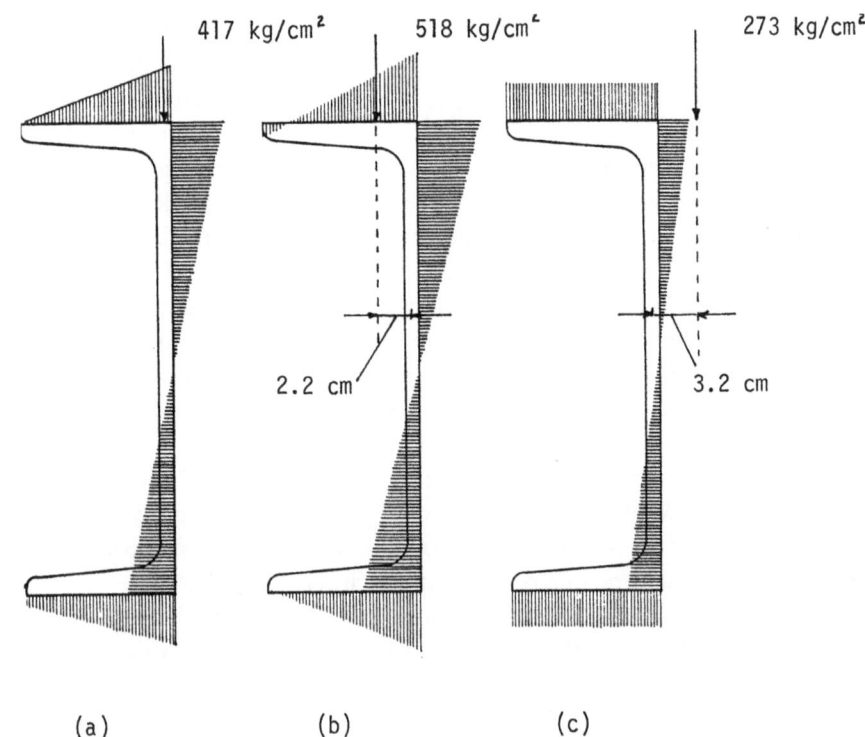

417 kg/cm²   518 kg/cm²   273 kg/cm²

2.2 cm   3.2 cm

Figure 79. Channel beam theory leading to Maillart's discovery of the shear center. (*Source:* Drawing by Mark Reed)

(a)   (b)   (c)

and would use later in his criticism of Rohn:[35]

the use of extensive formulas and number work for the solution of a technical problem will usually be honored as the highest achievement. Quite falsely. The fewer calculations the practicing engineer can get by with, the better off he will be.

This statement, which is not the writing of a typical engineering argument, is rare in any technical writing, let alone in that of the Swiss engineering profession. Engineers of the early twentieth century knew that it was easier and far more economical to use mathematical predictions for structural performance than it was to try and base design on physical tests and collected field experience. But Maillart recognized that such mathematical predictions were sometimes unreliable and often difficult to check.

Between 1920 and 1921, Maillart attacked the overemphasis on mathematical theory in a series of five brief reviews in the *Bauzeitung*, the essay on Bach's book being the most extensive.[36] Beginning

in 1907, Bach had carried out a series of tests on steel channel beams (see Fig. 78) that produced results that he was unable to interpret. The problem was that the stresses Bach measured as the channel beams bent under loading [Figs. 79(a) and 79(b)] were radically different from those stresses he calculated based on the by-then classical beam theory [Fig. 79(c)]. As a good experimentalist, Bach could not dismiss his test results; as a knowledgeable academic, he could not question the proven classical theory. So he took another way out.

The channel is not a symmetrical section (about a vertical line) as is, for example, the common I-beam, for which the beam theory gives the experimentally verified correct stresses. Thus, Bach concluded, the classical beam theory holds only for symmetrical sections and there must be some other theory, which he could not as yet determine, for unsymmetrical ones.[37] The situation remained unresolved for 12 years, partly because of World War I. In 1920, following publication of the eighth edition of Bach's book, Maillart took up the problem and solved it by invoking a simple physical idea that he called the *shear center*.[38]

THE SHEAR-CENTER CONTROVERSY

Coincidentally, one of Rohn's students, Adolph Eggenschwyler, who was working on this same problem, published an article on it in December 1920 after Maillart had submitted his own article but before it appeared in the *Bauzeitung*.[39] Unlike Eggenschwyler and others later, Maillart's approach was primarily physical, not mathematical, and it began with clear simplicity. He took Bach's results for two different tests, which showed "errors" of increasing sizes as the load was moved to the left by 2.2 cm [see Fig. 79(b)]. Maillart argued that the "error" would be reduced by moving the load to the right and that there would be one position at which the "error" disappears. How to find that position?

In effect, Maillart said the following: If I start at the correct point (unknown, at first) and move the load a distance x to a point over the web [the vertical part of the channel, Fig. 79(a)], then the "error" will be (the stress as found by Bach) 144 kg/cm^2 (417 − 273). If I move it 2.2 cm farther to the left, the "error" grows to 245 kg/cm^2 (518 − 273) as Bach found. Therefore, the distance x must be proportional to 144 as the distance $x + 2.2$ is proportional to 245, from which $x = 3.2$ cm. In this way, Maillart located the shear center in thin air 3.2 cm from the channel web [Fig. 79(c)].[40]

It remained for Maillart to develop a theoretical and general proof of this remarkably simple explanation that had eluded students of mechanics. Maillart provided this proof in 1921, but it would not be fully accepted for three more years, opposed vehemently by Rohn to the end. Maillart wrote a second article, in reaction to one by a German engineer, in which he drew the broader implications of overreliance on academic experts by an uncritical profession. He also stressed the need for tests to confirm his theory.[41]

The following year, Rohn, in his response to another Maillart criticism, conceded that Maillart had given "a clear-sighted treatment of the center-of-

shear,"[42] but he retracted that praise early in 1924 after publication of Maillart's detailed article reporting on the results from a testing program carried out at his suggestion. In that article, Maillart demonstrated that the channel-beam tests made by Mirko Ros, a Yugoslavian engineer living in Baden, Switzerland, confirmed his shear-center theory of 1920: The problems lay not with Bach's purported anomaly, but rather in his faulty interpretation of test results.[43] Maillart was not satisfied with a few new test results; he wanted to stress a more general idea. "One may ask oneself, whether basic errors found in the textbooks and also presented in most school courses will be adhered to in the future." He then proceeded to show how mistakes such as Bach's appeared also in another major German work.[44]

Such provocative statements finally brought forth a direct response from Rohn, who on March 22 insisted that the center of shear was not an inherent characteristic of the cross-section, which Maillart claimed it was. Rohn also declared that Maillart's ideas did not work for angle-section beams and that tests were needed to develop a new theory.[45] On April 12, Maillart challenged Rohn to present evidence rather than opinions, and demonstrated that his method did work for angle sections.[46] The debate had become as much an attack on, and defense of, the Zurich establishment as it was a technical argument.

By this time, enough literature had appeared in which Maillart's ideas were accepted even if not always credited to him. On the major technical questions, Maillart was correct. The center of shear is a characteristic of the section as is the center of gravity, the type of loading Maillart did not consider does not influence the general applicability of his ideas, and the angle section is easily solved for a center of shear.[47]

Rohn's weakness arose in part from his recent election as Rector of the ETH and partly with his lack of deep technical insight.[48] He was recognized as a good administrator, but not as an engineer in the same class with Wilhelm Ritter, Emil Mörsch, or Robert Maillart. Even Eggenschwyler found fault with his professor's arguments and on May 31 publicly agreed almost fully with Maillart.[49] In that same issue of the *Bauzeitung*, Maillart finally closed the shear-center debate on what was for him the central technical issue:

After all one may not award preeminence to the mathematical treatment over the experimental one. Rather both take their lead from the purely rational [*verständnismässige*] or even intuitive [*gefühlmässige*] consideration of the essentials of the thing, otherwise mathematical developments as well as test results will be superficial, groundless and without support. Only when a matter is thought through first, will it be comprehended by calculation and experiment.

The center of shear was the most historically significant Rohn–Maillart debate because of its influence on the profession. The leading twentieth-century scholar of the history of the strength of materials, S. Timoshenko, wrote in 1953 that "this question [of the shear center] was cleared up by R. Maillart who introduced the notion of the center of shear and showed how this point could be found."[50]

Eric Reissner, one of the outstanding theorists in late twentieth-century structural engineering, further noted that "What strikes a contemporary reader of Maillart's article is his *understanding* of the physical aspects of a problem and his ability to give an essentially correct simple mathematical analysis of this problem. This at a time when the 'authorities' were not yet prepared to undertake such an analysis."[51]

By the spring of 1924, Maillart had settled the shear-center issue at the cost of losing all rapport with the powerful Rohn and his new protégé, Max Ritter, who was soon to follow Rohn as professor in Zurich. By publicly chastising these authorities, by proving them wrong, and by drawing from the specific technical issues sweeping generalizations about academic shortcomings, Maillart put himself in permanent opposition to those engineers who controlled Swiss technical education and set the standards for structural theory after World War I.

Attacking the Establishment: Research and Practice, 1920–1924

ALPINE FAILURES

Maillart's concern with theory and education and the professional conflicts into which this concern led him were only a part of an activity in the early 1920s that included also a series of designs aimed once again at Rohn. They illustrate the insatiable inquisitiveness of Maillart's mind, and in each case, his technical originality presented a public affront to the academician's prestige. In this period, Rohn was a consultant, a researcher, and a designer, and Maillart published sharp criticisms of each aspect of his adversary's engineering actions. First were the consulting activities of Rohn on hydroelectric power plants.

In general, the rapidly increasing demand for electric power throughout Europe and, in particular, the natural fast-flowing Swiss rivers with high mountain lakes, all led to a focus on hydroelectric power-plant construction. Therefore, following the war, hydroelectric power was the major technical problem in Switzerland. The pages of the two Swiss building journals, the *Schweizerische Bauzeitung* and the *Bulletin technique de la suisse romande*, were filled with technical descriptions and scientific analyses centered around construction and operation of power plants.[52] Of all these plants, the most noted structural difficulty arose at the Ritom Lake high above the Tessin River just south of the Gotthard Pass. This difficulty along with Rohn's defective consulting report on it stimulated Maillart to study the problem of pressurized tunnels (Druckstollen).

The plant was built under the direction of the Swiss National Railways as one of two generating stations for electrification of the Gotthard rail line. The tunnel, carrying water from the high mountain lake down to the power turbines in the valley, was concreted late in 1919 and connected to the plant on February 20, 1920. On May 4, the first pressure test showed a loss of water through leakage. After emptying the tunnel, numerous cracks, mostly horizontal, were observed. Between May 7 and June 29, further tests only aggravated the cracking. Finally on July 1, massive leaking from the last test led to a washing away of trees on the slopes below the tunnel. Shaken and fearing a disaster, the railroad management called in Ferdinand Rothpletz, an expert on tunnel construction. After a site visit, Rothpletz in turn called in Arthur Rohn and Jakob Büchi of Zurich, an engineer with long experience in hydroelectric construction.[53] This three-man Ritom commission, after site visits and new tests, produced a detailed *Report* that was published in seven consecutive issues of the *Bulletin technique*.[54]

As soon as the first difficulty was spotted, the *Bauzeitung* began to run a series of articles on the plant.[55] Noted for his independence and his penchant for criticism, the veteran editor August Jegher (1843–1924) wrote an outspoken comment for the August 21 issue. Particularly repugnant to Jegher was the specific contentions that there had been no failure, that the commission had everything well in hand, and that it was making tests to find out what pressure the pipes should carry. This defense of the commission's work stated that such tests were necessary for every power plant using pressurized tunnels, a patent lie because the allowable pressure is always established by design. This idea is the same as one that would claim that a bridge is loaded after construction in order to determine what type of traffic it can carry.

The *Bauzeitung* editor identified the political nature of the Ritom commission and the questionable technical ideas it had produced. He ended his article with what he called an engineer's ethic:

These latest conditions bring us close to the seriousness of the times in which we live. These remind all our colleagues in the construction field of their obligations, in so far as it is possible, to fight against such frivolous political whitewashing in city or country and in every place and at every time to strive for the full, uncosmetized truth.[56]

Maillart stayed out of the political and ethical debate, preferring to focus on technical questions. He began with an analysis of the Ritom difficulties in the spring of 1921 after the commission's report had appeared.[57] The difference between Rohn's and Maillart's ideals is illustrated by comparing their respective solutions. Rohn's commission recommended a "herculean" solution consisting of a tunnel lining made of precast blocks that were then to be covered by a thin layer of concrete cast directly against the blocks and heavily reinforced with two ring layers of steel – one near the blocks and one near the inside face of the concrete lining. Finally, a thin layer of mortar would cover the concrete [Fig. 80(a)].[58]

By contrast, Maillart designed a concrete liner cast directly against the rock without any reinforcement, using reinforcing steel only in a thin layer of high-quality concrete cast over the liner [Fig. 80(b)]. This more practical design would require only simple calculations, cost less, and would control cracking more effectively.[59] Maillart's solution was later adopted for many projects both in Switzerland and in Austria.[60]

One such project was the power plant near Amsteg, for which the Rohn commission design proved so difficult that it was abandoned after 100 meters. Both the chief engineer at Amsteg and another member of the Rohn commission urged that Maillart's design be adopted and it was. The contrast between the Rohn commission design and Maillart's is dramatically illustrated by the economy and elegance of the latter. It is much thinner (32 vs. 57 cm), it has only one layer of reinforcement versus two, and it provides greater security against cracking.[61]

In late January 1921 for the Kloster-Küblis Plant, Maillart drew up a solution to the pressurized-tunnel problem similar to his design for Ritom and again very different from Rohn's approach.[62] Here the problem was to design a concrete lining to run more than 8 kilometers through a tunnel carrying water from a high mountain lake down into the valley to turbines.[63] Construction began by first drilling a circular tunnel and then placing an un-

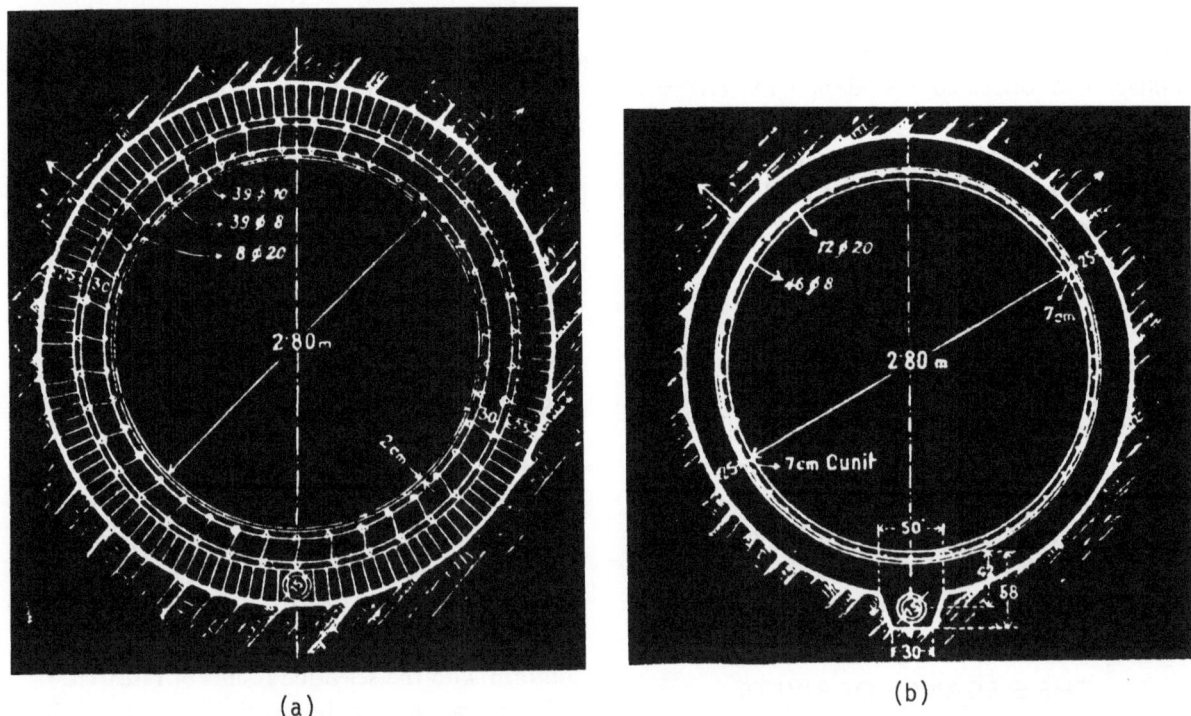

(a)   (b)

Figure 80. Pressurized tunnels: (a) Commission design; (b) Maillart design. (*Source: Bulletin technique de la suisse romande* 47 (5–11) (March 5 to May 28, 1921):104–5)

reinforced-concrete lining by spraying an inexpensive type of concrete over the rough rock surface. Reinforcement was then placed over the lining and covered by a high-quality sprayed concrete. The virtue of Maillart's design was its low cost and its assurance of being watertight.

The last part of Maillart's tunnel treatise appeared in the *Bulletin* early in 1923. Shortly thereafter, he published a paper in the *Bauzeitung* that was a direct result of that treatise. This article on the design of tunneling shows Maillart's willingness to expand his interests – in this case to geology – successfully criticizing the standard theory of rock pressure developed as early as 1878 by the noted Swiss geologist Albert Heim (1849–1937). Heim had worried about the safety of deep tunnels because a rock's internal vertical pressure increases as underground depth increases: At great depths, the roof of a tunnel would cave in. Maillart argued that

safety was not controlled by the weight above a tunnel, because at great depths there is also more horizontal pressure (like the pressure water exerts on the wall of a tank) that prevents the rock from caving in vertically as long as the tunnel is circular in shape.[64]

Maillart used his studies of concrete arch structures to explain natural phenomena. His grasp of the engineered structures became a means to better understand the natural world: science as applied engineering. Here he approached the work of Heim critically just as he had the work of Rohn's students, but it was not to discredit Heim: rather it was to modify the earlier work and to lay a basis for safe and economical design. Unlike the situation with Rohn, he made it clear that his results were "in principle not contradictory to the views of Prof. Heim." As soon as the article appeared, he sent a copy to Heim, who responded promptly: Although he did not fully agree with Maillart, he conceded that the engineer's basic approach was sound.[65] Maillart had discussed his ideas with Professor Charles Andreae (1874–1964) in Zurich that winter. In his own article on the same question, Andreae, who was to succeed Rohn as Rector of the ETH, observed that Maillart's work represented "the most satisfactory discussion of the problem so far."[66]

In these theoretical studies, Maillart was putting ideas together in new and surprisingly simple ways. He enjoyed the challenge of modifying the results of one of the best-known scientists in Switzerland. Although his theoretical work is now largely forgotten, it illuminates his approach. He could work creatively within the discipline of engineering – which he accepted – seeking to clarify analysis that would facilitate the practitioner's task.

THE ELEGANCE OF RIVETS

In his reactions to Rohn's research, Maillart could get carried away by ideas that were theoretically intriguing but practically unrealistic. When his mind probed incisively into areas in which he had little construction experience, the results were interesting intellectually, but were sometimes technically too complex. One such example appeared in his 1923 study of rivet connections for steel structures. Rohn had approached the same problem through a detailed laboratory program, whereas Maillart, with no such resources at his disposal, proposed new design ideas, which gave the riveted plates an elegant shape that would improve their performance. (Maillart had begun his rivet study in 1920, applied for a German patent on the idea in early 1921, applied for a Swiss patent in 1922, and wrote up the results for publication on July 28, 1923.)[67]

The standard method of connecting two steel plates was by inserting solid-steel cylinders – the rivets – into matching holes in each plate [Fig. 81(a)]. Maillart observed that a basic defect in usual practice is the neglect of fatigue, or changing stresses, as in a railroad bridge under locomotive loads. Steel has a much lower failure stress where loads come and go than it does where loads are permanent. Thus, if the design load (usually about half of the failure loads) results in high stresses in some rivets, those rivets will be more susceptible to fatigue and hence early failure.

Maillart's solution to this problem was to shape the plates connected by the rivets; he changed their width and their thickness in ways calculated to stress each rivet equally [Fig. 81(b)]. In this way, the plates are designed to give a calculated result: The designer becomes the active agent in determining how the structural detail behaves. By contrast, in the cases that Rohn was studying at the Technical Institute, the researcher accepted the connection details given by practice and simply studied how they worked. For Maillart, the designer was an active participant in setting the performance; for Rohn, he was a passive observer, who then had to conform with the scientific results of research.[68]

Rohn was so upset by Maillart's article on rivets that even though he was on vacation, he immediately sent a response to the *Bauzeitung*. He claimed Maillart's article could not go unanswered because "it gives the impression to the nonspecialist in steel structures that such an essential perspective has been up to now overlooked in structural steel design."[69] Without addressing the problem of changing loads and fatigue, Rohn insisted that overstressing in some rivets was insignificant. He concluded that Maillart's theory was faulty and his ideas useless in practice. Maillart's arguments about rivets were essentially correct, but his ideas for finely shaped rivet plates were not, in fact, practical in steel construction.[70]

The rivet debate illustrates Maillart's urge to create new forms in even the most obscure aspects of structure. Of equal significance was his insight into the problem of fatigue and here he was clearly ahead of his time. No writing on this problem predates Maillart's paper.[71] Probably because the *Bauzeitung* was not widely read in the United States (only three libraries have complete sets), Maillart's study never entered the mainstream literature on the fatigue of riveted joints. Beginning in 1924,

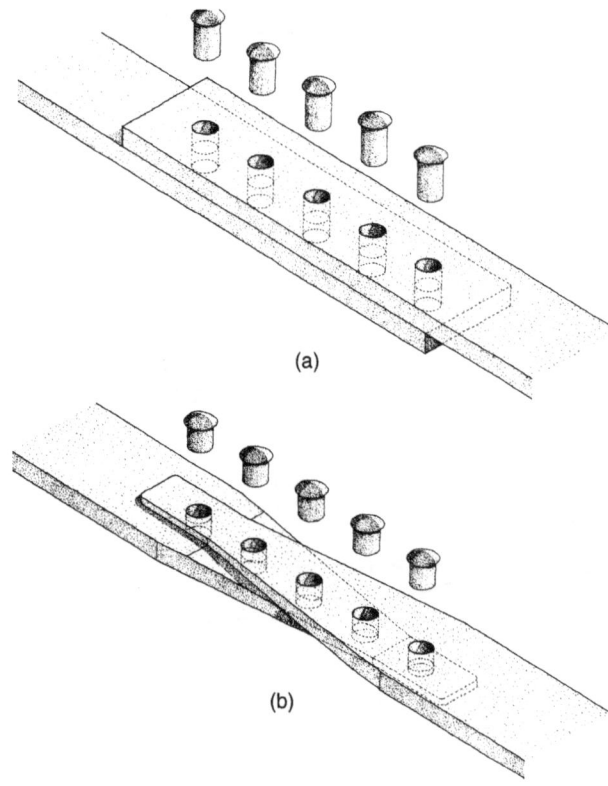

(a)

(b)

Figure 81. (a) Conventional rivet connection; (b) Maillart patented rivet connection. (*Source:* Drawings by Mark Reed)

when Maillart began finally to get numerous design commissions, these problems occupied him less. At the same time, the deepening gulf with Rohn began to have a direct impact on his design career.

ROHN CONTROLS THE ZÄHRINGER BRIDGE COMPETITION

Between 1920 and 1923, the *Bauzeitung* discussed six major bridge projects, four of which Rohn had played a central part. Nowhere are his ideas seen more clearly than in the role he played for the Zähringer Bridge competition, announced in 1920 but not reported on until April, 1923.[72] This was an unusual competition because the city of Fribourg commissioned two official designs. One was for a rebuilt suspension bridge of 271.6 meters free span to replace the suspension bridge that, when com-

pleted in 1835, had the longest span in the world (273 meters = 900 feet); the other was for a 56-meter concrete arch and a 190-meter viaduct on long concrete piers designed by Jaeger and Lusser, who had designed the Pérolles bridge nearby. Builders could make bids either on the official designs or on one of their own. There were twenty-seven submissions, including one by Maillart in early 1921, and the "expert commission" (not officially a jury because this was not a normal competition) met in early March to study them.

The commission was unhappy with all the entries and finally decided to make its own design – a seven-span concrete viaduct – which it then asked Jaeger and Lusser to detail [Fig. 82(a)]. In April 1922, having just completed construction of the Pérolles bridge, Züblin and Company was given the contract to build the new Zähringer Bridge. When Rohn's article on the bridge appeared in April 1923, construction was already well advanced. Rohn was clearly the leader of the expert commission and the choice of a final design was largely due to him.[73]

Rohn's article, presenting publicly the commission's arguments for rejecting other designs in favor of its own, characterizes the fundamental differences between himself and Maillart. Rohn faulted Maillart's design for fitting neither the valley nor the city profile. He then proceeded to justify his commission's stonelike viaduct because it had the form "of a classical Roman valley crossing." His solution did not grow out of modern engineering, but rather out of a superficial overview of the landscape coupled with a preference for ancient forms made of traditional materials. Rohn thereby picked up where Robert Moser had left off before the war.

Unlike the cases of the pressurized tunnels and the rivets, Maillart did not attack Rohn publically, rather he left that to his friend, Carl Jegher. The *Bauzeitung* editor recognized Rohn's close-minded attitude right away, noting that the reasons Rohn gave for the decisions were based "to an unusual degree upon subjective feelings." He made it clear that he did not agree with them.[74]

Even though Maillart's design represented an entirely different conception from Rohn's – forms ap-

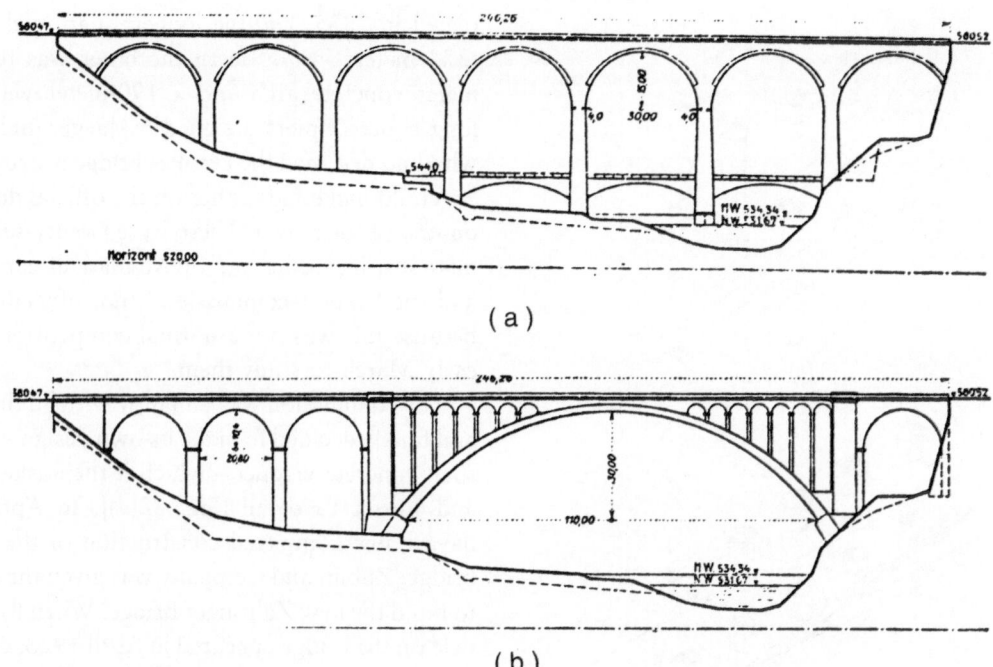

(a)

(b)

Figure 82. (a) Zähringer Bridge by the Consulting Commission (led by Arthur Rohn), 1923. (b) Zähringer Bridge competition design by Maillart, 1921. (*Source: Schweizerische Bauzeitung*)

propriate to concrete rather than to stone – Maillart's Zähringer arch of 1921 resembled more his Pérolles design of 1908 than it did his later designs. Maillart's main arch had a modern profile, but he used traditional Roman arch forms for the short spans and designed heavier pillars above the main arch springings [Fig. 82(b)]. Like the prewar Pérolles and Lorraine proposals, this design is daring in scale, but not yet characteristic of a new structural idea. That new idea was slowly developing in Maillart's mind in 1923, and, as usual with him, it would need careful testing on small projects, before it could become a full-scale realization.

In early July, Maillart received news that promised trouble for him within the Swiss engineering community: Rohn had been elected – by a large majority – Rector of the ETH. Maillart's chief antagonist had become Switzerland's chief engineer. It is not impossible that Maillart had hoped to become a professor in Zurich. Certainly his research

between 1920–3 merited a professorship and his concern about proper technical education implies it. With the election of Rohn as Rector that possibility was gone. More importantly, the bridge design competitions would now be even more securely under Rohn's control. Principally because of Rohn, Maillart lost each of the three major bridge competitions of 1923 – Zähringer, Baden-Wettingen, and Sitter – all rejected in favor of ideas that were essentially architectural, not engineering.[75] This criterion, that a bridge be "architectural," became more entrenched in Switzerland thanks to Rohn and to his increasingly influential ally, the German architect Paul Bonatz.[76]

From 1920 to 1924, Maillart's hostility to Rohn intensified. At every public occasion, it seemed, Maillart spoke out against Rohn.[77] But Maillart could not afford to let his quarrel with the establishment prey on his mind; he was too preoccupied with securing enough standard, large-scale work to

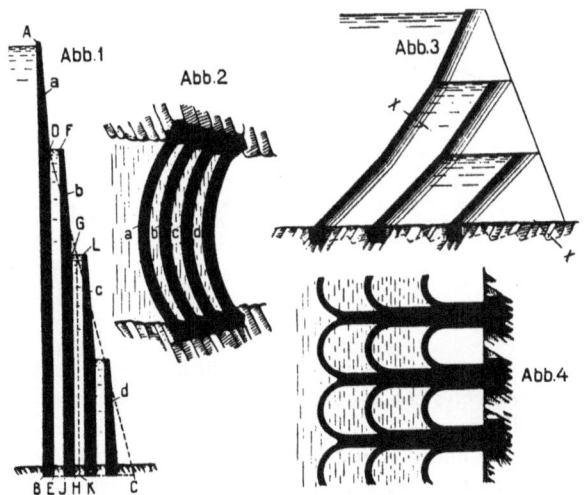

Figure 83. Dam design patent by Maillart. (*Source:* Swiss Patent Office, Bern)

pay off his debts and to reestablish a profitable business.

New Business, New Forms
1921–1924

REHABILITATION

Even within the more modest capacity of a consultant, for which large capital outlays are unnecessary, it took Maillart over 2 years to find a job in Zurich that would provide work for his small office staff. Only in mid-July 1921 did he receive the commission for a multistory office building for the Rentch Company on the Reitergasse. For it he decided to use his flat-slab design: He needed only to lay out the scheme; any one of his engineers with a draftsman could complete the design and drawings.

In February 1922, Maillart received his first important commission in Geneva: the Swiss National Bank building.[78] This, followed shortly by four other Geneva projects, allowed him to hire Albert Huber, an experienced engineer, to take on much of the calculation and supervision of the draftsman. His discouraging trip to Riga the following year and his decision in 1925 to withdraw from the Barcelona office would allow Maillart to shift his attention away from international expansion and to focus on his Swiss practice. In 1925, another office – in Bern – would also begin to function.

Maillart kept searching for design commissions by proposing new design ideas sometimes even for structural types that he had never built. One such case was a new Swiss hydroelectric plant in the Wägital in 1922. He had devised a design for arch dams, radically reducing materials by the use of hollow-barrel arches and very thin concrete surfaces (Fig. 83). Again Maillart was indirectly attacking Rohn and his research. The professor was a consultant for a dam at the Wägital and he was supervising a doctoral student on dam design.[79] Maillart unsuccessfully proposed this design for a dam at Rochemolles in September 1921 at the same time that he filed for a patent in Germany. The following year, he patented the idea in Switzerland.[80]

In 1927, Maillart would begin to rethink the problem of arch dam design when a *Bauzeitung* article on French arch dams prompted him to revise his patent essay slightly and publish it for the first time. Then in a second *Bauzeitung* article of 1928, Maillart showed that within certain limits the thinner the dam, the better the design, that is, the lower the stresses, just the goal that he always sought in his own designs.[81]

Like his rivet design, Maillart's dam idea was soundly based on the performance of structures under load but was judged impractical by the engineering profession, which never built any major dam following his model. Maillart had worked on two dam designs in the early 1920s and would continue to seek such commissions, but without success, up until 1931.[82] Here were examples of ideas proposed without the construction experience to verify their economy. When he could base his ideas on full-scale tests and structures in service, they proved practical and eventually influenced the design profession.[83]

A RADICALLY NEW FORM

Rarely, even in the life of a pioneer like Maillart, does a single event transform the overall vision. But high in the Wägital Valley in 1923 one bridge would represent a major change in twentieth-century design: Maillart's deck-stiffened arch over the Flienglibach. By late summer 1923, Maillart had completed seven bridge designs and numerous miscellaneous small structures for the new hydro-electric power plant above the Lake of Zurich at Wägital, where his friend Simon Simonette had contracted to build a road above the projected water line of the dammed river.[84] For the Flienglibach, Maillart reduced the arch thickness to a practical minimum. He designed the concrete parapet so that it would stiffen this thin arch, thereby achieving a much lighter arch without additional concrete anywhere else in the structure [Fig. 85(c)]. He realized that the necessary vertical connections between arch and deck forced those two parts to act together and thus allowed the deck to stiffen the arch. Maillart had seen the potential for a hollow box in the conventionally designed Stauffacher. In the standard design for the 1920 Marignier Bridge – given to him merely to calculate and draw up – where the parapet served only as a safety barrier, he now saw the possibility for a new form in which that parapet became an integral part of the structure.[85]

Maillart's insight, however, came primarily from three other experiences: his studies with Wilhelm Ritter, the performance of his 1912 Aarburg Bridge, and the ideas that informed his Zuoz–Tavanasa designs. Ritter had written in 1883 about the interaction between arch and deck in steel bridges, noting that arch bending could be reduced by deck stiffness [Fig. 84(a)]. Maillart's class notes from Ritter's courses show deck-stiffened arches in both wood and steel. One example, the typical American covered bridge, has a thin wooden arch stiffened by

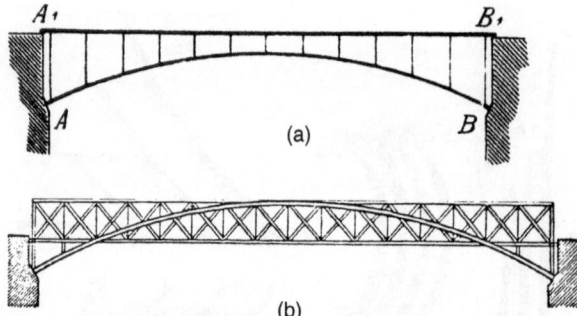

Figure 84. (a) Drawing from Wilhelm Ritter, 1883. (*Source: Schweizerische Bauzeitung* vol. 1, No. 1 January 6, 1883, pp. 6–7). (b) American truss-stiffened arch. (*Source:* Wilhelm Ritter, *Der Brückenbau in den Vereinigten Staaten Amerikas,* 1894)

a truss that serves as the sides to the covered road-way [Fig. 84(b)].[86]

The second stimulus came from cracks that had appeared in the shallow deck beams of the Aarburg Bridge after it had been in service.[87] As in Zuoz, these cracks illustrated a new idea – that the deck must act together with the arch and hence must be reinforced for that interaction. At Aarburg, the parapet looked strong, although it was actually weak, having little reinforcement, and the shallow deck beams below the parapet were not reinforced properly [Fig. 85(a)]. Maillart realized that he could have used the deep parapet, properly reinforced, as the deck beam and thereby, following Ritter's idea, he could drastically reduce the arch thickness.

Finally, Maillart's 1902 patent application gave his major bridge ideas and was a third stimulus to his thinking in 1923. In elevation, the solid wall is similar to Stauffacher (in appearance only), Zuoz, and Billwil; and in section, the deck and the arch are connected by transverse walls. Of course, in 1902, this did not mean that the transverse walls were the essential connectors of the deck and arch; Maillart stated in the patent that the connection was by the longitudinal walls. However, the patent drawings clearly suggest that the transverse walls could provide the main connection between deck

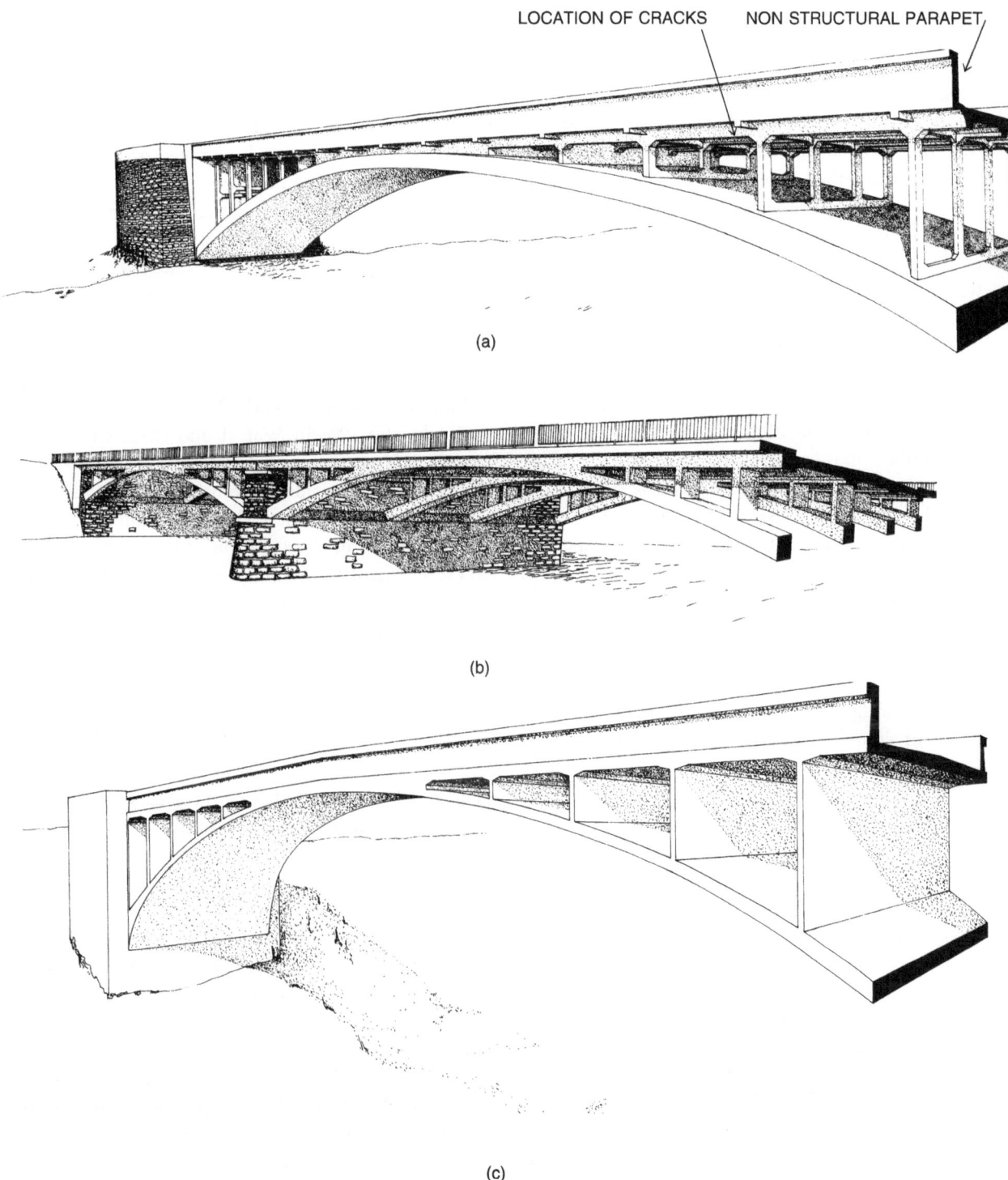

LOCATION OF CRACKS   NON STRUCTURAL PARAPET

(a)

(b)

(c)

Figure 85. (a) Aarburg Bridge, 1912; (b) Marignier Bridge; (c) Flienglibach Bridge, 1923. (*Source:* Drawings by Colin Ripley)

and slab just as the Marignier bridge implied a similar connection [Fig. 85(b)]. Maillart seemed even to imply that possibility in his patent statement: "The way in which the connecting steel bars are arranged is unimportant; what is essential is merely that the three major structural elements be well connected together. They can thus be disposed in other ways than that shown in the figure, or even built completely as a framework." Maillart had already considered the possibility of reducing the walls and leaving mainly the transverse walls as connectors. In sum, the hollow-box patent contained the seed for the deck-stiffened concept; first, by its technical rationale of connecting the deck and the arch together to save materials; and, second, by its visual suggestion of that connection by transverse rather than by longitudinal walls.

Starting with the Flienglibach Bridge [Fig. 85(c)], over the next 10 years Maillart would realize twelve deck-stiffened arch bridges ending with the spectacular Schwandbach of 1933 and the Töss of 1934. Ros's load tests confirmed their high quality and European artists marveled at their beauty, but the rancor at the ETH effectively banished Maillart's ideas from education. Max Ritter took charge of teaching bridges and embarked on a 20-year campaign of complex analytic research that ridiculed Maillart's simple calculations and effectively prevented students from learning about such new ideas.[88]

In the United States, where Maillart was almost totally unknown to engineers, academic attachment to analytic complexity seduced engineers away from new design ideas. In the 1920s, concrete-arch analysis gained a reputation for complexity through such publications as Charles Whitney's 1925 detailed mathematical elaborations of fixed arches (arches without hinges).[89] It is not surprising, therefore, that the American committee assembled to produce a design guide for arch bridges would view with trepidation the even more complex problem of fixed arches connected to spandrel walls and roadway deck. As they wrote in their final report:

The theoretical analysis of an arch with open spandrel and deck is so complicated that few engineers have ac-

quired the facility that justifies them in using it with confidence. Furthermore, the time required for such a theoretical analysis is excessive.[90]

Even America's best-known academic structural engineer, Nathan Newmark, had written his doctoral dissertation on deck-stiffened arches only to conclude in a paper published in 1938 that "The ordinary method of designing arch structure neglecting interaction is sound, in general, since structures so designed have apparently given good service."[91] He and his equally famous advisor, Hardy Cross, had recognized the possibilities for deck stiffening, but they did not see or appreciate the design potentials and they never referred to Maillart's completed works.[92]

Only after World War II and thanks to Pierre Lardy, a new teacher who replaced Ritter at the ETH, were these ideas from the early 1920s allowed to penetrate the ETH curriculum and began to influence a new generation of designers. Among these is Christian Menn, who designed numerous striking deck-stiffened arches beginning in the late 1950s. By 1972, the idea was taken up in the United States and it is now a fully accepted bridge form.[93]

THE CHIASSO SHED

Early in 1923, Maillart was to realize the same potential for integration in his design for a concrete water pipe that was to go over the Trebsenbach downstream from the main Wägital dam. He simply removed the separate (bridge) structure underneath and let the natural stiffness of the pipe bridge the stream and resist the internal water pressure (Fig. 86).[94] Also in the Wägital during the first 3 months of 1924, Maillart designed another deck-stiffened arch bridge, this one over the Schrähbach[95]; and he began to plan a warehouse and a shed roof at Chiasso. In each project, he sought to create new forms through a new integration of parts. The heavy loads of the five-story warehouse immediately suggested a flat-slab design, but the shed posed an entirely different problem that Maillart did not solve until the summer.

440
10.XII.23

Figure 86. Aqueduct over the Trebsenbach. (*Source:* Madame M.-C. Blumer-Maillart)

In the Chiasso shed, the unsupported span had to run the entire width of the warehouse (25 meters) rather than merely the 5-meter spans within the enclosed flat-slab building. The shed was to be open on three sides for the easy transfer of materials from the railway sidings and platform; it was also to be only one story in height with an overhang to line up with the overhang cantilevered from above the second floor of the adjacent warehouse. Maillart decided to use the required concrete shed roof slab in the same way that at Wägital he had used the required concrete parapets on the little bridges. He would integrate those elements with the standard load-carrying members. In each case, integration led to substantial reductions in concrete while producing a conserva-

tively stiff structure. Stimulated by his Flienglibach and Schrähbach designs, Maillart saw the main framing for this shed as a thin inverted arch connected to a stiff ribbed slab above it by means of slender vertical ties (Fig. 87).

No building design by Maillart has been subjected to more study and discussion than the Chiasso shed and yet the full origin of its form remains a mystery. There are three theories about how Maillart arrived at the truss-frame form: from modern architecture, or from purely functional requirements, or from earlier modern engineering. Some writers have noted the skeletonlike forms of certain plants, and have compared them with *Art Nouveau* decoration and with the works of the Bar-

Figure 87. The Magazzini Generali warehouse shed, Chiasso: standard view. (*Source:* Madame M.-C. Blumer-Maillart)

celona architect, Antonio Gaudí. Certainly Maillart was familiar with Gaudí's designs through visits to his Barcelona office, and he was aware of *Art Nouveau* through the *Bauzeitung,* which reported on architecture as well as engineering.[96] A close look at the shed form, however, will clearly dispel the first theory, which ascribes a direct architectural influence. Detailed analyses of structural mechanics and of the use for the shed itself show that no element departs from these purely functional requirements.[97] In spite of some visual similarities, bones and plants have fundamentally different functions and no first-rate structural engineer has ever created great works from such analogies.

But do the functions – properly understood – determine the form as assumed in the second theory? Was Maillart, by his superior grasp of structure and a careful study of use, able to discern the best possible solution to this seemingly simple problem of creating a single span, repeated many times, to cover a shed? No, Maillart did not evolve this form solely out of its functions; rather the functions provided the discipline within which he designed his form. The form is one of many which fit the functions, but for him function never dictated form.[98]

A third theory claims that its evolution came out of a new synthesis of existing forms within the tradition of structure itself. At least three suggestive structural forms already existed by August 1924 when Maillart completed the shed design. First there were the strengthened bridges on the Gotthard rail line: In the early 1920s, Maillart had often seen these bridges with their thin parabolic trusses.[99] Second was the 1897 idea of the Belgian, A. Vierendeel, for trusses without diagonals; by 1924, Vierendeel's ideas were well known throughout Europe and a bridge was in fact completed in Switzerland following those ideas in 1925.[100] The third structural form, and by far the most influential one for Maillart, was his own deck-stiffened arch bridge designs at Wägital.[101]

Unlike those bridges, the arch (inverted here) is not a curved slab in compression, but rather a concrete prism in tension. Also the verticals are thin struts rather than cross walls, as they are in the Wägital bridges. He has, therefore, retained in profile the thinness of those bridge elements while reducing them to an equal thinness horizontally. This extraordinary delicacy in the dimensions contrasts vividly with the concrete Vierendeel trusses for

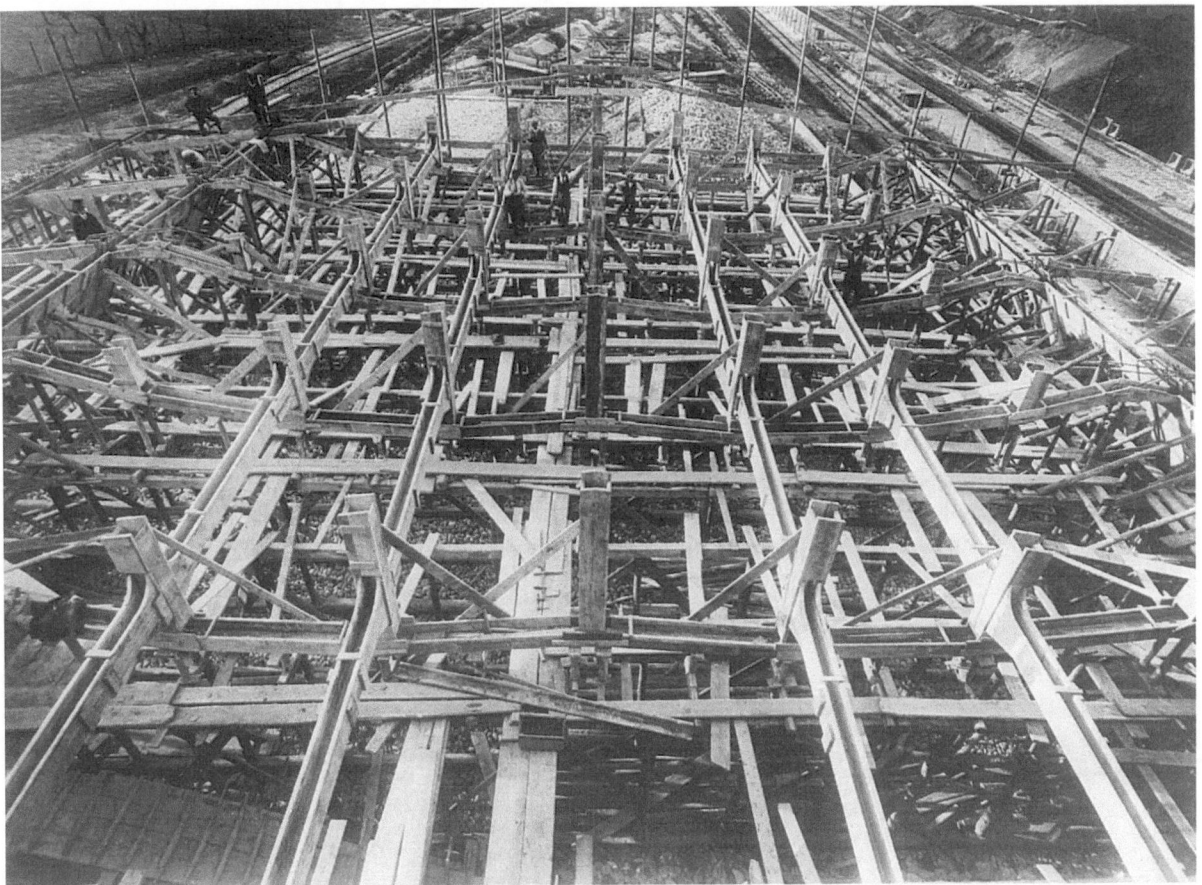

Figure 88. Construction of the Magazzini Generali warehouse shed, Chiasso. (*Source:* Madame M.-C. Blumer-Maillart)

which all elements appear heavier, especially at their junctions. The Vierendeel ideas – arising from metal truss forms – are quite different from Maillart's Chiasso concept, which is an outgrowth of the deck-stiffened idea.

The purely functional requirements of the Vierendeel and Maillart designs demonstrate that many other possible designs would be technically correct. Maillart preferred always a design giving a light appearance; his starting point was aesthetic, but his respect for function meant that each of the six frames be absolutely identical and symmetrical about its center line (Fig. 88). Within that restriction, Maillart was led by his visual imagination to a new solution for old problems. He neither copied

the past nor was influenced by the fashions of the present.[102]

THE FUNDAMENTAL IDEAS OF 1924

Both the Wägital bridges and the Chiasso shed roof are forms derived from the funicular polygon idea in which a string loaded with weights takes a shape naturally. The shape ensures that the structure (the string) will carry those vertical loads purely by tension along the string, because it cannot carry any compression (as can a column) or any bending (as can a beam). In the deck-stiffened arch, the arch form is simply the loaded string inverted so that the thin concrete slab carries the bridge dead load

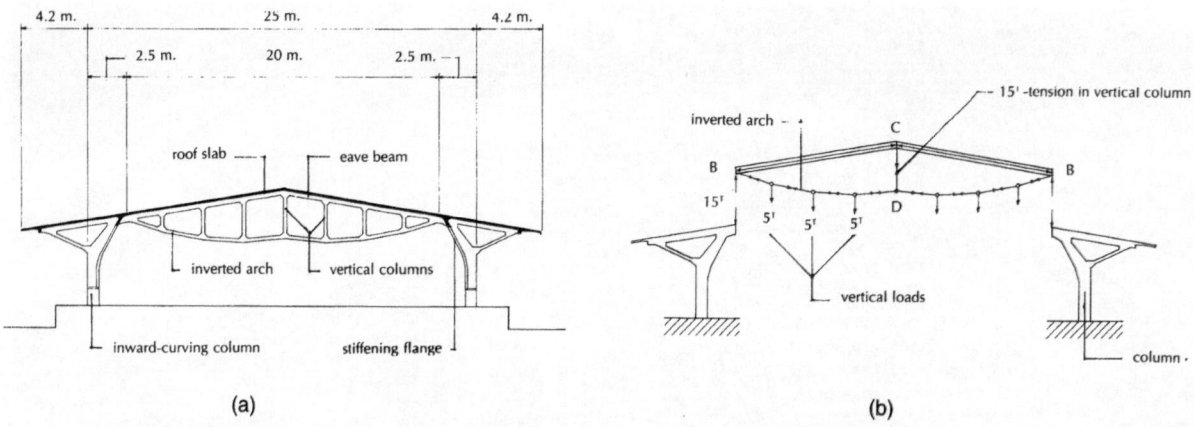

Figure 89. The Magazzini Generali warehouse shed roof, Chiasso: (a) span dimensions and structural elements; (b) imaginary separation of the 20-m center span from the column supports to show that the center structure carries vertical loads like a cable. (*Source:* Drawing by Mark Reed)

Figure 90. The Magazzini Generali warehouse shed, Chiasso, Switzerland, 1924, by Robert Maillart: profile view. (*Source:* Madame M.-C. Blumer-Maillart)

solely by compression (ideal for concrete). In the Chiasso shed roof, the string is the form hung directly from the supports and from the peaked midspan gable.[103] Maillart's first fundamental idea, therefore, is the development of a form from the string polygon (Fig. 89).

In both the bridges and the roof that form is correct only for the dead loads; the thin element below cannot carry the bending that would arise from live loads (trucks on the bridge or snow on the roof); therefore, the deck or gable serves as the beam to carry live loads by bending. They can only

do this because they interact with the thin elements and are much stiffer than those elements (Fig. 90). This is Maillart's second fundamental idea – the refinement of form from the interaction of its parts. Maillart was not the first engineer to discover or use these ideas, but he was the first to extend them to the limits of reinforced concrete and, in large part because of him, these two ideas have transformed the vision of designers after Maillart's death and the possibilities for new forms arising from his ideas have only begun to be explored by the 1990s.

Maillart had come through the first 5 years of

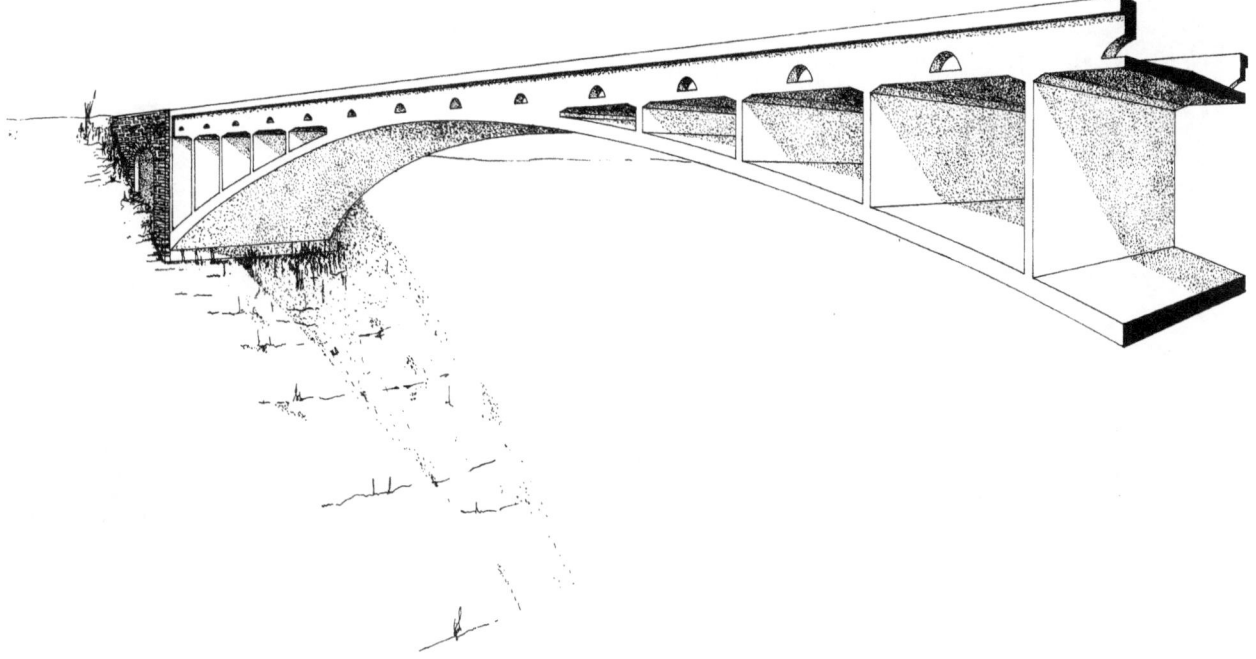

Figure 91. Valtschielbach Bridge. (*Source:* Drawing by Colin Ripley)

his impoverished reestablishment in Switzerland without much worldly profit: although his office had worked on over 100 projects between 1920 and 1924, only 20 had resulted in paying contracts. But he had found a rich source of inner strength in challenging the theories of Rohn and his colleagues in Zurich, and now as 1925 began, he found in his designs a means of expression that would create a new kind of practice freed from dependence on old forms and complex formulas.

Revival Through Bridges
1925–1927

THE BRIDGE AS POEM: VALTSCHIELBACH

Between the time he designed the Chiasso buildings and the last great innovations of his career, Maillart enjoyed a moment of lyrical creativity. From late 1924 and throughout 1925, three bridges dominated his imagination: the Valtschielbach, the Châtelard, and the Grand Fey. Unlike his major innovations, they did not specifically contribute to the technical development of structural engineering; but each one represented a unique aesthetic achievement.

The Valtschielbach design might be thought of as the result of a single burst of energy (Fig. 91): roughly analogous to the sonnet, the impromptu, the line drawing. Such was this wilderness crossing, whose thin arch seems to carry so much mass with so little material. Its impact is immediate. Further study might enrich our perspective, but ultimately it will only confirm what we sense at first sight – a new form. Its lyrical contrast of a light arch with a stout deck remains the primary delight (Fig. 92). Like Tavanasa before it, and Salginatobel after it, Valtschielbach invites further visits: Its simplicity is its virtue and deeper study will reveal how difficult that is to achieve.

The visual impact of the Valtschielbach Bridge

Figure 92. Valtschielbach Bridge. (*Source:* Kant. Tiefbauamt, Graubünden)

comes from the profile view where the full span, the arch thinness, and the depth of the ravine are all apprehended at once. It is just that view that stimulated Maillart's imagination in early 1925. He designed the bridge quickly, calculating its complex interaction of deck and arch in a matter of a few hours.[104] His design, submitted to the canton engineer of the Graubünden by a local builder, won the competition, and construction proceeded through the late summer of 1925 (Fig. 93). A load test, in September 1926, confirmed Maillart's simple calculation. As the first modern concrete arch to express extreme thinness, this bridge has become a landmark. Engineers would argue about its appearance throughout Maillart's lifetime,[105] but Maillart's own critique of it 8 years later would be the most telling and would lead to a new masterpiece.

THE BRIDGE AS PLAY: CHATELARD

The second major work that year was the aqueduct at Châtelard, a bridge designed to carry water and a narrow roadway across a highway and stream high up near the Forclaz Pass on the road between Martigny on the Rhône River and Chamonix in France. Maillart played with the design made by the Swiss railways (Fig. 94) to achieve a stronger and a more elegant structure. On October 3, 1924, he presented a new project that reduced the materials, lowered the cost by about 30 percent, and radically altered the appearance.[106] Maillart's form, submitted with the builder Simonette, greatly reduced the center span by tilting the two main supports to form a kind of arch (Fig. 95). Moreover, where his thin columns joined the hollow-box aqueduct girder, he created haunches to produce a stronger structure and to "give the structure a less severe appearance," as he wrote in his proposal (compare Fig. 96 to Fig. 97).[107] He completed the final design on February 9, 1925, and the aqueduct was built by Simonette that year.

Maillart worked the design into a more complex form than we see in the rigorously symmetrical single span at Valtschielbach, where the deep ravine is radically different from the flatter, highly unsymmetrical valley over which the Châtelard aqueduct passes. Water is carried in the hollow box, only the deep side wall of which appears in profile. Because of the constant water load, the girder profile does not change from end to end except for the light taper (haunching) over the legs. Maillart has care-

Figure 93. Valtschielbach Bridge construction. (*Source:* Kant. Tiefbauamt, Graubünden)

fully arranged the supports to make such a constant-section girder fully rational by keeping the main spans to a constant 16 meters.[108]

The girder is built into three types of structural forms: the continuous beam (approach spans), the cantilever (lengths on either side of the two girder joints), and the rigid frame (the main span combined with the slanted columns). This combination of forms, the rationale for which derives from the constant-depth girder required for an aqueduct, is resolved so simply that Maillart's talent for synthesis becomes evident only after the full length of the structure is studied. The top of the box serves as the roadway and the whole structure is so narrow and long that Maillart widened the approach supports at their base for lateral stiffness. He created

a surprising lightness by dividing each leg into two as it reaches the footing (at the far right of Fig. 97). Unlike Valtschielbach, here the drama comes from movement around the structure, rather than from a single view. The character of this bridge develops in time and is thus more like a play than a poem.

If, as students of this drama, we become more deeply involved in the mechanics of its plot, we would discover how Maillart chose a further structural symmetry, which even some of his most sympathetic students have criticized. The artist Max Bill, in his fine book on Maillart, finds disturbing the little stone pedestal that projects above ground to support one slanted leg (Fig. 98). Bill would have preferred a longer leg meeting the ground in the same way as the other leg and thus creating an

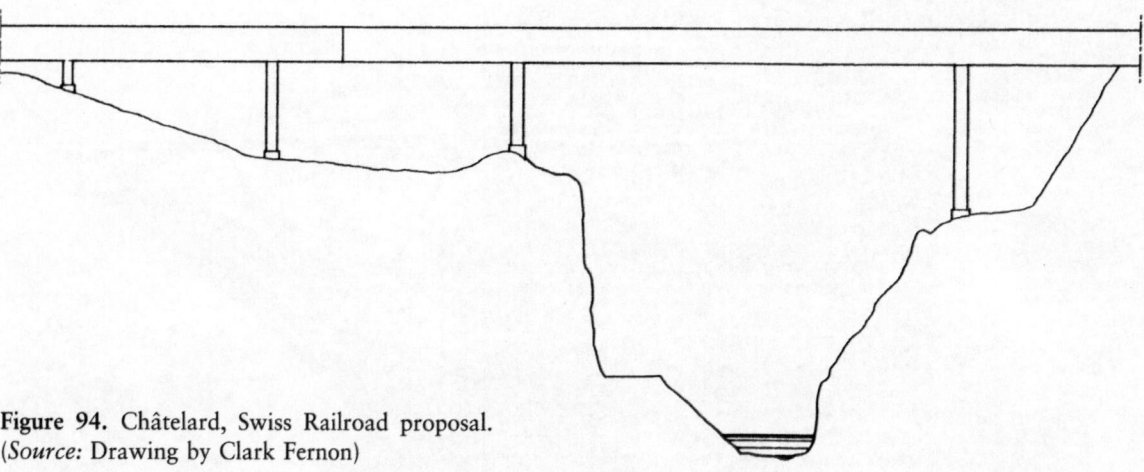

Figure 94. Châtelard, Swiss Railroad proposal.
(*Source:* Drawing by Clark Fernon)

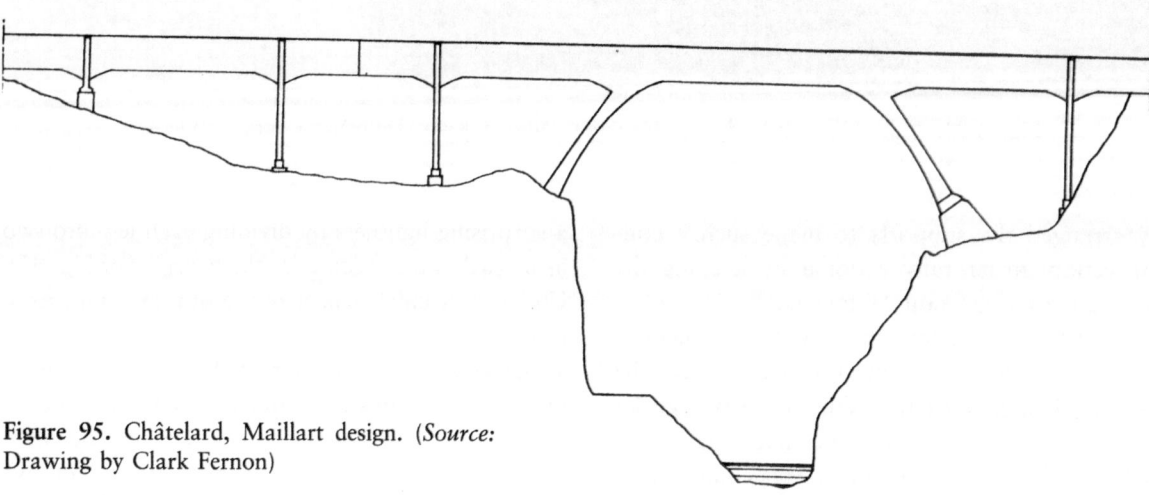

Figure 95. Châtelard, Maillart design. (*Source:*
Drawing by Clark Fernon)

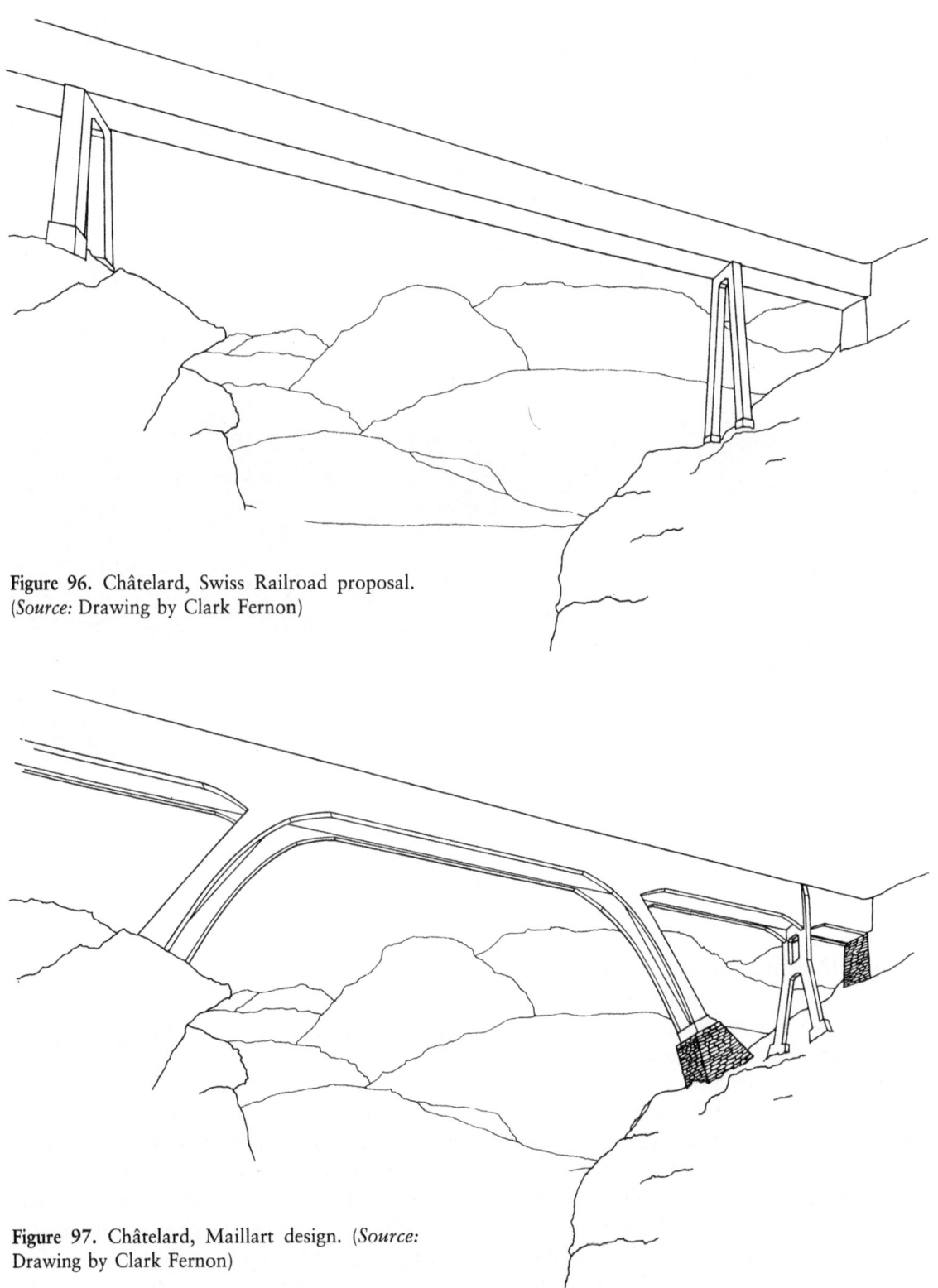

Figure 96. Châtelard, Swiss Railroad proposal. (*Source:* Drawing by Clark Fernon)

Figure 97. Châtelard, Maillart design. (*Source:* Drawing by Clark Fernon)

Figure 98. Châtelard. (*Source:* Madame M.-C. Blumer-Maillart)

asymmetrical frame to fit more the asymmetrical valley.[109]

But that pedestal reveals another aspect of Maillart's search for simplicity. With a symmetrical concrete frame, Maillart could calculate the frame forces on a single page of computations and thus render analysis subservient to design. His mind was freed from concern about forces to concentrate on the problem of proportioning the spans and especially the junctions of slanted legs and girder, slanted legs and supports, and girder to girder where the joints have to be watertight. Maillart's symmetrical frame reflects his concern for a trouble-free structure rather than for visual conformity to a given landscape. His structure was already so radically different from the materials and the terrain

that he apparently saw no reason for making a small concession to the latter.[110] Nevertheless, Bill's judgment is reasonable and Maillart's use of stone here is disturbing in much the same way as are the stone approaches to the Valtschielbach Bridge.

THE BRIDGE AS NOVEL: GRAND FEY

The Grand Fey Viaduct required yet another approach. Here Maillart took a problem that had far more constraints than at Valtschielbach or Châtelard, on a larger scale, with less opportunity for new form. Following World War I, the Swiss national railways intensified the program of rehabilitating its nineteenth-century iron bridges to carry heavier loads. The Grand Fey viaduct crossing the

Sarine River just north of Fribourg was the largest such project at this time and it required a completely new structure to be built at the same exact location without interrupting train traffic. These constraints led to the plan for a concrete viaduct within the old iron one, changing from flat trusses to barrel arches. Maillart was to advise both on the design of the superstructure and on its construction.[111]

In this case, the Swiss railway engineers set the overall form and Maillart was asked to study the dimensioning of the arches and vertical connections to the horizontal railway deck. The more rigorous constraints and greater scale at Grand Fey produced a new design, a series of light, new concrete arches between massive, tapering concrete piers that, because they contain the old iron pillars, extend up to the deck.[112] In the final form, the extra width of the pier extensions above the arch are unnecessary, but their existence represents the major building constraint of providing support for train traffic before the concrete superstructure was completed (Figs. 99 and 100). The arches and other verticals are so light by comparison to the pillars that in profile there is the same feeling of surprise as at Valtschielbach.[113] At Grand Fey, however, there are six spans that cross the wide valley and provide in their bare white concrete the sharpest possible contrast to the flat waterway and the deep forested banks on either side.

The viaduct provides another, radically different, experience when approached from the Grand Fey bank beneath the rail line and up into the bridge span itself. There, right below the tracks and resting on the crown of the arches, is a thin concrete slab running the full 383 meters of the viaduct that pro-

Figure 99. Grand Fey viaduct construction, Fribourg. (*Source:* Madame M.-C. Blumer-Maillart)

vides unparalleled views, including the Bernese Oberland in the distance (Fig. 101). Here the bridge takes on a completely different feeling – one of strength, confinement, and durability. Its arches are massive, its verticals powerful, and the metal guard rails provide a sense of security for the long walk across the valley. It is over this path in 1925 that Maillart loved to walk with Marie-Claire to show her the bridge under construction and the extent to which it challenged his technical skill.

By taking advantage of the heavy deck above the walkway and the numerous verticals, Maillart had suggested that the railway engineers make the arches light. This gives the bridge its distinctive appearance. But that lightness is relative to the profile view (seen best from the downstream roadway bridge) and not to the view either from the close banks or the walkway. Close up, the arches are obviously powerful, load-bearing elements; Maillart's

ideas have led to lightness within the context of large scale. Above all, like a long and complex work of fiction, this structure requires time and patience to understand. It does not burst forth like the Valtschielbach nor can it be taken in by a single observation like the Châtelard. Again, as in the other two works, stone form intrudes here in this case through the arcaded deck.

The challenges at Valtschielbach, Châtelard, and Grand Fey signified a renewed focus on design. The trips between Geneva, Bern, and Zurich became routine, whereas long trips abroad became fewer, shorter, and without the tension occasioned by large financial commitments. Trips such as one to Barcelona in February 1925 held little appeal for him. As he had previously observed about Paris, Barcelona in early 1925 was a huge city compared to which he anticipated that "Geneva will seem like a village."[114]

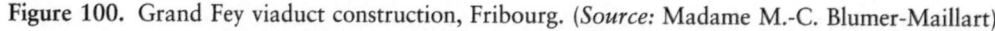

Figure 100. Grand Fey viaduct construction, Fribourg. (*Source:* Madame M.-C. Blumer-Maillart)

Figure 101. Postcard from Maillart to his mother. (*Source:* Madame M.-C. Blumer-Maillart)

Indeed, compared to Spain, France, and especially Russia, Switzerland itself was little more than a village where the scope of building could never be large. At just this time, his Swiss contemporaries, Othmar Ammann and Fred Nötzli, were beginning their independent careers in the United States, where the potential at this time for large-scale structures was greater than anywhere else in the world. Ammann began his George Washington Bridge design in 1924 and Nötzli his high-dam designs in California at about the same time. Ammann had come to the United States in 1904 and had been chief engineer on the Hellgate Bridge (designed by Gustav Lindenthal) completed in 1916 as the longest spanning arch bridge at 978 feet. By 1931, he would be the leading designer of steel bridges.

For his part, Maillart never considered leaving Switzerland, where he had to face design questions relating to more ordinary structures, under more stringent economic constraints, and where many of the best-known engineers were unsympathetic to his ideas. But he was no longer alone; three men within the profession had become his staunch allies and, with several others later, crucial to his success. Moreover, he liked living in close proximity to his family.

CHAPTER SIX

Solitary Designer

1927–1932

Colleagues and Family
1927–1930

THE PROFESSOR, THE EDITOR, AND THE BUILDER

On January 4, 1925, François Schüle died, having retired the previous year as director of the Federal Materials Testing Laboratory (*Eidgenössische Material Prüfungs Anstalt: EMPA*). His successor was Mirko Ros, the man who had directed the tests confirming Maillart's shear-center ideas. Now, in early 1925, he began to plan with Maillart new tests of unprecedented technical significance. The Zurich Alumni Association (*Gesellschaft Ehemalige Polytechniker: GEP*) had appointed in late 1924 a three-man committee that included Maillart to work with Ros on his new ideas. All during 1925 they met together to plan the testing of a wide variety of major concrete structures then under construction in Switzerland.[1]

Ros and Maillart formed at this time a fast, lifelong friendship. The two men became an easily recognizable pair at various testing sites: Ros, clad in his usual bizarre field garb, including puttees; Maillart always dressed conservatively, with his long winter coat in cold weather (Fig. 102). No one would do more than Ros to show the high technical quality of Maillart's designs as well as the correctness of his simple analyses for complex structures.

Equally significant was Ros's judgment of Maillart's completed designs as works of art.

Ros, the ebullient Yugoslav, was quick to see in Maillart the combination of discipline and play, which fit well with his own *joie de vivre* and sound technical understanding. More importantly, Ros unlike Rohn did not think of himself as a designer, and thus never felt competitive with Maillart. Ros was the natural inheritor of the Culmann–Ritter tradition of analyzing structures graphically, except now the need was for a systematic emphasis on the visual results of full-scale testing that is the natural complement to graphic statics. If graphic analysis was the language of engineering as Culmann had claimed, then full-scale testing is the speaking of that language. We may read Maillart's elegant computations with professional pleasure, but the realization of his ideas requires a full performance of the words in action. The bond between Ros and Maillart gave Maillart another outlet for his energies in opportunities to design convincing experiments. It was Ros who first publicized the Valtschielbach, the Flienglibach, the Schrähbach, and the Châtelard bridges in a major article published by the *Bauzeitung* in late 1927.[2]

Carl Jegher and Maillart had built a similar relationship. In complete control of the *Bauzeitung* since his father's death in 1924, Jegher began to encourage Maillart to write about his design ideas now that he had become more active. Jegher's respect for Maillart was stronger than for any other engineer largely because of the integrity Maillart

130

Figure 102. Maillart and Ros at Valtschielbach. (*Source:* Madame M.-C. Blumer-Maillart)

had shown through his articles in the *Bauzeitung,* beginning with the purloined Guggersbach design of 1908, the 1911 Lorraine Bridge competition scandal, and continuing through the debates with Rohn.

August Jegher's direct attack on the 1911 jury in Bern and on the 1920 Ritom Commission report set the example for his son, Carl, who was himself fiercely independent in his judgments, as shown in his criticisms of Rohn's behavior over the Zähringer Bridge. It was this independence of mind that attracted Maillart and linked him with Carl Jegher in the struggle to express new ideas about design. The

editor at this time hired Peter Meyer, an architect, to write regularly on engineering aesthetics, a novel idea for an engineering journal. Meyer's articles of the mid-1920s helped to stimulate an interest in reinforced-concrete structures as works of structural art, a theme subsequently taken up by Sigfried Giedion. Much of that later writing was made possible by Carl Jegher's steadfast view that such ideas were an essential part of engineering.

In early 1925, Jegher persuaded Maillart to write an article on Eugène Freyssinet (1879–1962), the best-known French designer in concrete, who had just completed a series of immense structures[3]; the next year, Jegher asked Maillart to write a reaction to the landmark exhibition at the Zurich Art Institute on public works and industrial structures viewed as works of art.[4] Jegher's insistence and encouragement led directly to the publication of Maillart's ideas on structure. These essays, of greater value to Maillart than the mere publicity they generated, made him rethink what he had designed, thereby reviewing past problems at the same time that he worked on new solutions.

It was particularly valuable for Maillart to reflect on his works in use – either under test by Ros or in print by Jegher – so that those in progress could take account of earlier problems. One example might be the heavy romanesque approaches to Valtschielbach that would soon give way to the open side spans at Salginatobel. Innovation for Maillart came through a steady improvement, not through a striving for novelty in each design. By this time, independence of mind, love of creative play, and a sense of discipline were major features of Maillart (Fig. 103) and his best friends reflected these characteristics.

The senses of discipline and play in Maillart were complementary rather than contradictory. Play, in its strictest meanings, signifies strict rules, rigorous practice, and aesthetic pleasure when taken in the context of a sport, game, stage production, or musical performance. Fantasy within well-defined limits characterizes these activities as it does bridge design in which the play with form must produce a safe structure that can be constructed economically.[5] After 1919, Maillart needed the help of a talented construction engineer to build his unusual designs for reasonable costs. Simon Simonette filled the role. It was through Simonette that Maillart had obtained the Wägital works beginning in 1922, and Simonette was also building Châtelard at the same time as Grand Fey.

Simonette was the contractor for the Grand Fey Bridge that was under construction throughout the summer of 1925. On October 4, Maillart and Nissen brought Marie-Claire for a detailed look at the great work nearing completion.[6] The engineer was pleased with its progress, largely due to his rapport with Simonette. He had difficulties before the war in communicating even with his own field men and now that he had no financial control over construction, it was more than ever crucial to find reliable engineers like Simonette who could realize his ideas. Simonette, like Jegher, was from the Graubünden. Even Simonette's two partners, Simon Menn and Florian Prader, were Bündners as was Lucien Meisser, whom Maillart had hired in 1924 to be his chief engineer in Geneva. Not only were Maillart's pioneering bridges in that sparse region, but many of his closest friends came from this same canton.

Sadly, on October 24, 1925, without warning, Simon Simonette died of apoplexy. Maillart was deeply affected.[7] Maillart's sense of loss is indicative of the role that a few talented builders would play in his life after 1920. Fortunately, Prader took over Simonette's firm and Menn went into partnership with a young Bernese engineer who would build some of Maillart's most significant bridges so that the Graubünden connection would continue.

ISOLATION AND ROUTINE

Paradoxically, at the same time that Maillart's business improved and he entered a period of new creativity, his personal isolation became more pronounced. In December 1926, Marie-Claire went to England to work. As summer approached, with his daughter away, and René and Edmond both pursuing careers abroad, Maillart was alone in Geneva. Marie-Claire was never far from his thoughts and when she failed to write for a whole month in late spring, he explained to her: "Do you not know that I think daily about you and that the feeling of emptiness around me grows painfully? Is it not sad enough that destiny separates us to make that separation greater by this prolonged silence?"[8] Yet when Marie-Claire again suggested that she live with him, he rejected the offer because, as he wrote her, "the order of things is that I leave this world long before you... and my greatest happiness would be to see you married! That is the natural solution and it takes away the logic of the idea of our living together."[9]

At the beginning of January 1927, Adolph Zarn's wife died of pneumonia. Barely a few days later, Zarn told Maillart that although he could keep a room there, it would no longer be possible for him to take meals with the family as he had.[10] Maillart began to spend more time in his office, where he had a sofa and the means for making simple meals. He developed the habit of resting in his office, at Rue du Marché 18 in the middle of the city; he ate either there, with family, or at restaurants. Within a year, Maillart decided to move into his office, thereby reducing his living expenses still further. Even though his three offices were busy, there was never money left over after expenses and payments on the large debt he had to assume to start up after the war.[11] Maria had motivated Maillart to earn money; he himself was indifferent to it. His Calvinist upbringing may well have been a factor, together with his singleness of purpose, in his willingness to accept a near monastic life.

With a slim bank account and no real home, Maillart nevertheless began 1928 in good spirits. It was a year of hope and prosperity throughout Western Europe. The inflation of the early 1920s had been overcome; there was a boom in the United States, whose economy was by now so crucial to Europe. Germany had recovered from the war and was prospering; and as a symbol of optimism, the Kellog-Briand Pact outlawing war was signed on August 27, to an outburst of popular enthusiasm.

Figure 103. Robert Maillart and his cigar, circa 1926. (*Source:* Madame M.-C. Blumer-Maillart)

Peace and prosperity seemed to have been ratified.[12]

Only Max von Müller's death, in early February, sounded a sad note.[13] But Maillart had interesting work and Marie-Claire had moved back to Zurich, having secured a job there with a firm of landscape architects. She came to live in the Nissen apartment.[14] Occasionally, she accompanied her father from Zurich to Geneva for the weekend. They then engaged in a bizarre ritual. Father and daughter would set up targets at the far end of the office's long narrow corridor, and with the engineer's elegant and powerful matched sporting pistols, the two would engage in a highly competitive shooting match. Gunshots echoed eerily in the empty building, with the older man and the young woman taking great delight in this unusual matching of skills.[15]

Already by the fall of 1926, the small Bern office – little more than a room for a draftsman's table and one engineer's desk – had sufficient work thanks to the Lorraine Bridge commission to justify the hiring of Ernst Stettler to run it.[16] At the same time, the series of large buildings Maillart began in Zurich in 1921 now allowed him to expand that fledgling office.[17] Geneva remained to the end his central office with the most people – usually three or four engineers and one or two draftsmen plus a secretary and an accountant.[18] Maillart made a regular weekly tour from Geneva to Bern to Zurich; in each office, he made all the important decisions himself. No structural engineer has ever before or since worked like Maillart did between the mid-1920s and 1940, personally directing three separate offices and doing all the innovative designs.

Often he would sketch out a new idea on graph paper or on used envelopes while riding the train between cities. At his Bern office, Maillart rarely checked his subordinate Stettler's calculations. When Stettler once asked for such a check, Maillart replied with good humor, "You studied at the ETH. I know you can do the work." It gradually became clear to the young engineer that Maillart knew ahead of time almost exactly what the calculation results should be.[19]

After a day in the office, often broken up by a visit to some agency like the city building department, Maillart seemed to lose interest in the work and would invite a group out to dinner that included Stettler and sometimes his wife, the architect Klauser (with whom he had designed the Lorraine Bridge), the city engineer Armin Reber, Eugen Losinger (the young engineer whose firm was building Lorraine), Simon Menn (Losinger's chief engineer), and others. Following a fine dinner, Maillart invited them to play Revoluzger, a card game he had devised in Russia to occupy the hours after the November Revolution when work dried up. Revoluzger was more of a challenge than Wist (Jass in German) and yet not as demanding as Bridge; thus, it held Maillart's interest without excluding those for whom Bridge was too taxing. Part of his lifelong effort to avoid small talk, this game provided a

means to socialize and also enough of an intellectual challenge to keep the evening interesting.[20]

The next morning Maillart would be on the express train to Zurich, where he spent afternoons with Alois Keller, his chief engineer there. In 1928, this office was barely larger than the Bern one, having at times only Keller and one draftsman, but it contained a small private room for Maillart, where he received visitors and where, on a meticulously cleared desk, he sometimes wrote letters, brief reports, or did a few calculations. After reviewing the Zurich work, Maillart would usually join Paul Nissen for dinner and have a long evening visit with his daughter (Fig. 104). On Saturday, Maillart had lunch always at the same, specially reserved table in the simple Schutzen Restaurant with Ros, Jegher, Prader, and often some of Prader's employees, including the vivacious Mlle. Oechslin Prader's private secretary. There Marie-Claire would join them and become part of the jovial and distinguished club of Swiss engineers, who were quite outside the circle of establishment engineers like Rohn, Ritter, Adolph Bühler of the Swiss railways, and their young protégés. These Saturday lunches, like the Bern evenings and those spent with Nissen, were remarkably free from shop talk. True to character, Maillart refused to sacrifice his evenings to potential clients. His routine suited him perfectly.

EDUARD BLUMER

Maillart held a few formal lunches to honor visiting dignitaries. One was in early summer 1928 for Fritz Emperger (he dropped the von after the war), the lively Viennese engineer who had published Maillart's pioneering works before the war. Emperger was in Zurich to meet with Ros and Rohn in connection with a forthcoming international conference in Vienna, which was to be the second such event projected by Rohn during a 1922 gathering in Zurich. Emperger had always been a great admirer and good friend of Maillart's; both had a lively sense of humor and a questioning attitude toward the "authorities."

Figure 104. Maillart and Marie-Claire, 1928. (*Source:* Madame M.-C. Blumer-Maillart)

The luncheon, for about twenty people, took place at the Gotthard Hotel, where Maillart habitually stayed in Zurich. As usual, Maillart came with Marie-Claire, and Ros was accompanied by several of his young engineers.[21] One of these was a dashing young man from the canton of Glarus, Eduard Blumer (1901–80). Blumer had completed his studies in 1924 at the ETH and after military service had accepted a job in 1927 with Ros. In late 1928, he had taken a new and promising position with Royal Dutch Shell in the Dutch East Indies (Indonesia), where he was scheduled to go the following year.[22] His reaction to Marie-Claire was not unlike Maillart's to Maria 27 years before. Impulsively casting aside their previous engagements for that day, the young couple spent the afternoon together after meeting at the midday festivities.

After a lively courtship, Marie-Claire announced her engagement on January 9, 1929, to Blumer. They were married in February in the high snows of the Glarner Alps and immediately sailed to Indonesia. Maillart was pleased to see his daughter happy even though he did not know Blumer well. He cautioned Marie-Claire that "the beginning will be hard but if you win the struggle you will have the great satisfaction of having done it without the aid of others and with your own efforts."[23] This view of marriage as one's first truly independent act reflected Maillart's own marital experience.

At the same time, Maillart's two sons had both begun their own careers outside Switzerland. Edmond was working for a British firm that specialized in steel welding.[24] In May, he was called to Switzerland to consult on a broken steel pipe for the Grimsel hydroelectric works, thus complementing his father's earlier work for power plants.[25] René, meanwhile, was gradually getting established in the hotel business with the help of his aunt's brother in Paris.[26] Both sons appeared in Switzerland from time to time and Maillart tried to keep in touch with them, but they were going their own ways and that appealed to him. This independence was particularly crucial for Edmond, who needed to separate himself from his well-known father. Maillart's expectations for his younger son were less than for Edmond, and as a consequence, there was rarely any tension between them, even though Maillart did not always approve of René's escapades.

On December 9, he received a telegram with the news that Marie-Claire had given birth to a daughter named also Marie-Claire. He dashed off a letter to the new parents in German expressing his delight. He signed it "Papa and Grosspapa [Papa and Grandpapa which can also be written as = (1 + grand)papa]."[27]

Maillart wanted to know the unfaçaded truth about his daughter's life in Indonesia just as he wanted to know about all the cracks in his bridges.[28] The defects and imperfections told him more about the Blumers' lives than all the rosy prose. The worldwide depression, triggered by the 1929 crash of the American stock market, brought rumors of major losses for Royal Dutch Shell and Eduard Blumer worried for his job. A sudden admission of dejection by Marie-Claire made Maillart suspicious of her glowing earlier reports on their tropical life. He also worried about the crisis and continually agonized in his letters about the dangers of a shrinking economy; but he always wrote directly about his own worries. He could put these aside once they had been expressed and focus his mind on a new design.

From Tavanasa to Salginatobel
1927–1930

RUIN ON THE RHINE

In September 1927, Carl Jegher was preparing to publish a long, illustrated article by Mirko Ros on Maillart's major bridges designed since the war: Valtschielbach, Flienglibach, Schrähbach, and Châtelard. This was the first publication to summarize Maillart's new ideas, which Ros characterized at the end of his article by "the aqueduct over the Eau-Noire (Châtelard) is a structure which embodies the technical-economical thinking and artistic design of the engineer; it has an unmistakable affinity with the structural ideas which formed the basis of the Tavanasa Bridge designed and built by Maillart and Co. as early as 1904."[29] As these words were being set in type, the Tavanasa was being destroyed by forces no bridge could have withstood.

On the night of Saturday, September 24, 1927, and during the morning of September 25, rain battered the villages along the Vorder Rhine river, and finally just before noon, with the river water rapidly rising, a large section of the hill 500 meters above Tavanasa gave way, bringing an avalanche down the right bank of the valley. Houses disappeared, seven people were killed, and Maillart's bridge was swept away, most of its ruins ending up on the left bank of the river (Fig. 105).[30] Following a study of the ruins, Professor Mirko Ros emphasized that this bridge was "a small but prototypical masterpiece of Swiss bridge art. Its ruins are now silent testimony to the skill and to the love of the creator for his work."[31]

Maillart used to counsel Maria never to blame events for your misfortunes, but rather to try and regulate your own actions so that they do not become the cause of your regret or suffering. When the Tavanasa was lost, his primary response was to reevaluate the original design rather than to lament its destruction. The first step in restudying it was provided by Mirko Ros, who carefully tested the ruined concrete to show its strength and endurance. Maillart never doubted the Tavanasa's quality, but he realized the form could still be improved. So he set out to design a replacement that would be better than the original.

By 1927, no one had a deeper understanding of reinforced concrete than Maillart. Basing his new Tavanasa design on that understanding, Maillart now took the final step to complete freedom from the constraints of a two-millennia tradition of stone masonry bridges. For the Tavanasa replacement, he started to explore first a deck-stiffened design like Valtschielbach and then a three-hinged one like the 1905 bridge itself.[32] He was beginning to see that the old Tavanasa had retained vestiges of the masonry past: heavy stone abutments and a solid block at the crown. Maillart determined to eliminate these anachronisms, thereby designing forms with little reference to older structures. Maillart gave his design to the builder Florian Prader to bid in a design-construction competition for the replacement. Maillart and Prader lost that bid, and a far less interesting bridge designed by a local engineer was built the following year. The double loss sparked a fundamental change in Maillart's ideas about bridge design.

THE CLIMB TO THE SALGINATOBEL

That summer, Maillart realized how he might apply some of his concepts for the rejected Tavanasa to a new and unique site high in a remote region of the Graubünden. There an obscure dirt road was being pushed slowly up into the Alps to connect the tiny village of Schuders to the valley town of Schiers. But the road faced a huge obstacle, the deep ravine of the Salgina brook, where a new bridge would be necessary.[33] The canton decided to hold a design-construction competition in which low cost would be essential. The federal government agreed to a

Figure 105. Tavanasa Bridge in ruins. (*Source:* Madame M.-C. Blumer-Maillart)

subsidy, and on July 10, the competition opened with a deadline fixed for September 15. Maillart joined once again with Prader. By August 31, 1928, Maillart had sketched out his entry and the Geneva office had made three drawings for the competition.[34]

These drawings show decisively that Maillart's vision had shifted away from stone images. With the Salginatobel design, reinforced-concrete structure emerged from its adolescence. Maillart at last achieved a design that belongs to a new world of form. At the same time, in England, the philosopher R. G. Collingwood used the example of reinforced concrete to describe "A philosophy of progress":

a particular age has the task of realizing beauty in a par-

ticular way. We, for example, have invented reinforced concrete, and our task is to discover how to make it beautiful. We shall not do it by pretending that reinforced concrete is stone or timber; for it is a material with a very decided nature of its own, and we shall only build beautifully in it by understanding this nature and expressing it in our designs.[35]

With his 1928 design, Maillart expressed the essence of the reinforced-concrete bridge and his three drawings, profile and cross-section, convey the entire form. The eight rough sketches Maillart made in developing the Salginatobel form are also of the profile and the sections, with no perspective drawings or views depicting the bridge as it would actually appear to an observer stationed at any point near it. Maillart was working out his vision, not three-dimensionally as would an architect, but two-dimensionally as would an engineer.[36] The reason for this fundamental difference in thinking lies in the distinction between the two art forms: structure and architecture.

The designs for the Tavanasa, the Valtschielbach, the Châtelard spans always portrayed the form in profile as did plans for the Chiasso shed structure. In the latter two, the cross-sectional view is also of visual interest. Maillart, like structural engineers in general, shaped the forms to control the forces, whereas architects in general shaped forms to control spaces. Even in his warehouse designs, Maillart kept the column spacing and all details as regular as possible, whereas architects saw such flat-slab ideas as allowing the free plan. As we shall see later, Maillart developed a series of deck-stiffened arch bridges that curved in plan and cannot be fully defined in profile and section, but even these grew directly out of planar forms.

The engineer has to visualize the action of forces due to gravity and to wind, and the sequence of construction using scaffold and wet concrete. These are overriding design constraints; the first defines the performance of the completed form in service and the second determines the way that form is brought into being. Both constraints are so complex in principle that the designer must represent them, or model them, in as simple a way as possible in order to be sure of the final result. The fundamental simplification is to study forces in one plane at a time and this method fits closely with the nature of the loads on bridges. Gravity acts vertically and wind acts mainly horizontally. These two primary loads act independently and can be thought of separately.

The bridge profile responds to the gravity loads of the entire structure; for example, the arch form provides a means of transferring to the abutments the entire weight of the structure and the traffic. In designing that transfer, designers can play with form as a means of controlling forces. For the Salginatobel, Maillart focused on the 90-meter main span and studied carefully the ratio of arch rise

(about 13 meters in the final design) to arch span, and the variation in arch depth throughout the span. Out of many possible economical and technically correct forms, he intended to choose a profile view that would be the most striking. If he had been at all concerned with how the bridge would appear from either bank or from any view other than profile, he would necessarily have been led away from the detailed visual study of rise to span and depth variation.

One has only to look at the foreshortened side view of the finished bridge to see in its appearance of strength a visual contradiction of Maillart's search for a light and delicate profile. This is why nearly all published photos of the Salginatobel are taken from the adjacent mountain where the profile is seen clearly in its entirety. It is this view that Maillart drew in the summer of 1928. To some critics, this limitation seems to be a fatal constraint of artistic creativity, but it is really the same type of restraint operating on the painter or photographer. The discipline of representing vision in a single plane never hinders the creation of an artistic form. Just as painting is understood differently from sculpture, so must engineering structure be distinguished from architecture.[37]

These drawings also show how Maillart fused the deck and arch into one structure with a straight parapet line from end to end and with slender verticals on the sides. It is a form that could not be derived from stone or timber. It comes naturally from a material cast in the field rather than fabricated from cut blocks or sawed cylinders. In the final Salginatobel Bridge, gone are the romanesque approaches of Valtschielbach, the heavy pillars of Grand Fey, and the stone abutments of the Tavanasa. There were two variations to the approach spans: one with romanesque stone arches and the one finally built with straight concrete beams. Maillart apparently hedged still on the issue of traditional aesthetics, but the actual bridge does not keep that ancient form. Maillart would come only slowly to these forms with surprising results, but in 1928, his actual bridge was radical enough to have eliminated the masonry heritage and to have followed rigorously the creative potential within the discipline of profile and section.

THE SALGINATOBEL BRIDGE

Prader submitted the Maillart design with its two approach-span variations on September 15, 1928. It received strong support in the Graubünden by late September from the district engineer. By October 19, it was clear that he and Prader had won the competition and that the canton wanted the less-expensive design without approach arches.[38] Maillart had chosen the three-hinged arch partly because it could move without stresses when the sandstone wall on the Schuders side of the ravine moved.[39] In part, he chose the Tavanasa-type form for the dramatic contrast it made with the ravine and the surrounding terrain.

The engineer's principal inspiration, however, came from his studies of graphic statics with Wilhelm Ritter (Fig. 106), his own past designs, especially the 1905 Tavanasa and the designs he had made in late 1927 and early 1928 in the vain attempt to win the new Tavanasa commission.[40] With the Salginatobel a certainty, Maillart turned to its detailed design, beginning with studies of the great arch profile and cross-sections. He made these studies himself and he provided the basis for the computations, but entrusted the detailed calculations to Ernst Stettler in Bern, his best bridge engineer.[41]

From these calculations, we can see how Maillart thought about the crucial relationship between forces and forms.[42] First he laid out the profile (Fig. 107) and divided it into roughly equal lengths. Next he computed the dead weight of each length, adding them together to get the total arch weight. He worked with only half the bridge because of its symmetry. Maillart then imagined that the half arch tried to rotate about the support hinges (at point A in Fig. 107), but was prevented from any motion by a horizontal force, H, at midspan (point C in Fig. 107). He proceeded to equate the rotational effect (moment) of the known dead-weight forces with the same effect of the force H, but in the opposite direction (counterclockwise about point A).

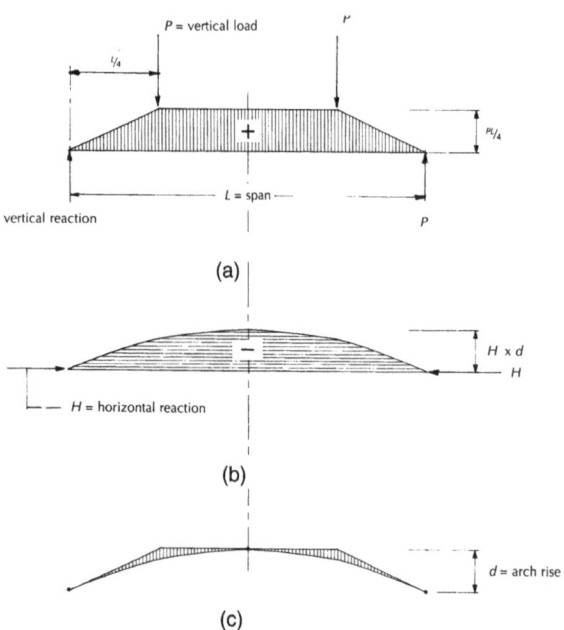

Figure 106. Drawing of moment diagram.

From this equation, Maillart calculated the force H and could now find the internal forces in the arch caused by the bridge weight.

In theory, these forces should be only compression with no bending, but surprisingly Maillart permitted some bending under dead load. The reason is that it is difficult to estimate a profile whose weight distribution will eliminate such bending. It requires much trial and error, and Maillart had little patience for lengthy calculations. He, therefore, settled on a form that was not theoretically perfect.[43] Apart from his drive for simple computations, he had a more powerful justification for an inexact dead-load solution, which was that the live load would unavoidably cause bending and the major problem was to prevent the combined dead and live loads from overstressing the concrete. His fusion of the deck and arch ensured low stresses throughout, especially because the profile was deepest between the hinges, where the live-load bending is greatest.

The final maximum stress in the arch of 790 psi

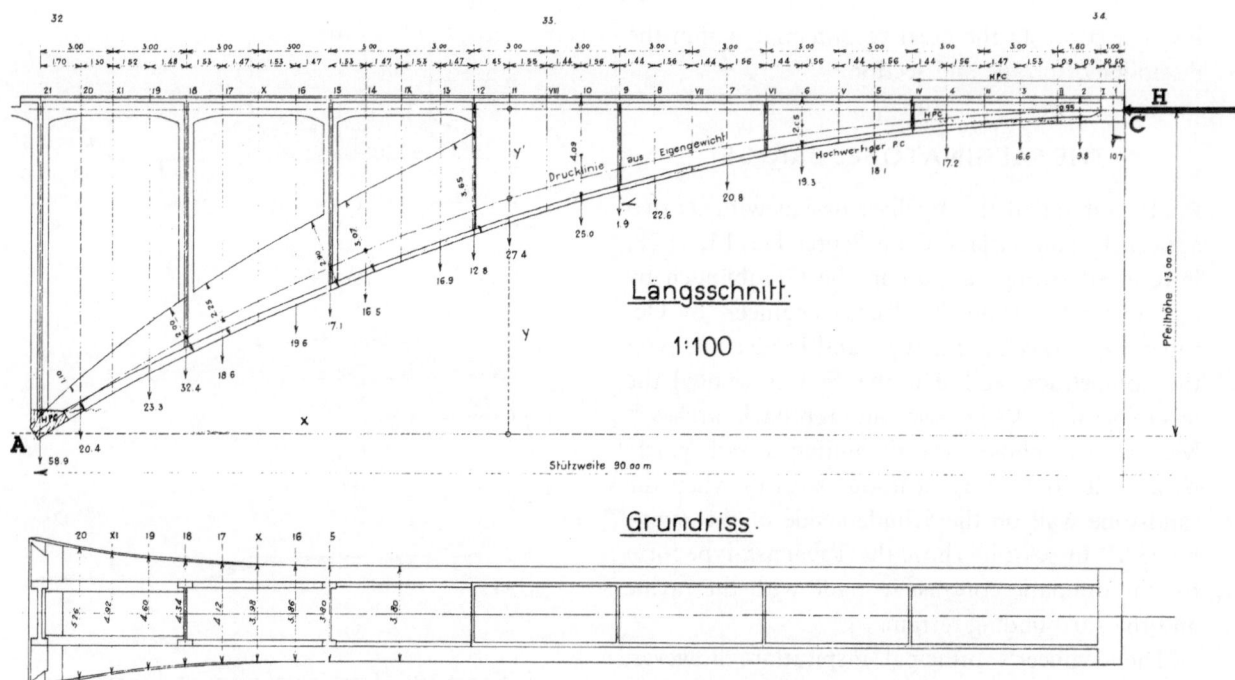

Figure 107. Salginatobel half profile with weights used in the calculations. (*Source:* Maillart and Co., Drawing No. 168/11, pp. 32–4, June 1, 1929)

(pounds per square inch) is well below an allowable stress of about 1,575 psi for the concrete used in the Salginatobel Bridge.[44] Had Maillart adjusted the form more carefully by laborious calculations, he could have reduced the maximum stress to about 600 psi, clearly of no advantage in such a conservative design. The more exact form would have been imperceptibly different in appearance.

By the spring of 1929, Maillart had completed his design, Stettler had finished the detailed calculations, and the drawings were done in the Geneva office. Between May 7 and June 5, the office finished the drawings and construction began.[45] Maillart traveled to the site at the end of June to oversee the critical work of preparing the rock walls to receive the arch foundations.[46] Then near disaster struck.

The scaffold designer and centering constructor, Richard Coray (1869–1946), on site to supervise the construction of this exceptionally high and light temporary wooden structure (Fig. 108), fell from the top about 120 feet to the base.[47] He lay for

weeks in the hospital. The scaffold's unusual lightness was due to Maillart's design that it carry only the weight of the thin curved slab forming the bottom of the hollow box. Once hardened, that slab could then support the walls and deck slab without help from the scaffold. Because of this lightness, it was essential to have Coray's close supervision during completion of the wooden structure. His absence meant that by midfall, the decision had to be made to hold off casting concrete until after the winter.[48] It was a disappointment for Maillart for whom the Salgina was by far the most interesting project he had in 1929.[49] But Coray recovered, the bridge was easily completed in 1930, and by summer it was ready to be inaugurated.

UNVEILING AT SALGINATOBEL

Early in the morning of August 18, 1930, Robert Maillart walked up the single-lane dirt road from Schiers to the Salgina Ravine. There he waited with Mirko Ros for the official party of about fifty peo-

Figure 108. Salginatobel scaffold. (*Source:* Madame M.-C. Blumer-Maillart)

ple who also walked up to where the new bridge stood gleaming white but still undergirded by its wooden scaffold. With the officials in place, Coray dropped the scaffold 3 millimeters, Ros' crew of young engineers read their meters, and the Salginatobel Bridge supported itself, as Maillart put it, "without any assistance and without any self-doubt whatsoever!"[50]

After a light lunch, Maillart gave a brief talk in which he "celebrated the Graubünden as a canton which has provided a pioneering service in bridge building."[51] This reference included his own pioneering concrete bridges: the first hollow box ever built (at Zuoz in 1901), the first slotted, hollow-box, three-hinged arch (at Tavanasa in 1905), the first major deck-stiffened arch (at the Valtschielbach in 1925), and the first horizontally curved deck-stiffened arch (at Klosters in 1930). But the high point of his Graubünden work was the Salginatobel Bridge. Ros and Carl Jegher recognized the bridge as a work of art and its designer as a great engineer and an artist.[52] Two other Swiss, Sigfried Giedion and Max Bill, soon began to write about Maillart as an artist. It was the beauty of the bridge that engaged other artists, architects, and engineers and has put Maillart's design in the center of more art museum exhibitions than any other comparable structure (Fig. 109).

Eight years after the opening of the Salginatobel Bridge, Maillart returned with some family, including Marie-Claire. (Fig. 174) Perhaps a necessary pilgrimage, because it summarizes his career in one powerful image.[53] The Salginatobel Bridge expresses in its history the ideal that all structures everywhere should be efficient, economical, and elegant. How that ideal takes form is always in part a mystery and the designer who can do it works a kind of magic.

The Salginatobel span is shown so often in books that it has led to the impression that Maillart's works are in such improbably dramatic settings that they have little meaning to ordinary bridging problems. A brief review of his major three-hinged arches will quickly show that Maillart developed the form over 40 years of designing at least eleven

such bridges of which the Salginatobel was the fifth, and that the settings were mostly those of ordinary bridges over the low banks of small rivers. Only the Salginatobel span has the dramatic setting of such a high "shapeless crag."

To what extent did Maillart make aesthetic choices in this bridge as opposed to technical ones? We have two definite statements by Maillart, first, "In the Salgina Bridge, the intrados [underside of the arch] has been rounded off under the crown of the arch in deference to traditional design,"[54] which he meant somewhat as a self-criticism and which he overcame in later three-hinged arches beginning with Felsegg. In a second statement, Maillart noted how he had widened the arch at its springing lines (supports) to avoid the feeling of lateral instability, that is, the visual sense of the narrow arch tipping sideways. The closer we look, the more we shall realize how Maillart played with the form to suit his vision while never violating the disciplines of efficiency and economy.

THE SALGINATOBEL AND THE LORRAINE BRIDGES

1930 marked the opening of Maillart's other largest bridge, the Lorraine. The difference between the two inaugurations was as decisive as the difference between the structures themselves. At Bern, the city of bridges, a full 7 months after its load test, the Lorraine opened on May 17 with a gala procession that was reported nationally and internationally. Maillart also spoke publically there (Fig. 110).[55] By contrast, the Salgina opening coincided with its load test and took place in front of a few local and cantonal officials. From that moment, however, the Salginatobel Bridge was recognized as "being in itself already an attraction and a spectacle."[56] It was even seen as "a symbol of the effort in favor of the isolated valleys" of Switzerland to connect them to the modern world and to avoid "the menace of their depopulation."[57] (Before the bridge, all goods had to be carried to Schuders on horseback over a rocky trail.[58]) Beside its novel form, Salgina was the first instance of the federal government in Bern pro-

Figure 109. Salginatobel completed. (*Source:* Madame M.-C. Blumer-Maillart)

Figure 110. Maillart speaking at the opening of the Lorraine Bridge, Bern, 1930. (*Source:* Madame M.-C. Blumer-Maillart)

viding a subsidy for a county road (including the bridge).[59] Thus, socially, the bridge's origin had national significance.

One commentator observed that although the bridge cost was well below all other bids, it was still expensive when compared to the number of people it served. The road and bridge cost 700,000 francs and served Schuders, a village with eighty inhabitants, which meant a cost of about 8,800 francs per person, whereas the Lorraine Bridge was 3,600,000 francs, which for each of the 200,000 people of Bern represented a mere 20 francs.[60] The bridge, therefore, symbolized the great difficulty of connecting little villages and at the same time, it demonstrated that the least costly design could be beautiful.

The final form of the Lorraine Bridge was less important than the process of its construction (Figs. 111 and 112). The bridge today is largely forgotten. The Salginatobel Bridge, on the other hand, ap-

peared prominently in the Swiss pavillion at the 1937 Paris World's Fair and in 1991 was the first concrete bridge to be named an International Historic Civil Engineering Landmark (Fig. 113).[61] It marked the beginning of the last and most creative decade of Maillart's life. By returning to his 1905 Tavanasa form, while avoiding Tavanasa's stone approaches, Maillart began in 1930 to explore new forms with almost no reference to his profession and its state of the art. He was mining his own past, dismissing the aesthetic baggage of traditional forms, and expressing a vision of new possibilities. Traditional Swiss urban leaders, however, were still not ready for Maillart's ideas. In particular, the Bern Bridge resulted in his further alienation from the "authorities."

The Lorraine Bridge provided profitable work in Bern, but at the same time incurred the enmity of Adolph Bühler, bridge engineer for the Swiss national railways in Bern, who had coveted that com-

Figure 111. Lorraine Bridge, construction. (*Source:* Madame M.-C. Blumer-Maillart)

Figure 112. New Lorraine Bridge. (*Source:* Madame M.-C. Blumer-Maillart)

mission. The bridge had its beginning well before the War, when Maillart lost the 1911 design competition for it. The *Bauzeitung's* strong support at that time helped him eventually to win the bridge. Partly thanks to that support, the city of Bern had asked Maillart in 1923 to restudy the design; early the following year, he had presented it to a professional meeting of engineers and architects in Bern.[62] Faced subsequently with a different design by Bühler, the Bern branch of the Swiss Society of Engineers and Architects set up their own commission to study the competing proposals. By the end of 1925, the city had chosen Maillart's design.[63]

Maillart's Lorraine design was finally approved, in typical Swiss fashion, by a citywide referendum in mid-1927.[64] The bridge was finished during the summer of 1929. Ros directed the load test on Sunday, October 6, 1929, and the bridge opened to pedestrians just before Christmas.[65]

The conflict between Maillart and Bühler would further handicap Maillart in getting major bridge designs either for the railways or for the canton of Bern. Consequently, Maillart was forced to work either on large buildings, where Bühler had little influence, or on small bridges, where local engineers with strict budgets were in charge. Just as Maillart's personal life was now more restricted, the parameters of his professional life were narrowing. But what was in one sense a constraint, in another was a release. The Bern wilderness would push his inventiveness to new limits for small-scale work as the Graubünden had done earlier.

Wilderness Bridges
1930–1931

THE CURVED BRIDGE AT KLOSTERS

For all of its acclaim, the Lorraine Bridge was a dead end for Maillart because, as the Salginatobel Bridge illustrated, by the time it opened, he had abandoned the very ideas on which the Bern design rested. He had derived pleasure from the project and especially in thinking out its design and construction between 1923 and 1926. But his concept of bridge design had been changing thanks to the challenges of the Wägital and Valtschielbach. Then, following the unsuccessful designs for a Tavanasa replacement and the award of the Salgina design, his interest had shifted away from expressions of mass toward the complete elimination of extraneous materials that culminated in the Salginatobel Bridge. Early in 1930, his mind became more engaged with the limits of lightness in bridges and again it was a Graubünden Bridge that stimulated Maillart.

Even before the landmark Salginatobel Bridge was completed in the spring of 1930, Maillart's imagination was engaged by the much less dramatic Klosters Bridge. The 30-meter span bridge had to support a rail line that followed a curved path as it crossed the Landquart River, thereby presenting Maillart with the new problem of integrating the vertical curve of an arch with a horizontally curved roadway deck. Moreover, the right-bank approach had to be open for a road below. This meant that even though the deck stiffening of Valtschielbach could be used here, that bridge's heavy stone approaches would not do at Klosters.[66]

In early 1930, Maillart had studied the official design, made by two engineers of the Rhätische Bahn, and in conjunction once more with Prader had proposed an alternative. Maillart proposed an arch that used only 39 percent of the concrete required by the official design, his principal idea being to reduce the mass of the arch without increasing the mass of anything else. This he did dramatically.

He had already achieved such things at Flienglibach and Valtschielbach; what was new here was the horizontal curvature of the deck.[67] The official design carried the arch straight across the river and then put horizontal kinks in the deck to allow the rail line to make its curve at the approaches (Fig. 114). It was an awkward solution that Maillart had in fact adopted at Valtschielbach and earlier in a small arch bridge over the Ziggenbach at Wägital.

Figure 113. Salginatobel on the occasion of its designation as an International Historic Civil Engineering Landmark, 1991. (*Source:* J. Wayman Williams)

The Salgina design of 1928 suggested the solution to Maillart. There he had widened the arch near the supports for lateral resistance to the wind. Maillart described that widening as desirable at Salgina both for technical and for visual reasons. Even though the unwidened arch could carry the lateral loads without danger, he explained, "nevertheless the arch slab is widened near the supports in order to take into account the lateral forces and the even somewhat terrifying narrowness of the bridge."[68] Numerical calculations alone could have justified at Salgina the even more slender solution of constant width; but the appearance of this already daring design would be improved, in his view, with no loss of lightness, by accounting for both the depth of the ravine and delicacy of the structure through a widening of the arch. Here he willingly added structurally unnecessary material for visual effect but without any loss of competitive economy. The en-

tire structure was still the least expensive of any of the nineteen competition designs. Furthermore, the increase in concrete by widening did reduce the stresses and, in theory at least, reduced the amount of reinforcement needed.

With this experience as a guide, in early 1930, Maillart saw the Klosters Bridge's horizontal curve as a means to a new idea for thin arch design, where, as he put it, "the sharp and ugly kink in the deck beams" could be avoided. Most of all, the arch itself could now be designed to widen at the supports in such a way that when looked at from above, one would see that horizontally, the arch followed the curved deck on the inside of its curvature (the concave side), but it widened substantially on the outside (the convex side). The arch width was therefore minimum at the crown or midspan and maximum at the supports (Fig. 115). Yet the visual effect was to permit a smooth view of

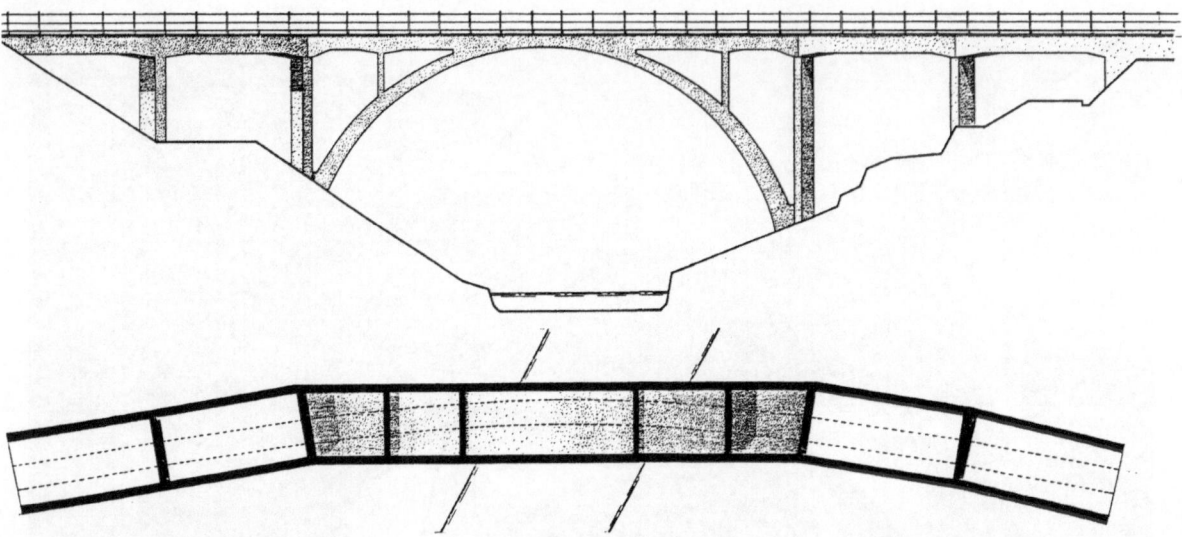

Figure 114. Landquart River Bridge at Klosters, official design, 1930. (*Source:* Drawing by Mark Reed)

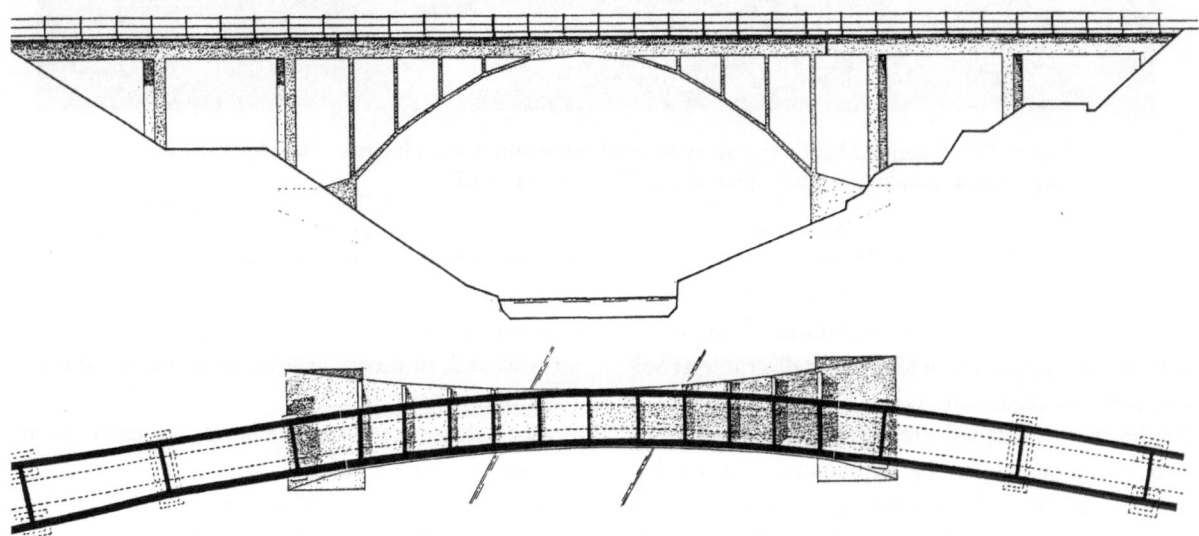

Figure 115. Landquart River Bridge at Klosters, Maillart design, 1930. (*Source:* Drawing by Mark Reed)

arch and deck on the inside curve and a striking view of the trapezoidal cross wall on the outside. These thin walls connecting deck and arch express visually, as Maillart put it, a "structural clarity" and at the same time they load the arch evenly in the lateral direction (as opposed to the several col- umns supporting the deck that are more common for arch bridges).[69]

Maillart's first thoughts on this design, expressed in February 1930, were to find "the most econom- ical and the most beautiful solutions."[70] For visual as well as load-carrying reasons, he designed the

Figure 116. Landquart River Bridge at Klosters, completed bridge, postcard from Maillart to Rosa, 1931. (*Source:* Madame M.-C. Blumer-Maillart)

deck girders to carry all the bending by making them deep. "This is desirable aesthetically, because the approach spans must also have deep girders." This necessity led Maillart to make the stiffening girders and the approach-span girders of the same depth, so that "the entire structure works together as a whole,"[71] as opposed to their less-attractive official solution (Fig. 116). In all his writing about this little bridge, he never separated his technical arguments from his aesthetic ones. He sought one solution to both problems. The last technical problem he noted at the end of his first report on the bridge concerned the foundations for the arch; these were satisfactory, but an unanticipated movement in the hillside would cause later difficulties with his structure.[72] Inspired by the problem of integrating the entire bridge into one coherent form, Maillart's

exploration of horizontal curves for the Klosters Bridge in 1930 further changed his view of design.

LOST BRIDGES AND PROFITABLE DRAWINGS

In November, a bridge competition in Basel awarded five prizes of from 10,000 to 15,000 francs but none to Maillart. "We have failed that splendidly," he wrote the Blumers, "I proposed an innovation, judged to be not extravagant and it is especially the architects [on the jury] who have found the silhouette 'impossible.' So it goes! . . . In these competitions it is necessary to try and follow the ideas of the jury rather than one's own!"[73] This he was not prepared to do anymore. Had his ambition been for material position and praise from the Swiss establishment, he would have hired an architect and made a silhouette that fit the prevailing mode.[74]

As 1930 ended, Maillart found that he had not suffered too much in the first year of the world economic crisis. Throughout the months directly following the stock market crash, Maillart had been making designs continually and was called on more and more to speak in public, to write about his ideas, and to serve on major Swiss professional commissions.[75] He had made a satisfactory income in 1929, and the following year, his offices in Zurich and Geneva were still busy. Unfortunately, the small bridges for which Stettler played an important role in Bern did not provide enough work to sustain that office profitably once the Lorraine Bridge was completed. The Depression had not come fully yet to Switzerland and, in any case, Maillart's reputation was by now strong enough to ride out any temporary lull in building activity. He ended the year with optimism characteristic of this time in his life. "We do not want to complain because [the year] has kept us in good health."[76]

In May 1931, he had been able to separate himself finally from Adolphe Zarn, his partner of nearly 30 years.[77] He was happy to be free of Zarn, whose humorless, militaristic personality never appealed to him. Maillart was now fully on his own with no financial obligation to a partner. Zarn's departure left the Geneva office with three engineers besides Maillart (Meisser, Humber, and Huber), a secretary, a bookkeeper, and a draftsman. Maillart's primary business worry was keeping these engineers busy with the type of work for which he could get a reasonable contract. Much of his day – too much for his taste – was taken up with consulting reports for which he could charge only his own time. Still, 1929 had been a good year for him largely because of the Lorraine Bridge, the Credit Suisse Bank building in Geneva, and the post office building in Zurich (the last two were commissions from architects).[78] These were large projects that required many structural calculations and drawings. (The Credit Suisse project necessitated ninety large drawings of the reinforced-concrete, mushroom-columned design.) Maillart could make a good profit on such buildings because his engineers had learned from him how to make simple calculations rapidly and he had trained his draftsman to draw the details with amazing speed. As Stettler recalled, Maillart "took works from architects because there he could make money – a mushroom slab floor could be calculated in a half day or a day and handed over to Daetweyler [the draftsman] who was his expert in drawing up all these details in accordance with my sketches."[79]

AMMANN AND MAILLART

As Maillart worked on a series of little bridges, none with spans over 100 feet, his countryman, Othmar Ammann (1879–1965), arrived in Switzerland to lecture on his recent designs for New York City: the George Washington suspension bridge of 3,500 feet in span and the Bayonne arch bridge of 1,652 feet in span. Both of these steel structures, in the last stages of construction by then, were the longest spanning works of their type in the world. On the immense scale of New York City's bridges, Ammann had already made innovations, even though his aesthetic sense, much like Maillart's before 1927, still carried traces of masonry architectural ideas. The occasion for Ammann's visit was the seventy-fifth anniversary of the ETH, celebrated by a symposium from November 6 to 8, 1930. Maillart gave as a reason for not attending that the cost of the affair was too high; although out of loyalty to his friend Ros, he attended the fiftieth anniversary celebration of Ros's EMPA (Federal Institute for Materials Testing) on September 10.

At Ros's urging, Maillart reluctantly wrote a brief article for Ros's celebration. In a few minutes, he dashed off what was to be his single most important statement on the generating idea behind his designs.[80] "Mass or Quality in Concrete Structures" reads today as convincingly as it did when Ros published it. But it did not reach the engineering profession at large until the following year. The densely written article contains fundamental themes. By placing mass in direct contrast to quality, Maillart

Figure 117. Othmar Ammann (1879–1965), circa 1930. (*Source: Fünf Schweizer Brückenbauer*, p. 31)

summarized the challenge that the art of structural engineering was presenting to the tradition of classical architecture, and to the clutter of mathematical complexity that he felt to be rampant in academic engineering circles of his day.

His technical arguments as usual reinforced his aesthetic ones. "There is no doubt that light, slender structures will one day be praised by the layman for being as beautiful or even more beautiful than massive ones. Only their greater efficiency will be treated in this speech." He argued that massive structures in reinforced concrete were potentially less safe because of higher foundation settlements and larger stresses due to temperature changes that arise either externally from the weather or internally from the chemical reaction by which cement and water form bonds.

Following the ETH symposium, there was a gala luncheon at the luxurious Grand Hotel Dolder, high above the city overlooking the lake. There Maillart sat with Ammann; Wilhelm Ritter's two most distinguished students conversed at some length. Ammann had just received an honorary degree and was at the very top of the profession (Fig. 117). Whereas the Frenchman Mesnager, speaking at Ros's celebration without naming Maillart, com-

plimented Switzerland for its small bridges, the whole Swiss profession was showering praise on Ammann for his record-setting designs. What were Maillart's feelings?

In a letter to Marie-Claire, he described his meeting with Ammann and he expressed his feeling that "In any case I envy none who have gone further outwardly and materially than I. I can say that I am perfectly content and wish only to continue my present life as long as possible."[81] He would have welcomed the commission for a large bridge, but he took instead his "nontranscendental" possibilities and made do. Indeed it was his serene inner life that permitted him a clarity of vision that could – when given the opportunity – make something transcendental out of ordinary and inconspicuous projects. His material returns were meager: For the design of the Klosters Bridge, he received 10,000 francs ($2,000 in 1930) – his largest bridge commission for that year.

MAILLART EXPANDS IN ZURICH

As Maillart pursued structural problems that would have seemed trivial to almost anyone else, some of his designs began to cause a stir outside of the engineering world. Two outsiders who entered his life at this time portrayed Maillart as an artist. The first was a young architect, just graduated from the ETH, named Hans Kruck; the second was a leading figure in the modern movement, Sigfried Giedion. In school, Kruck had heard of Maillart's flat-slab ideas; then, following graduation, he had gone to Frankfurt to work for the noted Dutch modern architect Mart Stam. From Germany, Kruck had written Maillart about how "the supporting structure of a building must be an integral part of its architecture, and not simply a frame to hang architecture on to."

Kruck visited Maillart in Geneva late in 1929. As Kruck recalled years later, "the first meeting of our eyes is still vividly in my memory. It must have been my first discovery that a man can smile with his eyes." After making a remark about the overly efficient Germans, Kruck felt he had gone too far, but Maillart laughed and invited him to be seated.[82] Maillart was busy, but as usual, his manner was unhurried; he decided to see how the young man would react to his newest bridge design. He got the plans for the Salginatobel and spread them in front of Kruck, looking at him with a little smile. Kruck studied them silently, afraid to say something foolish about the strange form that he had difficulty visualizing from the stark two-dimensionality of the engineering drawings. Maillart finally broke the silence by observing that "If you cannot now form an opinion, then surely you will be able to do so later when working in my Zurich office." Kruck was astonished that Maillart should hire him in such an offhand manner[83]; Maillart, in turn, was amused by Kruck's enthusiasm and had already decided that the young architect could be useful in Zurich, where he was trying to concentrate on large design contracts for buildings. He had hired Hans Kruck partly because his father, Gustave Kruck, was an influential political figure in Zurich, who had played, for example, a major role in the Wägital project.[84]

Small bridges and numerous consulting requests came to Maillart in 1931, most without his solicitation.[85] Although his income now allowed him to think about making a trip to Sumatra to visit the Blumers, he was too much in demand for that.[86] As late as midsummer 1931, he still had not suffered from the world economic crisis.[87] But the Depression increased the Blumers' worries about Eduard's employer, and in Germany, radical political elements were coming into prominence.[88] Maillart described Adolf Hitler to his daughter not prophetically as "an April fool's joke." On New Year's Day, Maillart sat alone in his office-apartment and reflected on his past year of 1930. He decided not to follow the pessimistic lead of the Swiss banks, but rather gave his employees raises and even a Christmas bonus, "in spite of the dim prospects for 1931."[89] He saw the economic crisis approaching Switzerland, but he did not prepare for it. Partly he was confident that new work would come and partly he simply did not wish to think about future difficulties. Typically, his mind responded readily to design and structural problems; as in Russia, it reacted less willingly to economic and political ones.

In the summer of 1931, Maillart needed to hire another engineer, but typically made no systematic search. In August, Kruck ran into an old school friend, Marcel Fornerod, who was looking for work. Kruck volunteered to set up an interview with Maillart in the Zurich office. On Saturday, October 3, he introduced Fornerod to Maillart, who was at the office for his regular visit. Kruck left right away and the interview proceeded, not unlike the earlier Kruck interview, for about 2 hours.[90] Fornerod noted that Maillart "was quiet, had something of a smile much of the time, and created a very relaxed atmosphere considering it was a job interview."

Fornerod had worked for 2 years in the United States and Maillart quizzed him intensely on his experiences, particularly those involving multistory concrete-frame structures and concrete-arch dams. Maillart pressed Fornerod to describe in detail the simplified analysis method developed in the United States for arch dams. Partly he was testing the young engineer's ability; but partly also he was always searching for simpler ways to solve complex problems. Fornerod was surprised by Maillart's keen interest in arch dams. He did not realize, of course, that the American procedure used the same ideas that Maillart had pioneered 20 years earlier for his St. Gallen gas tanks.[91] Nor did Fornerod know of the patent and the articles that Maillart had published in the 1920s on arch dam design. Maillart satisfied himself that Fornerod understood those new ideas and offered him the job for 600 francs per month, which Fornerod immediately accepted.

This expansion in staff coincided with the move to a new Zurich office at Bleicherweg during the week of October 5, when Fornerod began.[92] The new office, with a view of the lake and the Uetliberg Mountains to the south, now consisted of Alois Keller, engineer and manager; Hans Kruck, architect; Fornerod, engineer; and several draftsmen. The bulk of the work was reinforced-concrete buildings for which architects had made the design and Maillart simply calculated the structural elements, and made engineering drawings. Almost nothing of the structure could be seen in the final buildings.

Keller knew well the details of reinforced concrete, he could organize the calculations and direct Fornerod, and he had no pretensions himself to original design. Maillart could rely on Keller not only to do standard work without any supervision but also to carry out Maillart's ideas on unusual designs with little discussion. In this way, the office was by 1931 a financial success and one that caused Maillart few difficulties. On the other hand, by the end of that year, he worried about a lack of work in Geneva, where he had a larger staff. Of the three offices, the operation in his hometown of Bern was closest to his heart, even though it still consisted only of Stettler and one draftsman. Here the growing number of small-bridge projects, however unprofitable, still challenged him intellectually.

THE MICROSCOPIC BRIDGES OF BERN

The canton of Bern is almost as large as the Graubünden, with even higher mountains. There in 1931, Maillart had to content himself with his only bridge work after the challenging 1930 Graubünden bridges at Salgina and Klosters. "These microscopic bridges that they give me either to design or inspect: It would take 200 of them to equal the Lorraine," he wrote on September 1.[93] But there were no more Lorraine bridges in Bern, and the canton recognized that for small provincial projects where economy reigned, Maillart was unsurpassed in Switzerland.

From a business point of view, he should have been trying to curry favor with the canton architect Bösiger, with powerful figures like Bühler and Rohn, and with the architects of large city building projects. He rather did the reverse; he studied little

Figure 118. Engstligen River, Spital Bridge, 1931: profile. (*Source:* FBM Studio Ltd.)

Figure 119. Engstligen River, Spital Bridge, 1931: view from underneath. (*Source:* FBM Studio Ltd.)

bridges such as Spital and Bohlbach and stayed within the society of close friends and family.

The Spital Bridge near Adelboden allowed him to study a new problem for deck-stiffened arches. This 120-foot-span bridge crosses a small stream on a skew, that is, the roadway is not perpendicular to the stream. There Maillart designed two separate thin arches that are not side by side but themselves skewed.[94] Although the profile view is not unusual (Fig. 118), this design gives a strange and unique visual impression to the underside of the structure (Fig. 119).

Figure 120. Bohlbach Bridge, 1932. (*Source:* Ernst Stettler)

Recognition and Calamity

THE ARCHITECT AND THE HISTORIAN

The Bohlbach Bridge above Interlaken represented a preliminary and smaller-scale experiment for a much greater work. At Bohlbach, the road has to cross a little ravine and double back on itself, as at Valtschielbach, where in plan there are kinks in the roadway approach. However, as he had already done at Klosters in 1930, Maillart decided to design the Bohlbach Bridge to cross the stream with a smooth curve, thereby avoiding kinks in the roadway (Fig. 120). This also allowed him a new means of expression by integrating the horizontally curved stiffening deck to the vertical curve of a very thin arch.[95] This horizontal curve is much sharper at Bohlbach than at Klosters, and the result is a more striking expression of curvature, which served as the essential test for his Schwandbach design 2 years later. Maillart stored up the experience from these little works while trying to ward off the effects of the economic crisis by increasing his higher-profit work in Zurich on large, standard buildings.

As soon as Kruck had begun to work on March 1, 1930, he set up a darkroom in the office to pursue his passion for photography, and in early summer, after some objection, Maillart agreed to let Kruck take his picture on the balcony of his Zurich office: Maillart found himself on the balcony in front of Kruck's camera "before I could defend myself."[96] This photo was to become the most famous one of Maillart (see Fig. 121), frequently published and used by Max Bill as the frontispiece for his book, *Robert Maillart*. Maillart's family did not much care for the picture; they thought that it should have been retouched to remove the wrinkles.[97]

Maillart agreed it made him look "old and hairy."[98]

Kruck's photo caught something in Maillart's eyes that he had noticed during their first interview: a quiet amusement with the world around him. It is not the look of someone who is sentimental or easy going. Rather what Kruck caught was the look that made Maillart seem so appealing to some, and so intimidating to others. Kruck soon learned that Maillart could be curt and sharp, but he also found that he spent much of his time in quiet, highly focused observation, what Maillart once called "pure observation." While studying that Kruck picture, one has the distinct feeling that it is Maillart who is doing the observing.

On one of his Friday visits to Zurich, Maillart had another unusual visitor. A fellow Swiss, who had once studied mechanical engineering at the ETH, Sigfried Giedion, came to talk to him about his work. He had written an article on the buildings of Freyssinet and was putting together one on Maillart.[99] When it appeared in the *Cahiers d'Art* early in the summer of 1930,[100] Maillart did not take it too seriously, referring to it as "nothing at all, because the author is not a person of much importance. The rest of the 'cahier' is perhaps more interesting." Admitting to his daughter only that "perhaps this will augment your pride . . . ," for him it was worth no more comment.[101]

The article may not have interested Maillart much at that time, but it put him before a new audience, one that believed itself to be in the cultural vanguard. The *Cahiers d'Art* published in Paris was a lavishly illustrated monthly magazine that contained the newest ideas and images in art, architecture, music, cinema, and the stage. Like modernism itself, the *Cahiers d'Art* was looking hard for the novel, the sensational, the up-to-date in imagery. Its editor, Christian Zervos, had written already for the first issue in 1926 that "those spirits concerned with life's problems are asking if poetry has deserted the creation of the artist to take refuge in the work of the engineer."[102] Maillart's poetry was quietly waiting to be discovered in the summer of 1930, and it was Giedion who announced the discovery. For Maillart, however, the August load test

Figure 121. Robert Maillart by Hans Kruck. (*Source:* Madame M.-C. Blumer-Maillart)

on the Salgina and the news of some little bridges in the canton of Bern were of greater interest than avant-garde essays. He showed no interest in the modern movement in spite of its new found interest in him.

MAILLART IN ENGLISH

In addition to Giedion, other writers on architecture began to recognize Maillart's pioneering ideas. One of these, an English essayist and architectural critic named P. Morton Shand, visited Maillart's office shortly after completion of the Lorraine Bridge and wrote in 1931 the first article on his work in English, calling him "a reinforced concrete engineer of international celebrity" whose Lorraine Bridge has a "dazzlingly beautiful arch." Shand illustrated the article with the Klosters Bridge, the Salginatobel, the Tavanasa, the Châtelard Aqueduct, and the 1924 Chancy-Pougny Water Tower (Fig. 122).[103] In 1932, Maillart was to be mentioned in the *British Architects' Journal,* and Shand published his Chiasso shed in *The Concrete Way* along with an article decrying the backward state of concrete structures in England compared to "Switzerland and Germany and Scandinavia [where people] are getting so far more, such infinitely better value, out of modern life. How?" he asked and he answered "above all by building logically and economically with the building materials which our age has perfected," that is, reinforced concrete.[104] Shand began a friendship with Maillart that during the 1930s brought the engineer considerable honor in England.

In 1938, Shand would again publish a major article on Maillart in which he claimed that he "has certainly transformed [concrete bridge] structural design more profoundly, than any man yet born . . . a direct consequence of the far-reaching modifications he introduced [is] that a wholly new formal aesthetic in bridge-building is now emerging."[105] It is surprising, but at the same time explicable, that an amateur like Shand would see what it took the engineering profession another half century to grasp. What Shand saw in Maillart is not unlike what John Ruskin saw in Joseph Turner or what Maxwell Perkins saw in Ernest Hemingway, something that was radically new and someone who was revealing certain fundamental truths for the first time.

In addition to the bridges, Shand would focus on the two new Maillart forms for buildings: the flat-slab floor and the funicular-truss roof. He described the 1912 multistory federal granary at Altdorf by asking: "Surely, the abstract austerity with which these columns rise, the classic purity of their proportions, and the graceful fan-shaped corbelling of their capitals are as beautiful in their own way as those of any one of the Orders?" Shand gave four photos placed one on top of the other to show "where flights of octagonal columns are superim-

posed directly above one another floor by floor in progressively diminishing girth, massive as normal crypt pillars in the basement, and slender as Gothic clerestory groining in the attic." The photograph stack shows what can never actually be seen in one view since only one floor at a time is visible in such an enclosed warehouse. Shand is correct in identifying new orders in these smoothly flared-top columns, whose girth expresses the combined weight above, but he is stuck with a conventional vocabulary that Maillart had set about to reorder as well.

Maillart reacted favorably to such praise in spite of some small errors.[106] One of these perpetuated a common misunderstanding, notably by his statement that "Perhaps Maillart's greatest contribution to structural engineering is that he dared the risks of practical experiment beyond the limits then set by calculation. . . ."[107] Certainly, Maillart dared to

Figure 122. Water tower at Chancy-Pougny by Maillart, 1924. (*Source:* Madame M.-C. Blumer-Maillart)

design new forms for which standard calculations did not exist; but in his mind, these forms did not involve engineering risk. He enjoyed challenging peoples' set ideas of what is safe and usual. "Now we shall give them something to think about," he would say with a smile. But Maillart never took risks with the physical performance of his structures; he never compromised safety.

The internationalization of Maillart's reputation would continue. On January 11, 1934, the New York *Engineering News Record* published an illustrated article on his deck-stiffened arch bridges.[108] That March, the editors of *Cahiers d'Art* organized an exhibition devoted to modern building in their Paris headquarters. The *Neue Zürcher Zeitung* reported that "the photographs of Maillart's Swiss bridges justly captured the place of honor." This was high praise considering that the exhibit featured Swiss designers such as LeCorbusier and included "a large model of the grandiose but controversial project by the Englishman [*sic*] Paul Nelson [with structure by Maillart] for the city hospital of Lille."[109] Maillart had met this American architect, who was practicing in Paris, when he had visited that city about 2 years earlier. The architect and the engineer had decided on a collaborative effort when Nelson lectured in Zurich in April 1933. Primarily, the prominence given Maillart was due to another article by Giedion in *Cahiers d'Art* featuring two new bridges by Maillart (the Schwandbach, 1933; and the Töss footbridge, 1934). While Giedion was active in France, Shand continued to serve the same role in England where Herbert Read's *Art and Industry* appeared in 1934 featuring a photo of the Salginatobel supplied by Shand.[110] As far away as Australia, Maillart's bridges began to be discussed.[111] But when Kruck and Giedion first encountered Mailart in 1930, this international recognition was far from Maillart's mind.

A NEW GENERATION

When the Maillart family assembled in Bern for Christmas 1930, it seemed to be an oasis of well-being amid signs of global distress. Although frail,

Figure 123. Maillart family in the early 1930s: (*left to right*) Alfred, Rosa, Bertha, Robert, Paul, and Max. (*Source:* Madame M.-C. Blumer-Maillart)

the 86-year-old Bertha Maillart took part with joy. All five of her children were there with most of their children. Each of her children was in good health and prospering reasonably well: Robert at the height of his career, Alfred the dentist, Paul the furrier, and Max with his photography shop. This was the interwar bourgeoisie, in the best Swiss sense of that word. They were happy to have Switzerland as their home with Bertha, the symbolic anchor of the family (Fig. 123).

The next generation, a product of the 1920s, had a different outlook. For them, Switzerland was confining and offered little opportunity. Bertha's grandchildren wanted to move away, as had all three of Robert's children. Rosa's son and his family lived now in Australia, and Paul's daughter Ella was something of a roving journalist.

Maillart was surprised and pleased to see the change wrought in Edmond at this time. "Naturally it gave me great pleasure to see him . . . because he has become very kind and 'a gentleman' [Maillart used the English word]. Also his pocketbook is better stocked and he throws something willingly in the direction of his old father!"[112] Edmond had become, finally, not only a colleague but a friend. The essential feature of this change was his newly found independence based on his escape from Switzerland and his success in engineering work.

Ella Maillart had returned from Russia in December 1930 and immediately gave a lecture in Geneva on the Soviet Union. It was so crowded that 4 days later she repeated it. Ella was beginning to support herself as a journalist and writer; she was one of a small group who would establish an international reputation as a professional woman at this time. She was anxious to explore the novel and the exotic and she had immense energy and talent with which to pursue her interests. A high jumper, sailor, and skier (captain of the Swiss women's team), she gave a glowing account of the new Communist state that was lively and convincing to many. But she did not convince her Uncle Robert. "As a foreigner she has surely seen things that were made for propaganda and naively she believes all that. Needless to say the Bolshevik sympathizers were enchanted by her lecture and they came to interview her. She and her parents are enthusiastic and one must not contradict them. It is like a contagious disease."

Ella's talks were fully covered in the press and Maillart sent an example to Marie-Claire with the comment that "one wants to laugh upon reading this, but it is sad to see admiration for a group whose works are the result of the greatest banditry in history!"[113] Maillart had watched the Bolsheviks in action. He never lost affection for Ella, but he did not swallow the propaganda that had begun to build up in the West as the Depression seemed to discredit democratic capitalism.

As for Marie-Claire, she had been eager to leave Switzerland since her departure from Geneva. Her adventurous spirit, although worrisome to Maillart, had his admiration. But unlike Ella, she did not have political ideas; her chief concern late in 1930 was for the financial stability of Royal Dutch Shell and Eduard's job. Despite her nonconformity, she had chosen to marry a real Swiss-German from one of the mountain cantons, who nevertheless toler-ated his adventurous wife's hunting expeditions in the jungle with friends while he directed civil works.

Fearing that they would have to return to Switzerland jobless, Marie-Claire asked Maillart if he would consider hiring her husband. In response, Maillart expressed confidence that Eduard would easily find work on roadway design and construction and that working for his father-in-law would not be good for him.[114] What Maillart did not say was that Eduard's road building experience would have been of little value to the type of structural design office run by his father-in-law. In discussing this delicate question with his daughter, Maillart made clear that he valued the Blumers' independence from him despite his regret in having them so far away. He assured her that they would never be neglected by him.

THE GOOD SAMARITAN

On Wednesday, October 14, Maillart set out for Bern from Zurich on his usual weekly tour. Canton engineer Trechsel had asked Maillart to consult on the reconstruction of a concrete-arch bridge in Adelboden over the Gilbach. On Thursday morning, October 15, Maillart, Stettler, and Trechsel left Bern by car. They drove south along the Aare River to Spiez; there they turned right to follow the Kander valley up into the Bernese Oberland, where Maillart had just completed two bridges over the narrow Engstligen torrent.

The road ends 1,356 meters above sea level, at Abelboden, the resort to which the Maillarts had often come before the war. There the engineers inspected the reconstruction work in progress with the builder O. E. Kastli. Early that evening, they began the return trip to Bern. Kastli invited Maillart to drive with him, but Maillart preferred to go back, as he had come, with his client the canton engineer, who had a big car and driver. They set out with Maillart and Trechsel in the back and Stettler in front beside the chauffeur.[115] Kastli followed in his car. The trip went well until they passed Thun going north.

Near the little town of Heimberg, the chauffeur spotted a cap and an overturned bicycle by the side of the road. Assuming an accident, he backed up the car and then turned it at an angle, allowing the headlights to shine on the bicycle. There they could see a man lying in the ditch. To their surprise, he immediately got up and began berating them for wakening him. He was clearly drunk. The engineering group, having left the car to help the victim, returned to their vehicle. Just as Maillart opened the back left door at that corner of the slanted car that jutted into the middle of the roadway, a car appeared speeding south. Stettler waved wildly in warning, and Maillart jumped onto the running board of the canton car, but they were too late, and the onrushing car collided with the back of the parked car. Maillart was thrown to the other side of the road with a force that he thought would be his "last leap. . . . I found myself at the edge of the road not knowing exactly if I were still in one piece!" The others realized immediately that he was hurt. Kastli's car appeared just then and they managed to get Maillart into it, while the speeding car continued south without stopping. (Later they discovered that it was driven by drunken soldiers returning to their base at Thun.) After waiting for the police, Kastli with Trechsel took Maillart to Bern, leaving the chauffeur and Stettler with the damaged car.

In the morning, a doctor came, examined the engineer, and gave the opinion that nothing was broken, but he did have a sprained foot. He lay there all day Friday, his leg burning with pain, writing his daughter and calling others by phone. He realized that his injuries were not negligible and that he would now need considerable help at least in the near future. The next morning things were not better, so the doctor ordered the engineer to go to the Engeried Clinic, where he put a plaster cast on Maillart's leg, still insisting that nothing was broken. Maillart found the clinic unpleasant and was happy when Edmond, Nissen, and his niece Marcelle came Sunday morning to drive him to Geneva. They gave Maillart crutches on which he could get around unaided. The day was splendid, the company joyous, and the release from Engeried welcomed.

Just north of Lausanne at Le Chalet à Gobet, they clambered out to have a festive high tea, after which they drove along the lake to Geneva. Arriving there about 7:00 P.M., they went directly to the Globe Restaurant, where practically all the Geneva Maillarts met them for a reunion supper. Edmond and Nissen then delivered Maillart to his office-apartment and put him to bed. They ended the evening there with Maillart playing cards.[116] But Maillart was in pain and it was therefore with considerable relief that he welcomed Amélie, his sister Rosa's maid, the next morning to take care of him and prepare his meals. She arrived on Monday, October 19, 1931; she was to be with him for the rest of his life.

AGING AND A DEATH

The next 9 weeks Maillart spent largely in bed. In spite of the pain, especially at night, and his inability to leave his room, Maillart did not miss a single day of work. He also had a stream of visitors: George Kitts, with his mother in Montreux, Nissen, Edmond, and the Geneva family.[117] Two weeks after the accident, Maillart was told that he had two small cracks in his ankle (the prolongation of the tibia), where the car door had slammed against him. This new discovery meant that he would be immobilized longer than originally thought.[118] By the end of December, Maillart was again able to travel,[119] but he was in continual pain. On February 6, Maillart turned 60 and thus became, as he put it, "more or less venerable . . . especially by still not walking very well, everyone must take me for an old man. . . ."[120]

On May 26, Rosa called from Bern to tell the family that Bertha had died peacefully at 7:00 P.M.[121] Bertha's unexpected death, funeral, and burial deeply affected Maillart. "In spite of saying that this must be and that in short it could not have been better, one begins to sense a huge void. The center of the family has disappeared and with her all of the preceding generation, so that definitely it is we who are now the old ones, myself and my brothers and sister."[122] The accident, turning 60, and now the disappearance of his only remaining parent came together in mid-1932 to remind Maillart of his own mortality.

Finally, in June, Rosa convinced him to see her doctor.[123] Upon returning to Geneva that Saturday, he found a note from the doctor ordering him to begin a strict diet with medication. "All that is good is forbidden!" Maillart protested. Despite his distaste for it, he agreed to the new regime because for the first time a doctor had recognized that the pain he suffered from originated not in his legs but rather from an internal cause – probably an injury to the liver and the kidneys, sustained in the accident.[124] By mid-1932, these combined problems made Maillart physically old before his time.

But he was still young intellectually, and still Spartan. He actually made his disability into a kind of new discipline for himself; the extra time it gave him was as liberating mentally as it was confining physically. Not only did his designs now take on new vigor, but his conflicts with the authorities became more public and revealed more technical insight. Between 1922 and 1932, the battleground had spread from Switzerland to Europe through the medium of a new type of organization: the international engineering congress, founded at the same time as the League of Nations and, like the political ideal of that organization, begun with the desire to help peace through cooperation between nations. Maillart's reaction to these meetings reveals the degree to which this conservative Swiss was a revolutionary within his profession.

CHAPTER SEVEN

Images from Within
1932–1934

The Interwar Congresses

ROHN AND THE CONGRESS IDEA, 1922–1928

Between 1926 and 1932, Maillart participated in a series of international conferences, presenting major papers in each one, but thereafter his interest in such meetings diminished as his own ideas became more unconventional. A first meeting occurred on September 29 and 30, 1922, when Maillart attended an international gathering of engineers convened by Arthur Rohn in the Zurich concert hall. Ostensibly, the group was to discuss the design of bridges and steel buildings. Rohn, however, had another agenda. He wanted to form an international society of structural engineers devoted to furthering contacts between countries and especially to bring together the French and Germans.[1] Surely, neutral trilingual Switzerland was the place to found such a society for which Rohn's organizational talent could be put to good use.

The early 1920s were a difficult time in postwar Europe, and Rohn did not succeed in organizing another meeting until 1926, when he, along with Ros, chaired a meeting in Zurich entitled "Swiss Engineering Structures, Theory and Practice: International Congress for Bridge and Structural Engineering." The last seven words were to be his term for an international society of engineers that would hold meetings in 1928 in Vienna and its first official Congress in Paris in 1932. It remains a thriving or-ganization in 1997, with its headquarters still at the ETH in Zurich.

At the 1926 meeting, devoted almost exclusively to Swiss bridges, the only paper on buildings was by Maillart.[2] It is a measure of Maillart's falling out with Rohn that his talk was on his flat-slab designs. Had the conference been organized by almost anyone else, he would undoubtedly have been asked to speak on bridges because he was by then Switzerland's most famous bridge designer.

In September of the following year, Maillart attended the first International Congress for Testing Materials in Amsterdam. There he presented results of concrete-beam research that he had carried out before the war.[3] Like the detailed description he presented in his 1926 Zurich paper, he again made public some tests that he had conducted while he had been a building contractor. But whereas the flat-slab tests were pioneering works of enduring value, the beam tests were not special, although they did provide Maillart with a firsthand knowledge of research that allowed him in the 1930s to propose radical new design methods for the Swiss code.

These Congresses were new to the engineering profession; they reflected a desire to put the subject on a scientific basis in which engineers would share ideas on structures the way scientists were sharing their research discoveries. The first truly international Congress on structures was in Vienna in 1928, home of the most prolific publicist for concrete structures, Fritz Emperger (Fig. 124). It was during Emperger's visit to Zurich to plan this meeting that Maillart and Ros hosted him at a luncheon at which Marie-Claire had met her future husband.

Figure 124. Congress at Vienna, 1928, showing Maillart, Max Ritter, Rudolph Saliger, and in the front row, the bearded Emperger. (*Source:* J. Aichhorn, Linz)

THE STRUCTURAL CONGRESS OF VIENNA, 1928

In late September 1928, Maillart and Ros prepared to go to Vienna as official representatives of Switzerland, and Maillart as a correspondent for the *Bauzeitung*. Maillart collected Marie-Claire from Zurich and the two headed off on the Arlberg Orient Express for the former Hapsburg capital. For the engineer, the Congress was a vacation from the intense activity of the summer culminating in the Salginatobel competition and a chance to be with his daughter. They arrived in the Austrian capital on Sunday, September 23, and took rooms in the famous Sacher Hotel. Ros was in a nearby room and the three of them were together throughout that week. Ros and Maillart jokingly predicted that their antagonist Rohn, of course, the leader of the Swiss delegation, would surely use the cliché "bridges between countries, between ideas, etc. etc."[4] And indeed, on the opening day, Rohn proclaimed that "Especially it is the engineer who is called to create such a work [the Congress] of understanding; and it is he who has built material bridges who knows the value of spiritual bridges from nation to nation."[5]

Maillart did not object to this idea, but rather to its constant repetition and to the pretensions of engineers who inflated their technical ideas into pseudopolitical ones. He knew that engineers, however friendly across national borders, had negligible influence on political events. Maillart was not at all attracted to the idea of visualizing international relations in engineering terms, just as he resisted the facile notion of the overdesigned bridge as a political monument. For Maillart, this Congress was a purely engineering event (Fig. 125) focused on new designs and old friends. It was also a chance to take Marie-Claire to gala events, including a performance of *The Rosenkavalier* at the opera, and a sumptuous reception at the Schönbrunn Palace.

Ros recovered from a cold sufficiently to play a major role in the Congress: running one session, giving a paper, and talking incessantly. Even while sick in bed, he urged Maillart to give a presentation on arch bridges.[6] So on September 25, Maillart spoke about the 1912 Laufenburg Bridge as an introduction to the Lorraine Bridge and the general idea of arches built out of concrete blocks.[7] The editor of the major German language journal on reinforced concrete requested Maillart to let him publish it.[8] It did not appear there, but rather in the proceedings of the Congress, in another German journal, and of course in the *Bauzeitung*.

But the professional highlight for Maillart was the chance to hear Eugène Freyssinet speak on the construction of his Plougastel Bridge, illustrated by a movie showing the floating of a huge scaffold for one of the three arch spans.[9] For Maillart, this presentation, as he later wrote in the *Bauzeitung,* "will remain for all the participants indelibly fixed in their memories." He spoke of the "most impressive and extraordinary daring as much in design as in the well-thought-out construction" of the French engineers Caquot and especially Freyssinet.[10]

Maillart was rarely given to such strong statements; this praise therefore reflects great enthusiasm, and Maillart was not alone. Freyssinet's Plougastel Bridge consisted of three hollow-box concrete arches, each spanning 610 feet. It represented by far the longest spanning concrete arches in the world in 1928 and one of the most brilliant construction devices – building a wooden form for one arch, floating it out into the estuary, and casting the first arch onto it. Once the concrete was strong enough, he removed the wooden form, floated it out to the adjacent location, and cast the second arch on it. The process was repeated for the third arch to complete the bridge spans. Such an operation at such a scale had never before been attempted; it was especially daring in the still relatively new material, reinforced concrete. For civil engineers all around the world at this time, Freyssinet's work was inspirational.[11]

FROM VIENNA TO LIEGE, 1930

Just 2 years after the Vienna Congress, Maillart again set off, this time to Liège with Edmond, Marie-Claire now being in Indonesia. In Vienna, the focus had been on international cooperation, expressed by the Swiss-inspired idea (not shared by Maillart) of bridges between nations. The charm of Vienna, with its ever-present music, provided a pleasant setting for technical presentations of general structural interest beginning with a major ses-

Figure 125. Congress at Vienna, 1928, showing Marie-Claire, Maillart, and Ros. (*Source:* Madame M.-C. Blumer-Maillart)

sion on aesthetics, in which the keynote address had centered on the engineer as designer rather than on the architect as aesthetician.[12]

In Liège, by contrast, architectural ideas of façade, decoration, and formal analysis prevailed and lost was the focus on engineering ideas as the means out of which an aesthetic could emerge.[13] A major aspect of the congress was visiting factories. There was very little lip service to international issues; Rohn was present, but less prominently so than in Vienna, as this congress was not organized by his new society in Zurich. More than anything else, the differences between Liège and Vienna lay in the focus in Liège on concrete as a building material.

Against this trend, Maillart's Liège paper was almost alone in expressing ideas about structure as an art form with a focus on full-scale observations of completed works. Whereas at Vienna his short talk centered exclusively on the concrete-block bridges at Laufenburg and Bern, at Liège, beginning with the 1890 Wildegg and ending with his own 1930 designs, he reviewed the best-known Swiss arch bridges. In spite of its broadly Swiss content, the major focus was on his own designs in his first reflective writing on the meaning of his work. It gave him a chance to trace his development from the Stauffacher in 1898 to the Salginatobel and Klosters, both then under construction, embodying his shift from the massive to the light and from imitative forms to something radically new. He ended the report with a repetition of his conviction that direct observation of bridge behavior is of paramount importance.[14]

His presentation was once again overshadowed by that of Freyssinet, whose Plougastel still dominated the profession. But even Freyssinet had by then gone beyond these great arches and was trying to make a commercial success out of his 1928 patent for prestressing, although he would not present his new ideas to the profession until after the 1932 Paris Congress.

The Vienna Congress was in the spirit of the return to normalcy that longed for the stability of prewar Europe. At Liège, there began the tradition of specialized international meetings focused on a typical twentieth-century technology. Gone was any sense of aesthetics (except for an isolated paper like Maillart's) or of history; all eyes were on recently completed works, on academic research, on new theories modeled after those in the natural sciences, and above all on the future. The reality of the Depression had yet to be felt by the engineering profession in 1930 as it would at the next major congress 2 years later in Paris. Maillart felt bound to stay the entire week in Liège, but without much enthusiasm.[15] He had been excited by Freyssinet and Caquot in Vienna, but for him, neither presented anything new at Liège.

THE PARIS CONGRESS, 1932

Even though he was at odds with most Swiss engineering, Maillart was so respected internationally that the organizing committee of the new International Association of Bridge and Structural Engineering invited him to be one of the three vice presidents of the second working session for the Paris Congress. Also, despite Rohn's prominence in

the association, it had made Maillart a member of the central committee.[16] Once again, Maillart met some old friends (Fig. 126) and saw his son René.

On May 19, 1932, the Congress opened officially in the Grand Ampitheatre of the Sorbonne, and the next day, following a series of papers on slabs,[17] Maillart presented his own "note on slabs without beams, *système Maillart*."[18] He spoke of two ideas that led him to his flat-slab designs: first was his concept of the slab as an integral element rather than as a connected set of flat strips; second was his idea that the column capitols by widening at their tops could greatly reduce bending in the slab. Behind these design concepts lay two more fundamental ideas: first, that as a single unit, the reinforced-concrete slab could carry loads as no previous structural element had done before. Maillart rejected all analogies to wood and metal. His second idea expressed his continuing empiricism: He would not "seek to establish a [mathematical] theory, because of the difficulties resulting from the multiplicity of factors one would need to consider. I will only cite here the influence of the capitols on the elasticity of the columns."[19]

Figure 126. Congress at Paris, 1932, showing Maillart and Emperger during a visit to Fontainbleau: card sent to Marie-Claire. (*Source:* Madame M.-C. Blumer-Maillart)

Overall, the 1932 Paris Congress focused less on design and on completed structures and more on the analysis of structures.[20] Maillart had fought against this approach since Schüle's 1906 Code Committee meeting and he continued to challenge academic engineers, now not only Rohn, but Rohn's protegé, Max Ritter, who in his position as Secretary was becoming prominent in the association. Maillart considered the scientific conference no model for engineering, and from 1932 on, he would proclaim that heretical view more publicly and less diplomatically. At the same time, in private, he continued to search for better designs by a process of revision not unlike that of a poet.

Maillart's Method
1932

THE GIUBIASCO BRIDGE

Maillart's ideas were potentially more widely applicable than those inherent in Ammann's big bridges or Freyssinet's arches because Maillart designed the kind of relatively modest bridge spans that became more common as automobile traffic demanded new and better roadways. As one example, following his accident in 1931, Maillart began to design a relatively long bridge over the Tessin River just south of Bellinzona between the towns of Giubiasco and Sementina.[21] First, he made a sketch of the entire 262-meter-long bridge with a 70-meter arched river span, and while still in bed he worked on this drawing, changing dimensions and making a few rough calculations. His draftsman completed it on November 27. Within a couple of weeks, he had written a report that included brief but careful calculations and a slightly revised drawing of the main span [Fig. 127(a)].

Maillart continued to work on the design in early 1932, perfecting details, none of which made much difference to the bridge performance or cost,

but all of which were significant visually. From the two figures, we can see the three changes Maillart made between December 14, 1931 [Fig. 127(b)], and February 2, 1932 [Fig. 127(c)]. First, he removed column A [Fig. 127(b)] and made the span 14.6 meters between the column over the arch springing [Fig. 127(c)] and the first viaduct column B. This change clearly separated the deck-stiffened arch with closely spaced columns from the approach viaduct with widely spaced columns. In the final design, he made the deck girder equally deep over both viaduct and arch, giving a visual integration to the whole. It was jarring to Maillart to see column A so close to the arch where it was not needed. Second, Maillart lowered the viaduct column pedestals so that they were underground and hence not seen. The approach viaduct looked more slender and uncluttered this way.[22]

Finally, in a third change, Maillart raised the deck slightly so that the deck girders would not be interrupted over the crown of the arch and thus the arch would become a more clearly articulated thin slab from end to end. In the earlier design, the arch merged with the deck girder over the central part of the span, just as it had previously done at Valtschielbach. Maillart's first design was, as usual for him, a paraphrase of recently completed works.

The idea of the first viaduct column being close to the last deck-stiffened column came directly from the 1930 Landquart design, which lacked visual integration between the main span and the approach viaduct. Having the viaduct column pedestals visible above ground in his first design came from the Salginatobel just as the deck girder merged visually with the arch slab came from Valtschielbach and later deck-stiffened designs such as the 1931 Spital Bridge. In this pivotal but unbuilt project over the Tessin River, Maillart felt his way to a bridge form that was visually more satisfactory to him.

In 1931 Giubiasco represented a design that was not yet fully satisfactory because of the change in span length between the main span columns of 3.72 meters and the viaduct columns of 14.6 meters. Having recognized that defect, Maillart would soon incorporate the aesthetic ideas of his Giubiasco de-

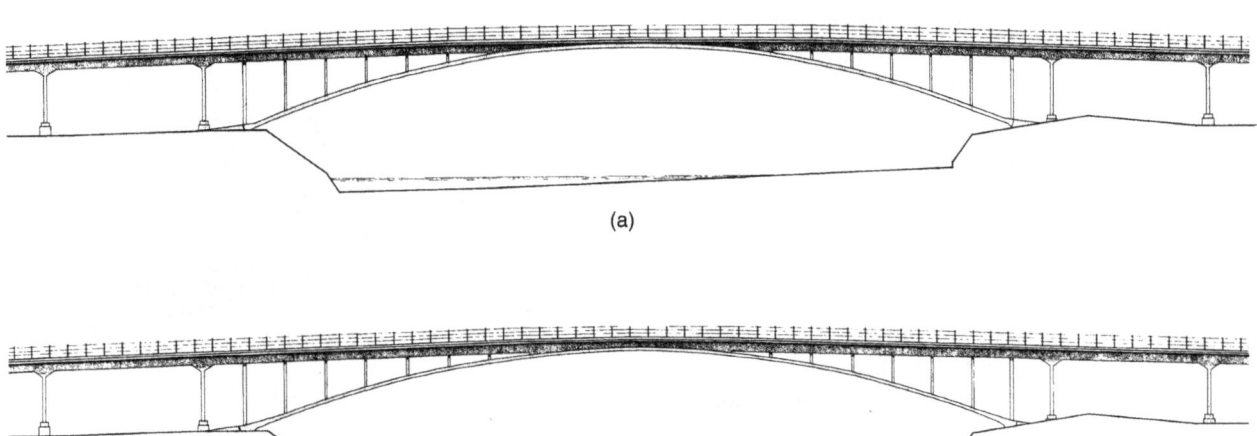

(a)

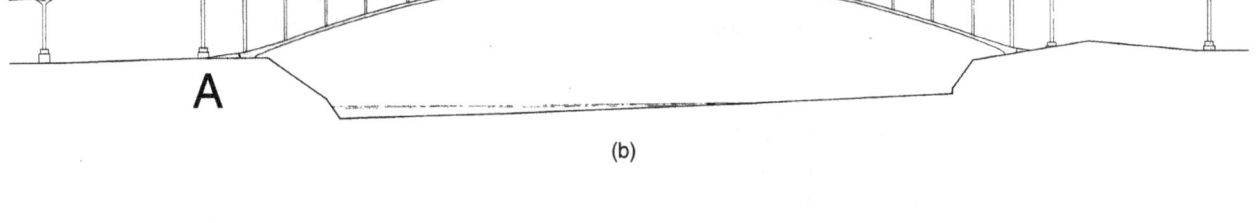

A

(b)

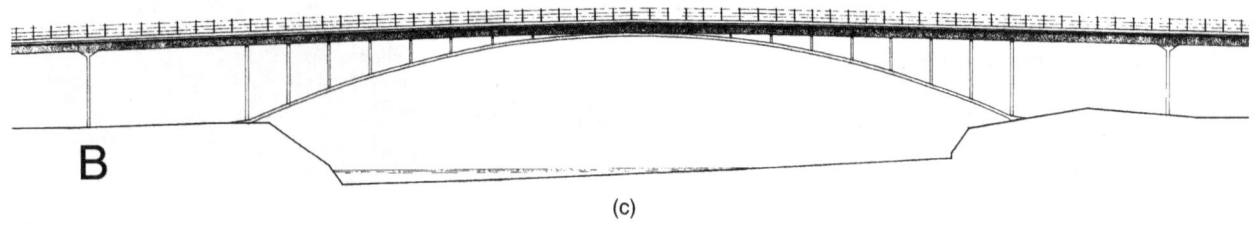

B

(c)

Figure 127. Giubiasco Bridge Project, 1932: (a) Maillart's design of November 27, 1931; (b) Maillart's design of December 14, 1931; (c) Maillart's design of February 2, 1932. (*Source:* Drawings by Mark Reed)

sign into the more sophisticated project over the Schwandbach in the canton of Bern.[23]

THE TÖSS BRIDGE

By the spring of 1932, Maillart had begun to feel the economic crisis. He was upset that the city of Geneva still gave him no work and he again contemplated changing his official domicile to Zurich.[24] He began to bring work from Bern and Zurich to the Geneva office just to keep the staff there busy. One of those projects was a small footbridge over the Töss River near Winterthur. Walter Pfeiffer, a

local engineer, had made a preliminary deck-stiffened arch design that he clearly identified as a Maillart-type structure (Bauart Maillart) and that he sent Maillart in April 1932.[25]

Pfeiffer's design has an arch thickness of 10 cm and a deck stiffener 90 cm deep. There are three cross walls over the arch on either side, and over the central half of the span, the arch and stiffener merge to give a slightly curved beam 1 meter deep. Maillart immediately redesigned the bridge and sent it back to Pfeiffer in late April. Maillart reduced the stiffener depth radically to 40 cm and increased the arch thickness to 14 cm. He also decreased the length of the merged arch-stiffener beam over the central part of the bridge by adding an additional cross wall on either side to give a total of eight openings (rather than six). With these changes, the bridge becomes lighter and more graceful. Here is the structural artist at work.[26] As with the Giubiasco design, he would keep this bridge in his head and modify it until it satisfied him.

His earliest sketch shows a very thin arch, a deeper deck, and the two merged over the central third of the span [see Figure 128(a)].[27] Picking up the project again in mid-June, Maillart retained the basic form but made two typical changes. First, he continued the thin arch farther out toward the abutments, thus slightly increasing its span and emphasizing more its visual lightness [compare Figs. 128(a) and 128(b)]. But the move also served to place the deck end right below the highest cross wall. He thereby removed the offset appearance of the cross wall and arch end, visually accentuating the delicate intersection of arch, wall, and abutment.[28] As usual for Maillart, such a change influenced the bridge performance and cost negligibly, but it did improve its appearance.

Maillart continually searched for such changes and he found a second one of even greater benefit. An April sketch showed the deck curb descending in a straight line to the river bank. In mid-June, he changed that by putting a kink in the deck curb in order to meet the river bank more nearly on a horizontal. This smoother transition looked better to Maillart and again he knew the change would not influence performance or cost. But that change still did not satisfy him, so that in late July, studying the bridge a third time, he decided to remove the kink and make a smoothly curved deck curb. It is this final change [Fig. 128(c)] that gives the little bridge its unusually graceful countercurvature.[29] Maillart's aesthetic ideas matured as he studied these minor details, which he had time for because the initial choice of form had reduced the problem of analysis to a minimum. Pfeiffer completed detailed drawings during the summer according to Maillart's redesign and the bridge was built in 1933 (Fig. 129).

The June 24 calculations for this 38-meter-span

(a)

(b)

(c)

Figure 128. Töss River Bridge near Winterthur, 1932: (a) perspective of April 21, 1932, design; (b) perspective of June 24, 1932, design; (c) final design with graceful countercurvature. (*Source:* Drawings by Mark Reed)

Figure 129. Töss Footbridge, Maillart design. (*Source:* Photograph by D. P. Billington)

bridge cover a mere ten pages, their simple formulas reflecting Maillart's long experience and revolutionary ideas.[30] (A comparable calculation made several years later by Max Ritter covered over 100 pages and was less complete than Maillart's; see Chapter 8.) Indeed, at the beginning of these calculations, he introduced his revolutionary methods' recently published in the *Bauzeitung.*[31] Maillart's light Töss footbridge design contrasts strongly with a much heavier and more costly footbridge designed by P. E. Soutter and built in 1931 also over the Töss River (Fig. 130).[32]

ROSSGRABEN

Construction of three more bridges in the canton of Bern went forward during the summer of 1932: the Traubach, designed in 1930; the Allenbach, for which he was only a consultant; and, finally, the Rossgraben, carrying a small mountain road over

the Schwarzwasser and north from Schwarzenburg. The previous summer, Maillart had submitted two designs patterned after the Salginatobel Bridge, which contrasted with an earlier official design for a straight-girder bridge of three spans with a total opening of only 50 meters between heavy looking, solid abutments [Fig. 131(a)].

Figure 130. Töss Footbridge by P. E. Soutter, 1931. (*Source:* Drawing by Mark Reed)

In the summer of 1931, Albert Binggeli, a builder in Schwarzenburg, proposed a heavy, three-hinged arch with a span of about 60 meters in competition with which Maillart made his designs of September 16.[33] One was for the same span and the other for an 80-meter span [Fig. 131(b)]. He considered the longer span better "for technical and aesthetic reasons." The local authorities agreed with Maillart who made a nearly final drawing of the 80-meter-span design in January 1932 after which Stettler made the detailed calculations. Maillart officially received the commission on February 18. The contract was awarded jointly to Losinger and Binggeli as builders.[34] Stettler produced the detailed forming plan in March and the final calculations in May while Maillart was in Paris.[35]

Construction proceeded with great speed over the Schwarzwasser; by early September, the arch scaffold was up and the bridge was completed by late October in a mere 3 months. Although not so grandiose as the Lorraine opening in Bern 2 years earlier, the Rossgraben inauguration was far more significant technically: first, because as Maillart observed, it was built in the world record time of 3 months, a statement quoted by all the local newspapers even though such a record could never be proven.[36] Second, it was built for the incredibly low cost of 82,840 francs for a bridge that spanned 82 meters with one hollow-box, three-hinged arch of even lighter appearance than the Salginatobel (Figs. 132 and 133). In the photographs of the Rossgraben celebration, Maillart stands near the center surrounded by girls dressed in the costumes of old Schwarzenburg; he is leaning on his cane, his beard white, but his head held high (Fig. 134).[37] About this time, he made another trip to the Salginatobel to think about his design (Fig. 135). He was about to change his view in subtle but substantial ways.

SUCCESS IN ST. GALLEN

As the Rossgraben progressed rapidly, Maillart initiated discussions with the canton engineer at St. Gallen about a new Thur River Bridge at Felsegg, not far from his Billwil design of 1904. The canton itself had made, in September, a design of four heavy romanesque arches, each spanning about 20 meters [Fig. 136(a)]; Maillart proposed in October to replace this with a single 72-meter-span arch al-

Figure 131. Rossgraben Bridge: (a) official design; (b) Maillart design. (*Source:* Drawings by Mark Reed)

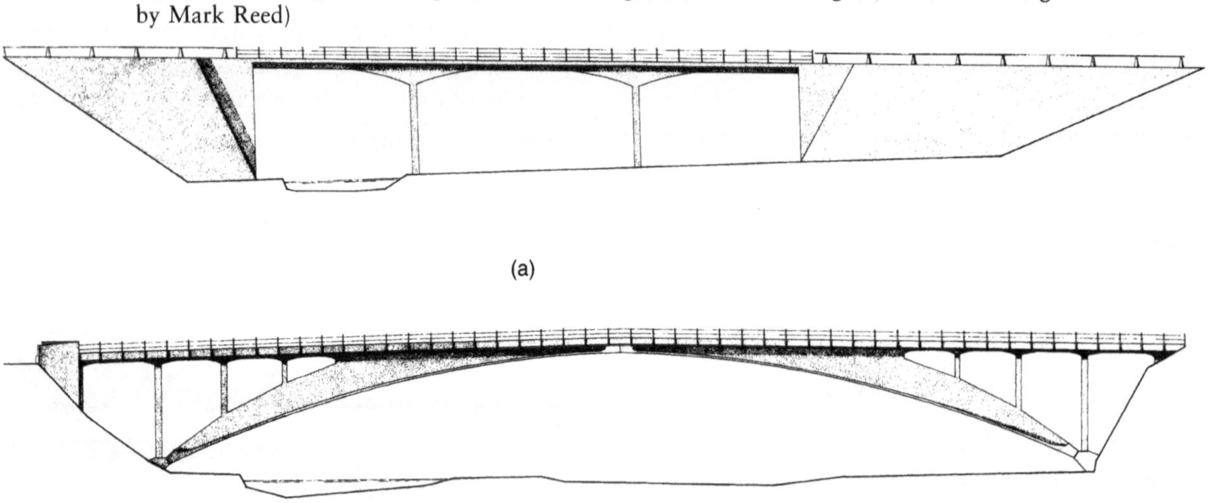

(a)

(b)

Figure 132. Rossgraben Bridge at Schwarzenburg, 1932. (*Source:* Losinger and Company, A. G., Bern)

Figure 133. Rossgraben Bridge, profile. (*Source:* Losinger and Company, St. Gallen)

Figure 134. Inauguration of the Rossgraben Bridge, November 19, 1932: Maillart at center, Stettler third from the right. (*Source:* Losinger and Company A. G., Bern)

most identical to Rossgraben [Fig. 136(b)]. The canton engineer was averse to Maillart's proposal because it challenged his own design. Maillart was not at all optimistic, but needing work in Zurich, he persisted.[38]

On Christmas Day, Maillart boarded the train for St. Gallen where a decision about the Thur Bridge design was reaching a climax. He got out his stationery and began a letter to Marie-Claire. It was a long trip and he had much on his mind as he reflected on past times in St. Gallen, when almost 30 years ago he had set out by train with Maria to go to the inauguration of Billwil, the first bridge he was to build by the *Système Maillart*. He recalled the former canton engineer of St. Gallen, who had tried to charm Maria.

Then he thought of the present canton engineer, whose name, Altwegg sounds like "old way" and who was not at all so charming, thought Maillart; "he does not like modern bridges and wants nothing to do with mine." It saddened him to think of

the problems associated with even such a relatively modest commission. For this bridge, he had already invested a lot of his own time as well as the time of young Fornerod. As the train rolled over the fertile low crescent of the Swiss farming lands of Bern, Maillart thought about his life in this small country where everything was organized so tightly that many little things – and mean ones at that – seemed to pursue him with insistent pettiness.[39]

On this Christmas, though, Maillart had no cause for gloom. A high-ranking political figure there had overridden the canton engineer's opposition, and on his arrival, Maillart was offered the design of the new Thur Bridge.[40] Thanks to the new Thur Bridge commission, Maillart returned in good spirits to Geneva and celebrated both his recent success and New Year's Eve of 1932 with the entire Geneva family at a festive dinner.[41] Unfortunately, the good news in St. Gallen was an isolated exception. Lack of work spread now to the Zurich as well as the Geneva offices. Even the coveted Thur

Figure 135. Maillart surveying his bridge, early 1930s. (*Source:* Madame M.-C. Blumer-Maillart)

Figure 136. (a) Thur Bridge at Felsegg, official design, 1932. (*Source:* Zurich Maillart Archive). (b) Thur Bridge at Felsegg, Maillart design, 1933. (*Source:* Maillart and Company, Zurich)

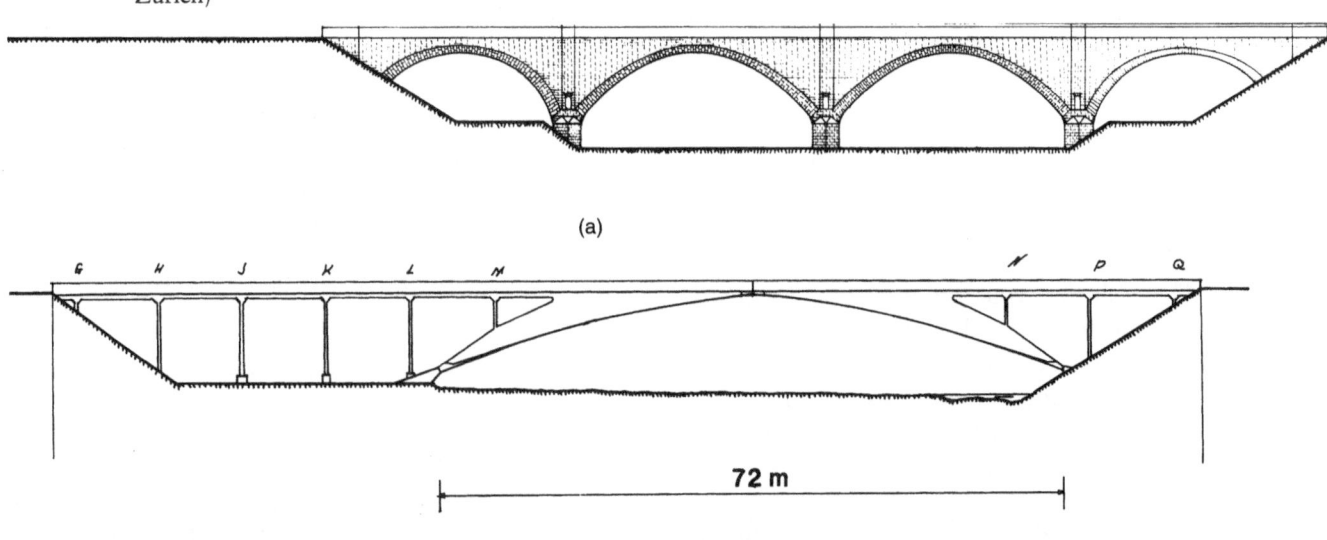

(a)

(b)

Figure 137. Thur Bridge at Felsegg, 1933. (*Source:* Madame M.-C. Blumer-Maillart)

Bridge commission provided barely a few month's production work. First, the calculations were formally written up and then detailed drawings were made by the draftsman following sketches by Fornerod to whom Maillart entrusted this major work.[42]

In the Thur Bridge Maillart modified his ideas in two ways in contrast to the Salgina and Rossgraben designs. First, he emphasized the crown hinge by creating a broken arch, and, second, he opened up the crosswalk to form a strongly articulated frame. The arch thus could be deeper at the quarterspan and the frames easily carried the wider roadway. This bridge carried a main two-lane highway rather than the single-lane, secondary-road traffic of the two earlier hollow-box, three-hinged arches at Salgina and Rossgraben (Fig. 137).

Maillart reviewed the design process each Friday, and on the last Friday in March, he took the full set of thirty-four drawings to St. Gallen, where the city council accepted them. Later there would be revisions and a few new drawings for construction purposes, but by April 1, production was finished, again leaving the Zurich office idle.[43] By mid-June, business was so bad in Geneva that Maillart put his entire staff onto a half-time schedule, and even then he could not keep them all busy.

The Schwandbach Bridge
1933

SCHWANDBACH DESIGN

In spite of their low profitability, Maillart devoted far more time to his small bridges than to large production projects. In retrospect, many of his design ideas seem so simple that we fail to realize how much thought they required. The little countercurvature in the deck of the Töss footbridge came to him only after over 3 months of intermittent study. If the Töss seems effortless and yet required careful thought, the Schwandbach required even deeper reflection (Fig. 138). Just as for the Rossgraben Bridge, Albert Binggeli proposed in October 1932 a bridge design to cross the small mountain valley of the Schwandbach, a little stream at a spot with no name, just a few kilometers from the Rossgraben. Binggeli designed a straight bridge with horizontally curved roadways at either approach (Fig. 139). The secondary road authority chose the design of Maillart's with Losinger as builder.

Maillart had been thinking about bridges curved in plan since at least 1923, when he designed the awkward, polygonal arch over the Ziggenbach in the Wägital. He had avoided the problem in 1925 at Valtschielbach by using the traditional kinked transitions between straight bridge and curving mountain roadways. Only with the 1930 Klosters Bridge and the 1932 Bohlbach Bridge did he arrive at the form of a smoothly curved roadway over the arch. At Schwandbach, as he thought about it in June 1933, the problem was more difficult because the span was much greater than at Bohlbach and the curvature much sharper than at Klosters.[44]

Maillart's report for the Schwandbach Bridge is only two pages long and it discusses mainly the complex layout problem that he proposed to re-

Figure 138. Schwandbach Bridge, 1933. (*Source:* Losinger and Co., Bern)

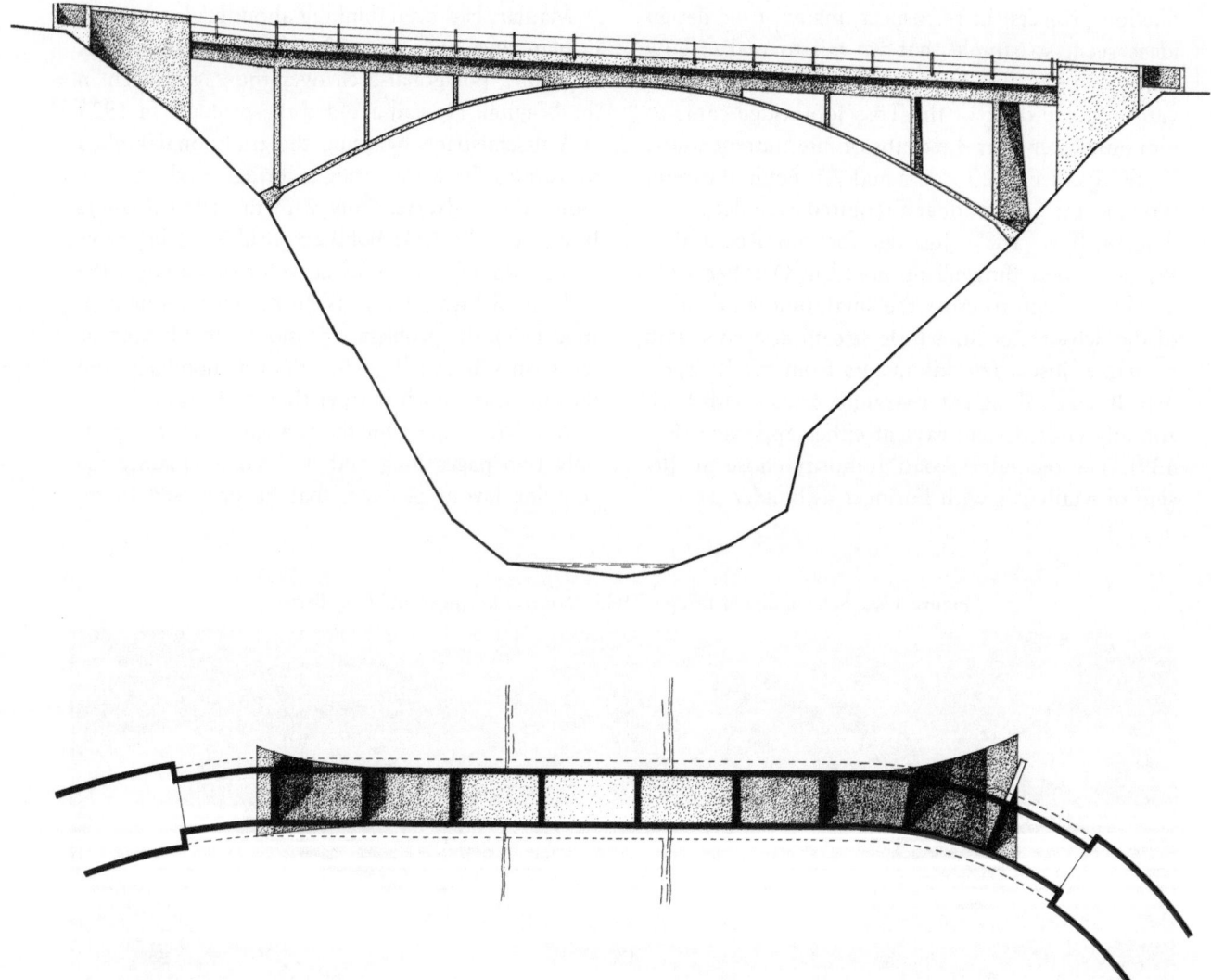

Figure 139. Schwandbach Bridge, 1933: competitor's design, profile and plan. (*Source:* Drawing by Mark Reed)

solve by using an elliptical curve for the roadway (Fig. 140). He devoted only one sentence to structure, merely to note that "the most economical bridge system here is the deck-stiffened polygonal arch [*der versteifte Stabbogen*], which also best fits the special plan layout [of the roadway]."[45] Maillart's design fee for this bridge, including site supervision, was only 5,300 francs (about $27,000 in 1996) or 10 percent of the total design-construction cost to the canton of 53,000 (about

$270,000).[46] Such works could hardly support a design office.[47]

On October 24, Losinger completed casting the concrete for the Schwandbach Bridge, and 4 days later, Werner Jegher published an article on it in the *Bauzeitung*. Not only did Jegher describe Maillart's technical work on curved bridges, but he praised Maillart's aesthetic while attacking a criticism of Valtschielbach presented in a book by Hermann Rukwied called *Bridge Aesthetics,* recently

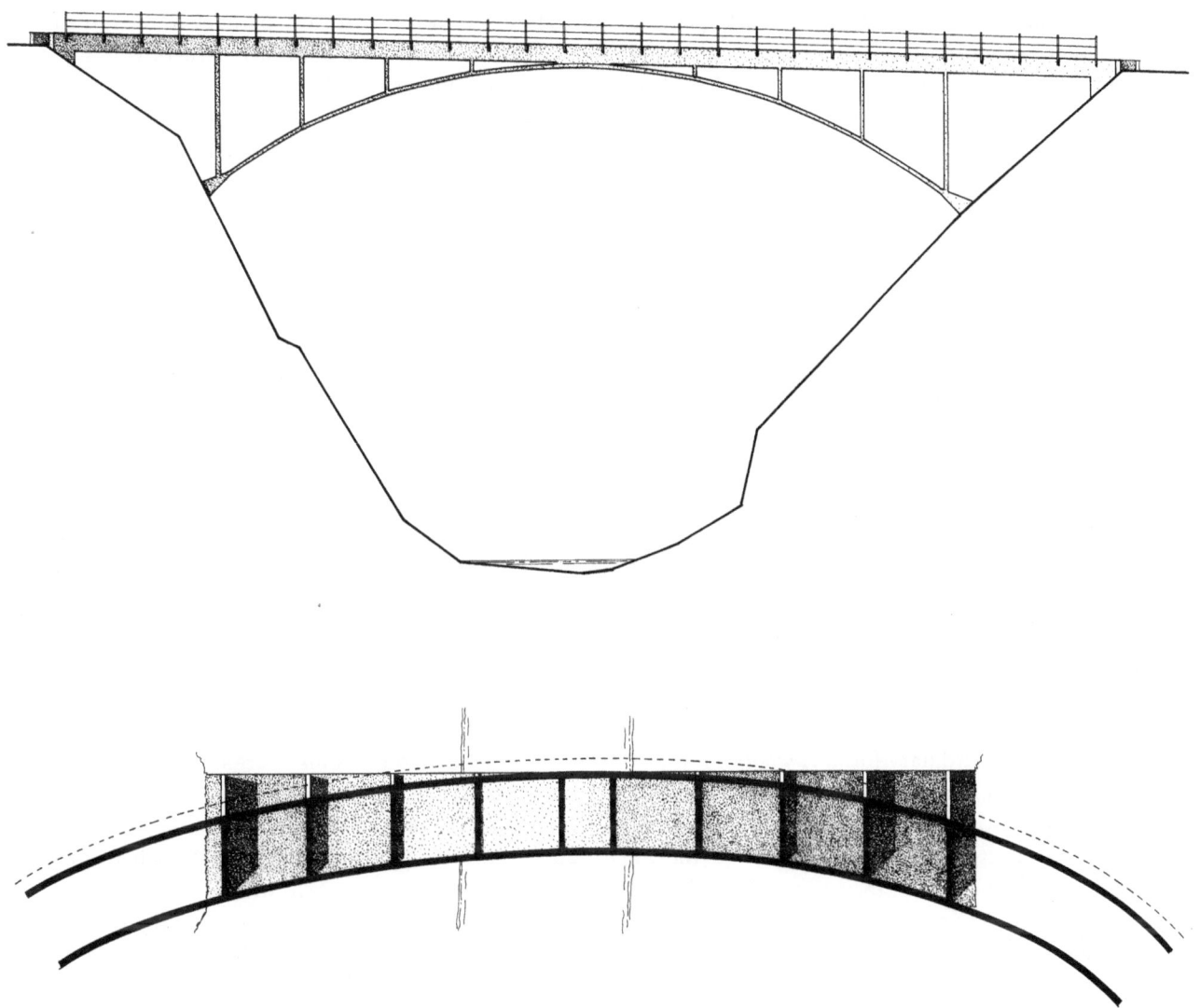

Figure 140. Schwandbach Bridge, 1933: Maillart design, profile and plan. (*Source:* Drawing by Mark Reed)

published in Germany. The young Jegher had already reviewed the book unfavorably and now he attacked the author's view that Maillart's deck-stiffened arches were ugly compared to heavy, stone-like, concrete bridges.[48]

At about this time, Ros spoke to Maillart about testing the Schwandbach Bridge. The secondary road agency in Bern had no interest in sponsoring a test and Ros therefore went to the Federal Foundation for National Economy requesting 6,000

francs for an unusually complicated load test with numerous measurements to study the effect of horizontal curvature. Unfortunately, a leading member of the Foundation Board was Max Ritter, who immediately vetoed the proposal on the grounds that there should be a mathematical theory to determine the effect of horizontal curvature in general. Ritter dismissed out of hand the costly study of a unique form and instead asked one of his assistants, Karl Hofacker, to study the bridge. Three weeks later, after making two models and about thirty pages of preliminary calculations, Hofacker reported that the problem could be solved mathematically. However, he did not take the time to make complete calculations,[49] and because Ritter was unwilling to reach a decision on the basis of an incomplete report, his objections stood.

Nevertheless, the Foundation finally agreed to sponsor the test in 1935, well over a year after the initial request. Ros made the tests and Fornerod, now an independent consultant, wrote up the results, which showed the accuracy of Maillart's simplified analysis and the excellent overall performance of the bridge[50]. The Schwandbach Bridge continued to appear in the literature up to the end of Maillart's life, but even more significantly, it began also to receive recognition as a work of art, beginning that spring in Paris. It would become one of the half dozen great bridges of the century.[51]

It is a masterpiece because it adheres fully to the discipline of structural art, using little material and constructed for a low price. Of the many bridges like it, the Schwandbach stands almost alone because of its extraordinary form. A smooth horizontally curving deck, integrated to a smooth vertically curving arch by means of trapezoidal cross walls (Fig. 138). These walls at their tops run the full width of the deck; and at their bases, the full width of the arch. This is a surprising integration because most such walls rest on only part of an arch width.

The Schwandbach Bridge shows how Maillart's ideas matured over the 8-year period since he completed the Valtschielbach design. The two figures illustrate an artistic progression: The 1925 Valtschielbach was a culmination of technical experience, but it is not an aesthetic masterpiece. At Valtschielbach, a U curve across a ravine is accomplished by a straight roadway deck and arch combined with sharp transition approach curves (Fig. 141). At Schwandbach, the same U curve is achieved by one smooth elliptical curved deck supported by an arch whose concave side is curved in plan and whose convex side is straight in plan (Fig. 142). At Valtschielbach, the approaches are heavy stone Romanesque arches (Fig. 143). The arch is

curved and merges with the parapet at the crown. At Schwandbach, the approach structure is light and open and has a girder of the same depth as the deck stiffener. The arch is polygonal and is separated from the lighter deck girder up to the crown (Fig. 144).

The most dramatic difference between the two bridges lies in how Maillart connected the arch to the deck. In the 1925 bridge, the arch and the deck were exactly parallel and rectangular cross walls easily connected them. At Schwandbach, however, the arch was wider than the deck. Maillart introduced trapezoidal cross walls that provided not just a technically correct transition of forces, but also gave a visually striking transition in form.

Maillart defended his Schwandbach design in the *Schweizerische Bauzeitung* in early 1934 by showing how alternative solutions proposed by an engineer, F. Bohny from Sterkrade (Rheinland), would have been inferior both technically and aesthetically. Figure 145 illustrates Maillart's defense of his design.[52] The first drawing shows the bridge as built, the second as proposed by Bohny, and the third as suggested by Bohny's argument. Of the second, Maillart noted that the constant-width arch (following the plan of the curved roadway) gave the appearance of tipping toward the right, whereas his design by having the arch widen at the support "gave the appearance of stability and repose." As for the third design, where the arch would be straight in plan, Maillart commented that in addition to its greater cost and use of more materials, "its aesthetic result is barely worth even discussing." In profile, the arch and the cross walls of the Schwandbach Bridge are, to an engineer, daringly thin, and to an aesthetically sensitive observer, they are made to seem even thinner by contrast to the strong horizontal deck girder (Fig. 144).

ANALYSIS IN THE SERVICE OF DESIGN

Maillart's analysis for the Schwandbach Bridge characterizes his design view as opposed to Max Ritter's applied science approach. Maillart designed the arch to carry its dead weight by compression

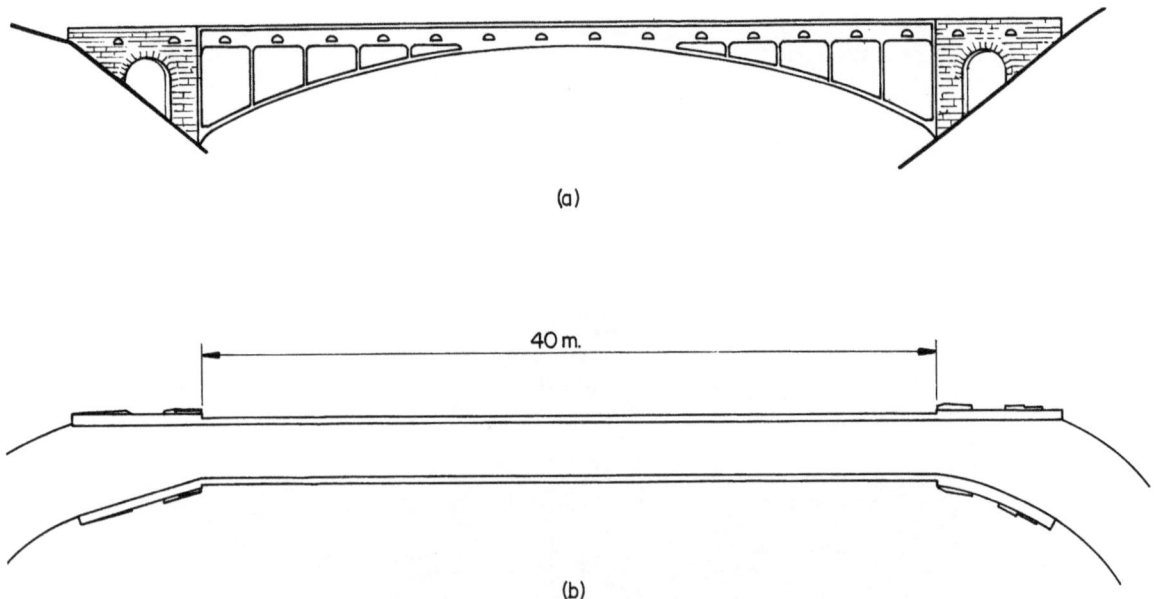

Figure 141. Elevation and plan of the Valtschielbach Bridge, 1925. (*Source:* Drawing by T. Agans and A. Evans)

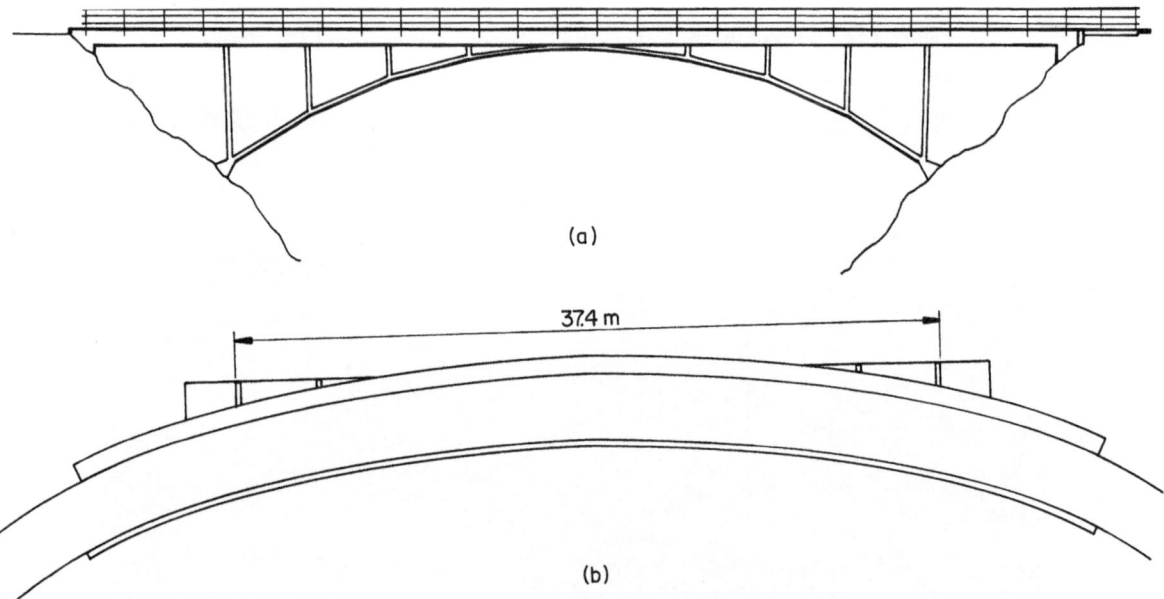

Figure 142. Elevation and plan of the Schwandbach Bridge, 1933. (*Source:* Drawing by T. Agans and A. Evans)

Figure 143. Valtschielbach Bridge near Donath by Maillart, 1925. (*Source:* R. Guler, Thusis)

Figure 144. Schwandbach Bridge, 1933. (*Source:* Losinger and Co., Bern)

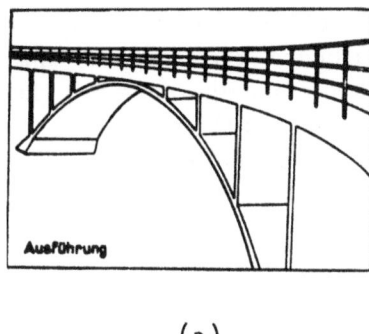

(a)

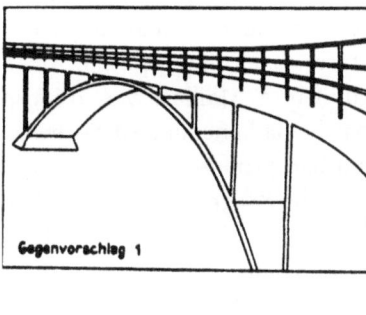

(b)

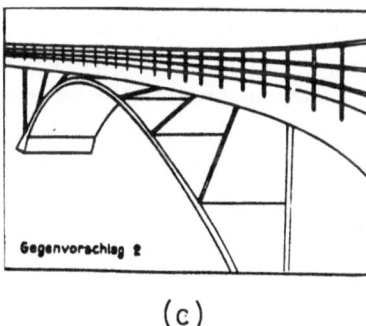

(c)

Figure 145. Schwandbach Bridge and alternate proposals: (a) Maillart design; (b) Bohny Proposal I; (c) Bohny Proposal II. (*Source:* Schweizerische Bauzeitung)

forces just as did the Romans. The reinforced-concrete arch is thin because it has only an axial compression. For dead load alone, the arch is an inverted cable. Weights hung from a cable give a form that puts the cable into pure tension. That same form when inverted gives an arch with pure

compression under that same load. That design idea was already well known and widely used in the nineteenth century.

What was new was Maillart's approach to the live loads. Once the engineer has fixed the arch form to fit the dead load, it cannot be changed and thus the live loads lead to bending. For example, when the live load is only over half the bridge length, the shape of a cable under that load would be half curved and half straight. Superimposing that on the arch shape gives a diagram of bending. Were the arch alone carrying that load, its bending would be substantial and its thickness would need to be greatly increased over Maillart's designs.

Maillart's solution, first used in his Flienglibach Bridge design of 1923, was to connect the arch to the horizontal deck and to make that deck much stiffer than the arch. Because both the arch and the deck must bend together, the actual bending force (called the bending moment) is distributed between the arch and deck in proportion to their relative stiffnesses. For the Schwandbach Bridge, the total moment would be 32.8 meter-tons with the very thin arch taking only about 5 percent of that value and the rest carried by the deck.[53] Thus, the live-load moment in the arch is essentially negligible and the arch can be extraordinarily thin (Fig. 144).

Maillart's design view led him to create a form in which the complex analytic problem of live-load bending is rendered unnecessary. Only the very simplified analyses sketched here were needed to ensure the validity of the design and to provide confirming evidence to the corroborating results obtained by the tests of Prof. Ros.

Meanwhile, throughout the 1930s, Max Ritter had worked on the problem of arch-bridge analysis, which considered the interaction of deck and arch. His applied science view dictated to him the need for a general or "scientific" theory covering all types of arch-deck systems; he then would be able to apply that theory to any set of dimensions for the system and obtain rigorously accurate results with which to proportion the elements (arch, verticals, and deck). His view was literally blind because he saw no specific designs, only abstract diagrams and algebraic formulation. His starting point was formula not form. For the Schwandbach Bridge, for example, Ritter needed thirty-six simultaneous equations to get a "proper" solution and even then he had to neglect the horizontal curvature. In the days before the fast digital computer, it is no wonder that his assistant only reported that the problem could be solved but that he could not possibly do it in 3 weeks. Maillart's solution took less than one afternoon.

In Maillart's mind, there was a direct connection between the authority of general theories and the authoritarian personality. He sometimes referred to Germans in that way, although he had good friends there who did not fit that pejorative label. Ritter was the clearest symbol of this tendency, and as the Schwandbach Bridge neared completion, his biggest confrontation with that professor began over the design of the Sihlhölzli Gymnasium (see Chapter 8). But before that controversy erupted into a legal battle, Maillart had to face the consequences of the deepening economic Depression.

DEPRESSION AND COLLAPSE

As the Depression finally spread into Switzerland, Maillart's business withered, but his design ideas seemed to flower. Amid the struggle to keep three offices in work, he had responded to an ordinary bridge problem with the stunningly new solution at Schwandbach for which the thinking on Giubiasco and Töss had prepared him. But along with that came the pain of retrenchment. Late in February, on learning that he had lost a bid for the Etzelwerk power plant, Maillart was obliged to dismiss two draftsmen in Zurich.[54] This left only five people (Keller, Kruck, Fornerod, and two draftsmen) with insufficient work even for them.

On the last Friday in April, he received in Zurich the good news that he had won the commission to design the new Amtshaus V (Zurich municipal building). Even though the structure presented little intellectual challenge,[55] he welcomed this big pro-

duction job. It lasted until early summer 1934. But this work required little engineering and could be done by Keller and the draftsmen. Engineering work languished as 1933 wore on.

The lack of work in Zurich culminated dramatically for Maillart that fall: On October 31, he was forced to discharge Marcel Fornerod.[56] It was a sad experience for Maillart because he knew this meant the end of the office as it had been. He saw in Fornerod someone who was unafraid of new ideas and who was sympathetic to his own ideals. Just a few months before, Fornerod had shown Maillart a new and simpler method of calculating complex concrete structures that came from the American professor Hardy Cross. Maillart was so intrigued with the idea that he had encouraged Fornerod to write about it and had promised personally to take it to Carl Jegher at the *Bauzeitung*. The result was a long article in that journal, which appeared 4 days after Fornerod lost his job.[57]

On January 28, 1933, the two feature articles in the *Bauzeitung* both described his works: the newly completed Swiss Credit Bank in Geneva[58] and the Quai in Vevey, which Maillart designed and Losinger was building. Its design was the first of its type in Europe.[59] The quai's uniqueness lay in the floating concrete caisson (or box) on which, when anchored, part of the quai rested. As with so many of Maillart's designs, its technical interest alone merited a major article in the *Bauzeitung*.[60] Begun at the end of 1932, it provided some production work for the Geneva office.[61] It was an unusual combination of profitability and challenge, but it alone could not sustain his staff for long.

On Wednesday evening, March 22, 1933, after visiting Stettler in Bern, Maillart sat down for dinner and a game of Jass with Losinger. At a crucial point in the game, the builder was called to the telephone; he returned shaken and pale to tell Maillart that the Vevey quai had collapsed during construction. Early the next morning, Losinger raced with Maillart by car to Vevey, where they discovered that only that part of the new Quai that was attached to the 1877 construction had given way.

Fortunately, no one was hurt. Earlier, Maillart had warned the city to remove the old structure before beginning to build, but officials had insisted on conserving the old work. The collapse was one more headache for the consulting engineer; it would take more of his time without the rewards of new production work.[62]

These were the risks of large-scale, custom-designed structures where construction problems could consume engineering time that was not easily compensated. The profit from design could be wiped out by such an accident. Maillart was continually torn between the need for routine production work to keep his offices intact and for projects that could satisfy his creativity. The former required salesmanship and contacts because so many of his competitors could do the same work; the latter, by contrast, usually required only Maillart because he alone could think out the design that often did not entail much production, as with the Töss Bridge. The collapsed Quai at Vevey was, in fact, repaired by Losinger and the entire project completed by September 1934.[63] On October 20, the canton president officially cut the ribbon; credit was given publicly to "M. le Professor Maillart," and the townspeople crowded out onto the new structure.[64] Maillart was surely amused to be classified with the academic world.

The Acrobat of Reinforced Concrete
1933–1934

GROPING FOR NEW FORMS

On June 26, 1933, in Geneva, Maillart met with a class of students from the Lausanne engineering school to show them his design for the Schwand-

bach bridge and other works. A friend, city engineer Robert Pesson, introduced him with a flourish as the "acrobat of reinforced concrete."[65] By this time, Maillart's reputation for graceful and daring designs had penetrated the profession and even the classroom. This was not true, however, at his own school in Zurich, dominated as it was by Max Ritter. It was not until after Ritter's death in 1945 (and 5 years after Maillart's own death) that a young assistant of Ritter's, Pierre Lardy, would bring the engineer's acrobatics into Zurich education, and help create a new generation of Swiss structural artists.

In 1933, however, Maillart's attention had moved away from education to the realities of his practice. As Losinger and Binggeli started construction on the Schwandbach, he began to work out the design for a new bridge over the Aare River near Wangen. He made two designs, quite novel for him, with flat elliptical arches that had three hinges like Rossgraben but with the hinges hidden as at Zuoz. Here Maillart began to face a design problem that was becoming more and more common, that of low highway crossings where arches were no longer suitable. The Wangen Bridge stimulated him to depart from the Rossgraben-type design and study even flatter three-hinged forms.[66] In early September, he began another Bern canton bridge, this one over the Aare River above Interlaken in the small town of Innertkirchen. Here he took the same form sketched for Wangen and worked it out in detail for this flat 30-meter span.[67] Maillart was struggling to find a flat form that would look light; he only partly succeeded.

In mid-September, he reflected on these problems in a letter to an American graduate student, Milo Ketchum. He noted that the advantages of deck-stiffened arches disappeared when they were too flat (as, for example, at Innertkirchen) and that hollow-box girders introduced by him in 1901 were better suited.[68] Maillart was thinking again about his Zuoz-type experience, when he had made a three-hinged arch in which the hinges were invisible as at Innertkirchen. In 1901, this struggle with form

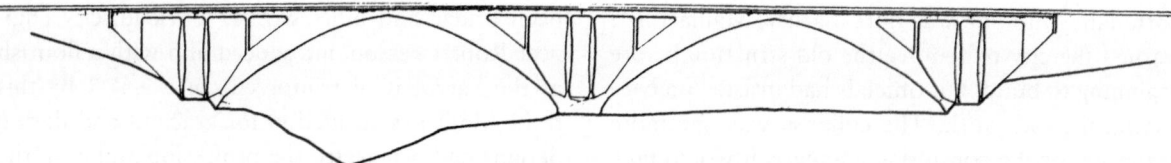

Figure 146. Development of the three-hinged arch form: Drawing of the Sirakovo Viaduct. (*Source:* Maillart and Co.).

had led him toward the Tavanasa design; now 32 years later, it was leading him in the new direction of continuous beams.

After the Innertkirchen Bridge, Maillart turned once again to Wangen, and on October 10, he presented a completely new design; this time not an arch, but a three-span beam bridge. It is an elegant concept, especially for the time, with a center span of 52 meters, but when Bösiger, the canton director of public works, saw it, he flatly said "no, not concrete" and although he paid for the design, he refused to have it built.[69]

The move from flat arches at Innertkirchen to the continuous beam at Wangen marked a decisive change in Maillart's thinking. He never completed another work like Innertkirchen, but he continually studied beam bridges after 1933 and five of his last eleven bridges were so designed. Maillart began his exploration of beam bridges in 1932 with a preliminary design for the Weissensteinstrasse in Bern; it was an ungainly design and he did not study the problem again seriously until 1934. Late that year, he returned to the beam form with the Liesberg Bridge and made a design that satisfied him (see Chapter 8).

Toward the end of 1933, Maillart had begun to think about the consequences of carrying his two arch forms to new extremes of scale and daring. The first of these new designs came in early 1934 for a bridge in Sirakovo, Yugoslavia, where he made a series of studies for a two-span, hollow-box, three-hinged arch structure.[70] Here he played with form in a startlingly original way, using columns hinged top and bottom and widened at mid-height as well as arches with hinges well out into the spans (Fig. 146). The compositions show Maillart playing with a form that he now under-

stood well thanks to his experience at Salginatobel, Rossgraben, and Felsegg. The two-span arch poses special visual problems because of the prominence of the central support. Long before, Wilhelm Ritter had argued for the Stauffacher Bridge that one long center span is always aesthetically superior to two spans.

FINDING A NEW DIRECTION

In mid-September 1934, Maillart designed a spectacular 200-meter-span thin polygonal arch for the Sitter Valley, in the canton of St. Gallen.[71] At the same time, he designed an 80-meter-span, three-hinged arch for the Russeinertobel near Disentis about 15 kilometers above Tavanasa; it would have the same form as his 1904 design but modified for the longer span to give a more polygonal profile.[72] Moreover, he began two new bridge designs for Geneva; again one in each of his mature styles: a thin polygonal, deck-stiffened arch at Lancy and a hollow-box, three-hinged arch at Vessy.[73]

In all these designs, Maillart expanded his ideas on form to larger scales and to more angular profiles. His sense of play was increasing, as sketches for the Sirakovo viaduct had already shown in May and July of 1934 (Fig. 146). The two remarkable facts in this evolution of style were Maillart's self-criticism while adhering to the same two forms. He saw the Salgina as too classical, too smoothly curved for the problem of longitudinal, wall-stiffened hollow boxes. He also saw that in the deck-stiffened, thin-arch designs of Valtschielbach and Traubach, the parapets were visually too heavy. These self-criticisms led him to make polygonal profiles for both types of bridges. The Sitter Bridge clearly shows Maillart's idea for an arch of

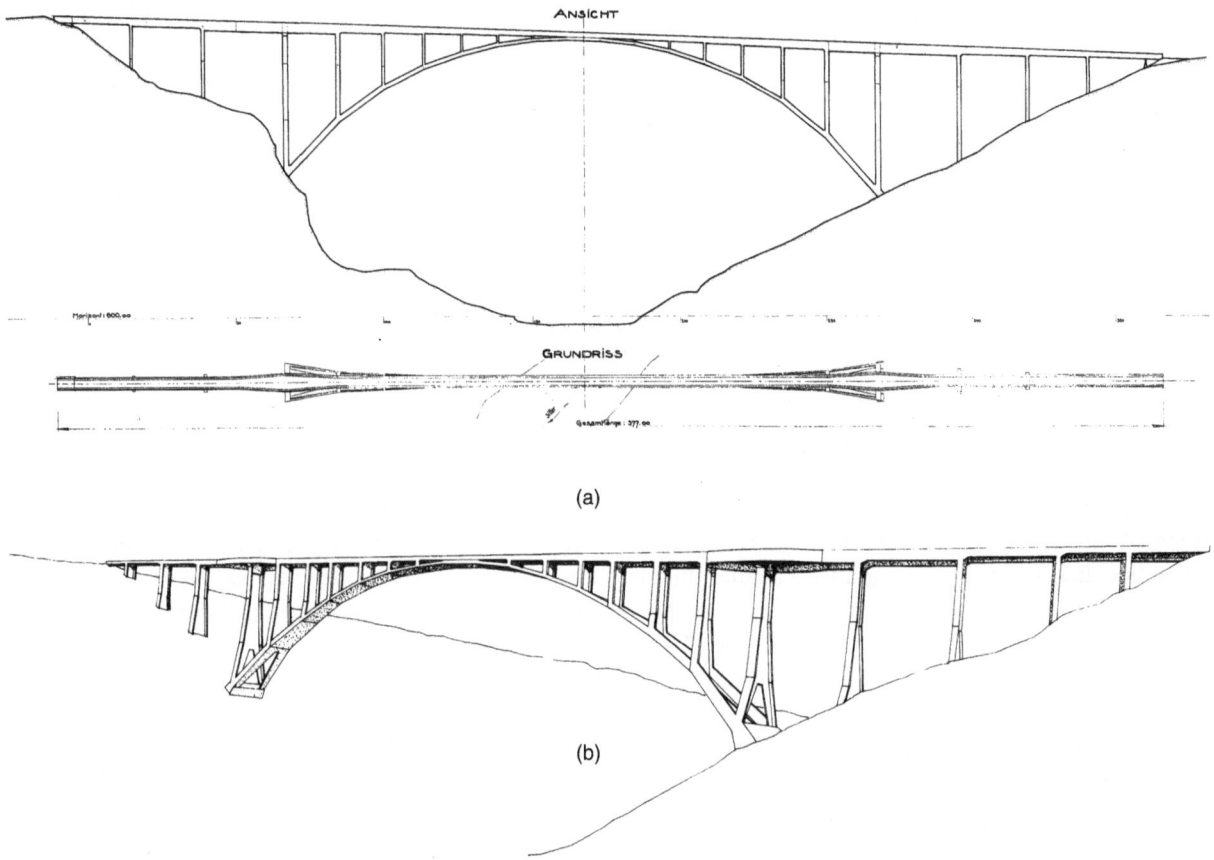

(a)

(b)

Figure 147. Sitter Bridge at Haggen-Stein: (a) Unbuilt design, October 1934. (*Source:* Maillart and Co.). (b) Perspective. (*Source:* Drawing by Colin Ripley)

great span – 20 meters more than Freyssinet's world record of 180 meters at Plougastel in 1930 [Fig. 147(a)]. A visual comparison of these concepts reveals Maillart's intense concern with lightness, angularity, expression of structure, and integration of overall form.

By 1934, Maillart had proven that thin polygonal arches were both technically sound and practical to build. But his designs had not gone beyond 40 meters in span. How could such a form be extended to 200 meters and retain the visual lightness demanded by Maillart's aesthetic sensitivity? His answer lay in forming the arch as a channel section that, near the supports, separated into two slanting splayed legs [Fig. 147(b)]. The visual effect of the arch slab and flanges (making up the channel section) is to integrate the main structure while per-

mitting it to be seen as made of thin elements. Were it merely two ribs (the flanges) without the slab in between, there would be the clutter effect of the two seen with different spacings from different angles.

Freyssinet's hollow-box design for Plougastel is, by contrast, a single smooth arch that is almost unvarying from support to support. Similar perspective views show the Maillart design to be much lighter looking, but even more striking is the integration of form in the Sitter design. The columns and the deck are also made up of flanges connected by slabs, and the arch polygon breaks slightly at each column connection, whereas in Plougastel, the three columns merely rest on the visually heavy arches with no reflection in the arch of the columns' influence.[74]

Maillart also introduced into the Sitter design the

new idea of decreasing the depth of the horizontal deck members (the flanges of the deck section) to match the decreasing height of the columns. The same decrease occurs in the flanges of the columns and of the arch. He also decreased the same members as the column height diminishes up the banks in the approach spans. These subtle plays of form are missing from the Plougastel as well as from Maillart's 70-meter-span proposal for Giubiasco in 1931. His 200-meter-span design of 1934 represents a critique of his own work as much as it does a contrast with his contemporaries. Indeed, Maillart was not reacting to the designs of others but rather to his earlier works. He was continuously turning inward toward images already impressed on his mind by having studied them in sketches, field visits, and simplified calculations. It was this inner dialogue that accounted for the spectacular works he imagined in the last half of 1934.

In spite of poor health and slow business, he felt comparatively well off. His work had just been published even in Germany, where he had yet to build anything. He still had more business than most Swiss designers,[75] and best of all, his Indonesian family was due home on their first home leave.

POLITICS AND PLAY

The peaceful Swiss environment in which Maillart was able to create the Schwandbach Bridge was an exception to the mounting political tumult of continental Europe. On January 30, 1933, President Hindenburg appointed Adolf Hitler Chancellor of Germany. Hitler's new position allowed him to begin his reign of terror, directed first against the Communists,[76] and within a few months, encompassing the Jewish population. Maillart was made acutely aware of the Nazi persecution through his closest friend in Germany, a Mr. Alexandre, a Jew, who despite his Swiss citizenship, no longer felt safe there. He arrived in Bern on April 1, ostensibly for the annual meeting of the Association of Swiss with Financial Interests in Russia (SECRUSSE) in which Maillart played a leading role. In reality, Alexandre

was trying to find a place to settle permanently either in Vevey or Lugano.[77] He remained in Switzerland throughout the war.

At the opposite end of the political spectrum, Stalin was creating an equally totalitarian state. In 1932–3 alone, Soviet collectivization killed several million people and placed millions more in forced-labor camps.[78] Maillart was horrified by both radical movements.[79] The ominous political situation, compounded by his own business disappointments in 1933, was brightened by his anticipation of the Blumers' first visit home in 4 years. By June, business in Zurich was better; Paul Nissen was away, and Maillart saw only Mlle. Oechslin and ate alone with her. She was so often with him during these years, that even his office workers assumed she and he were more than just friends. She was vivacious, full of fun, and intelligent; Maillart enjoyed her company and she seemed to prefer the company of older men and especially that of Maillart and Nissen.[80]

His attitude toward Mlle. Oechslin was both proper and unconventional, that is to say, it was a direct reflection of his personality. What was probably nothing more than a mild flirtation helped to fill the emotional void left by Maria's death, then by Marie-Claire's absence, and most recently by Bertha's death. In each of his three cities, he now had a young woman whose company he enjoyed, Mlle. Oechslin in Zurich, his niece Adrienne in Bern, and Simone Perrot in Geneva (who had been a close friend of his niece, Edith Wicky).

Maillart used the routine of card playing as a means of organizing his time with Rosa, Marguerite, and Marie. He was highly amused by the confusion of the card world at this time over the change in rules caused by the transition from an older (auction) to the newer form of bridge (contract).[81] In describing this confusion to Marie-Claire, Maillart observed that he needed neither company nor cards, because "bridge" puzzles could be solved by logic.[82] But his game of bridge eschewed fixed conventions, just as he repudiated the pseudo-discipline of scientific calculations professed by Max Ritter. Ritter and the others complained that Maillart's calculation methods were *tanzboden statik* (dance floor statics), that is to say, not serious enough for the risky job of making public structures.[83] Their pejorative name fits Maillart better than they imagined because he really did dance (in his mind) as he calculated.

Maillart expressed the connection between discipline and play for bridge by observing that he was not "at the pinnacle necessary for understanding all the rules and all the calculations that the science of bridge imposes upon us. If in spite of that Aunt Daggie [Paul's wife, who followed these rules religiously] always loses [to me], it is obviously because she had bad luck and partners who do not understand her bidding."[84] These words touched with irony obviously relate also to structure. The designer who knows all the rules and the "scientific" (read complex) calculations will invariably fail to attain the best designs and will blame that failure on bad clients (who insist on extreme economy) or dull sites (small standard overpasses). This failure arises more often because the committee rules replace the laws of nature in the designer's mind just as complex calculations stifle imagination. Without this attitude, Maillart could not have designed new forms such as the Schwandbach Bridge.

The summer brought Maillart no new work. He had begun to be aware of the Swiss Nazis who were intent, as he put it, on "smashing things up,"[85] encouraged by Hitler's domestic successes. With the failure in July 1933 of the International Economic Conference in London, Britain and America pursued more nationalist policies to relieve their countries' economic depression.[86] To the south in Italy, Mussolini tightened his fascist rule.

In August, Maillart visited René on the Isle of Jersey and Edmond in London; it was his first time in England.[87] As usual, when he left his native land, Maillart was indifferent to his new surroundings. But at least these stays with his sons cheered him and he also took heart in the news that the Blumers' trip home would be advanced. In early September, he learned that they would leave on October 7 from Sumatra and thus be in Zurich a few weeks later.[88] They actually arrived in November for 6 months'

leave. New Years Eve 1933 saw sixteen members of the Maillart family gathered together in Geneva for a celebration of the Blumers' homecoming. The family had a bigger and happier party than at any time since the couple left in 1929 despite the troubled times.

A JOURNEY OF THE MIND

While in Zurich in early 1932, Maillart had decided to rent an inexpensive two-room apartment with a kitchen in the Schwarzen Bären (Black Bear), a little old hotel, just off the Parade Platz in the center of the city (Fig. 148).[89] The purpose of the rental

Figure 148. Schwarzen Bären Hotel, Maillart's Home in Zurich in the 1930s: postcard from Maillart to the Blumers, April 3, 1936. (*Source:* Madame M.-C. Blumer-Maillart)

was primarily to have a place large enough for his children to visit.[90] Within the context of the engineer's frugal style of living, the rental represented an indulgence. But with no elevator, central heating, or hot water, it was indeed a minor indulgence.

During the winter of 1934, Maillart followed his weekly schedule except for more time in Zurich than his light work load there justified, because his apartment in the Bärengasse was now the Blumers' headquarters. When the 6 months were over and Eduard had to go back, Marie-Claire decided to remain behind for a few more months with her 4-year-old Marie-Claire. She wanted to spend more time with her father, and so, following Eduard's departure, father and daughter made a series of trips together, often to see Maillart's works. Her recollections give a picture of his last years at a time of physical debility and undiminished imagination.

One memorable trip took them to Billwil, where Maillart stood silently looking intently at his 1904 bridge. They noticed someone staring at them, and after a time, the stranger came closer and asked what interested them. Maillart answered simply that he had built this bridge in 1903. The man's face lit up, as he explained that he was a local farmer and that he recalled the time and especially the lady who came to the site and would visit his own family. The lady was of course Maria.[91]

As Marie-Claire stayed mostly in Zurich while her daughter spent time in Engi with the Blumers, she could observe firsthand her father's routine as she had during her weekend visits in 1928. But there were differences. First of all, she was now actually living with him, whereas before he had stayed at the Hotel Gotthard alone. Also, she was older and more observant now, and she could see that he looked older and moved more slowly since his accident.

One morning that spring Marie-Claire awoke at about 6 A.M. to see the door to Maillart's bedroom ajar. From her bed, she could see her father, in his bed supported by two large pillows. He was smoking his pipe, staring straight ahead, obviously lost in thought. The only motion came from occasional puffs on his pipe. After observing him for some time, she went back to sleep. At about 8 A.M., he called to her, as he usually did, to come have breakfast, and over coffee, she asked him what he had been meditating on for so long at six o'clock in the morning. Maillart smiled and said that he had made in his head the design for a bridge. In the spring of 1934, this could have been the Sirakovo Viaduct, a new three-hinged arch form that he was exploring in detail that year.[92]

But Marie-Claire's time, too, was coming to an end with the summer. On Friday, October 26, Maillart, his daughter and grand-daughter took the train from Zurich to Milan and then to Genoa, where they spent the night. The following morning, father and daughter took a long walk along the harbor. For Maillart, walking slowly on his heavy legs, it was a sad time. They had a splendid year together in Switzerland, but it was now over and she was returning to southeast Asia while he would return to little Switzerland. They talked of his plans to visit Indonesia the following year. Finally, Marie-Claire moved alone up the gangplank and onto the boat.

Maillart was too overcome to write her during the 3-hour train ride to Milan. Upon arriving in Geneva at 2:45 P.M., he met Rosa. Two days later, after work, Maillart returned to Rosa's apartment for his regular bridge with the ladies, now including the lively Simone Perrot. The old routine had once again taken over,[93] but in Zurich, it had become more unpleasant because of a major dispute between Maillart and the authorities.

CHAPTER EIGHT

The Great Debate
1934–1938

The Sihlhölzli Gymnasium Controversy
1933–1935

MAILLART VERSUS MAX RITTER

In May 1933, Maillart found a letter in Zurich that would open the most bitter controversy of his career. The case – involving a structural defect in the roof of his Sihlhölzli gymnasium – reveals the engineer under rare circumstances in which an error in his office led to a series of confrontations with the highest building authorities in Switzerland's largest city. Maillart never allowed the episode, which dragged on for over 2 years, to enter his family life; he never mentioned it in his letters and none of his family or associates ever referred to it.[1] But more than any of his other disagreements with Swiss officialdom, this incident was painful for Maillart because it appeared to give his enemies an opportunity to discredit him publicly.

In early March, the *Bauzeitung* had published a two-part article on the Sihlhölzli gymnasium (Figs. 149 and 150) and music pavilion designed in 1929. The article included a drawing that showed the reinforcing steel in the roof frame, a structure that Maillart conceived in the same way as he had the Chiasso shed.[2] Assistant city engineer Josef Killer had discovered an error in the reinforcement shown in this drawing. Killer and his associates aligned themselves with those who fought against Maillart and appear to have been delighted to have "caught"

the seemingly irreproachable engineer in what they considered to be a serious mistake: The amounts of steel in the central hanger and the ones next to it had been reversed on the drawings, so that the central one under the gable had less than half as much steel embedded in the concrete as required by the code (Fig. 151). Thus, by law and the contract Maillart had signed for the design work, he was required somehow to add that additional steel. Had the structure been only of steel (rather than steel-reinforced concrete), it would have been relatively easy to add a steel column, but concrete cannot be strengthened so easily.

Maillart spent considerable time on calculations to see if the roof was really unsafe. On April 26, he completed a five-page report in which he concluded that although there had been a draughting mistake, "Doubts [about the structure's safety] or even strengthening are uncalled for here."[3] Even though some steel would be overstressed, he noted that the increased stresses were still below values allowed in the new Swiss steel code of 1933. He suggested a load test as the best means of proving his conclusions. The city's initial willingness to consider this idea was quickly squelched by powerful forces led by Max Ritter at the ETH.[4] Ritter had not recovered from Maillart's stinging criticisms of research at the ETH and of the new reinforced-concrete code (eventually published in 1935), and he was only too happy to have an opportunity to discredit his adversary.[5]

On September 25, Maillart met with the authorities who proposed hiring Ritter to make calculations that would decide on the roof's safety. At the

Figure 149. Gymnasium at the Sihlhölzli, Zurich, 1929: main floor showing the slender frame span above which is the attic roof frame. (*Source:* Madame M.-C. Blumer-Maillart)

same time, they invited Maillart to appoint his own outside expert to make a similar study. Over the next 10 days, Maillart reflected on their ideas and found them faulty on three counts. First, he believed strongly that a load test was the only way to determine at this stage the true behavior of the structure, and hence to judge its safety. Second, he objected to the city's proposal of hiring a consultant because it would allow Ritter to set the criteria and ask the questions that he would then answer. For Maillart, the leading question was not "had a mistake been made and how can it be corrected?" but rather "is the structure unsafe and if so why?" The difference between these two questions reveals Maillart's approach to engineering as well as any professional issue could.

Maillart's third objection was to the suggestion that he hire a consultant of his own, because he realized that would merely produce two experts of opposite opinions as in a legal dispute. He therefore came up with the surprising proposal that Ritter serve as consultant both for him and for the city and that he (Maillart) be given the responsibility to determine the safety of the structure and the burden of any costs for repair.[6] Behind that apparently disarming offer lay Maillart's conviction that there was no one in Switzerland, or anywhere else for that matter, who was qualified to consult for him. By this time, it was he who was considered the prime engineering consultant in Switzerland. He was sure that the city did not understand how his structure worked and he suspected that Ritter did not either.[7]

Previously, Ritter had shrewdly avoided all public confrontation with Maillart. The latter's series of successes intimidated even his most resolute adversaries. Now, as the engineer hoped, he could use the gymnasium controversy to force Ritter into a

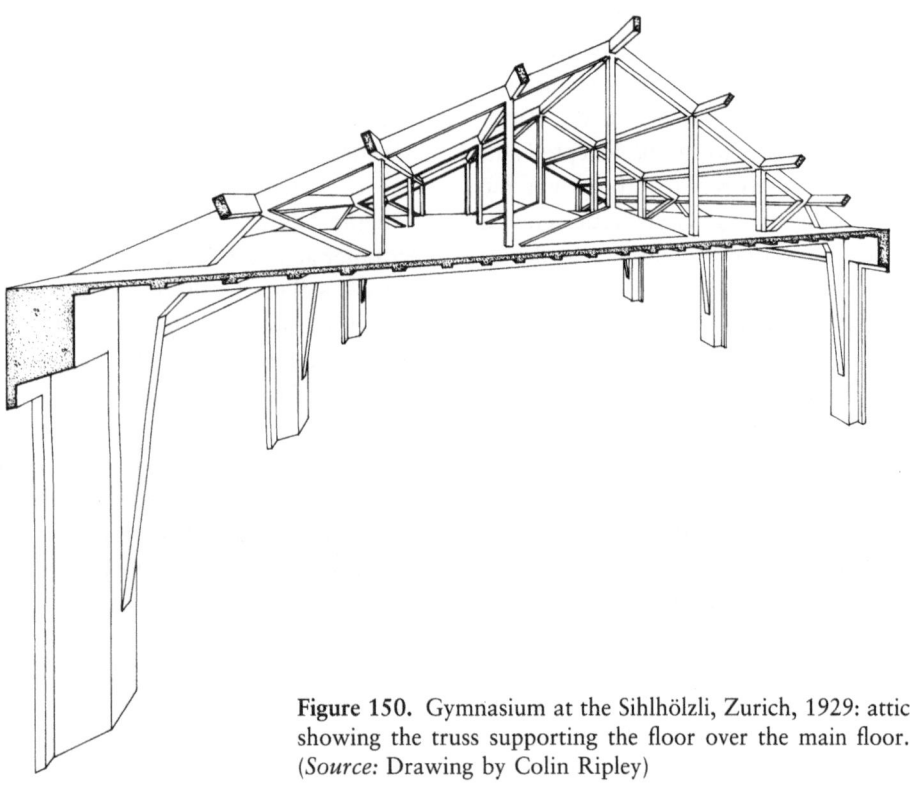

Figure 150. Gymnasium at the Sihlhölzli, Zurich, 1929: attic showing the truss supporting the floor over the main floor. (*Source:* Drawing by Colin Ripley)

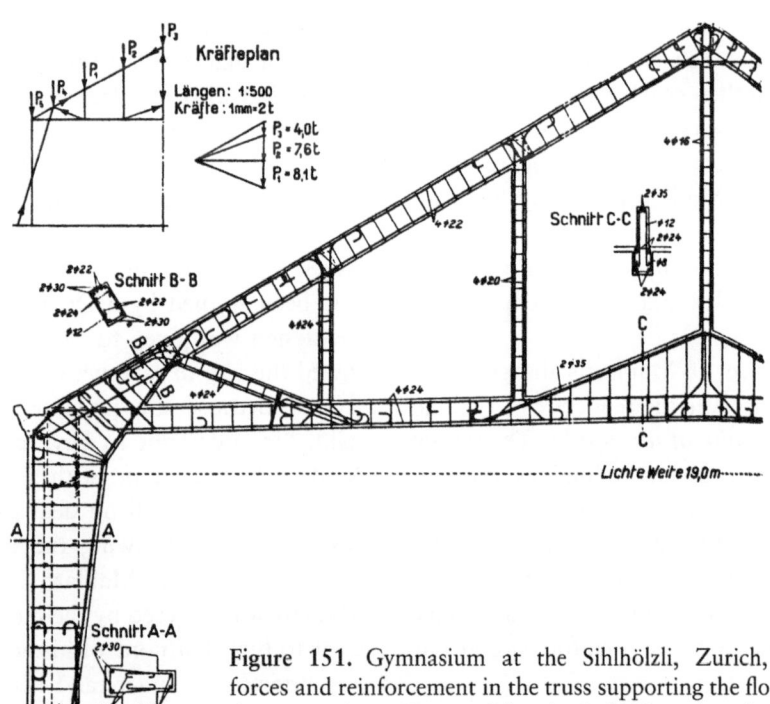

Figure 151. Gymnasium at the Sihlhölzli, Zurich, 1929: forces and reinforcement in the truss supporting the floor over the gymnasium. (*Source:* Schweizerische Bauzeitung)

public acknowledgment of their dispute. Ritter insisted that the new Swiss code should be the only criterion applied to the roof construction. Because of course it did not conform to the code, Ritter could be sure that his report would be accepted and that Maillart would be forced to strengthen the roof.[8] In anticipation of the city's requirement, Maillart designed a unique strengthening device for it.

As he had predicted, Ritter answered his own question: The structure did not satisfy the code and therefore had to be strengthened. Delivered to the city administration in February 1934, it was, for Maillart, both ridiculous and damaging – ridiculous because Ritter had produced over one hundred pages of neatly worked out calculations to justify a conclusion that had been obvious to everyone for a year, damaging because he saw right away that he would have to strengthen the structure at his own expense.[9]

Maillart responded on March 2 with a brief note to the city in which he agreed to add a horizontal strut but refused to accept any blame.[10] He took the position that the consultant's report bound him to add unnecessary reinforcement. Maillart's legal position was weak, but his technical arguments (not made public for another year) were sound. Once again, as in the controversy then reaching its peak over the new code, Ritter emerged victorious without having had to face Maillart publically. Well advised of his adversary's strengths, he relied on bureaucratic technicalities to win his arguments, rather than the complex realities of engineering practice.

Indeed, the essence of Maillart's disagreement with Max Ritter was over the tendency of calculations to set limits *in spite of test results*. The reason Ritter refused to advise the city to load test the Sihlhölzli gym was that the results would not have accorded with his lengthy but conceptually limited calculations. Maillart's relied on experiment as the best way to avoid the risks inherent in an undue reliance on calculations for reinforced-concrete structures.

IN DEFENSE OF DESIGN

On Tuesday, June 19, 1934, at 9 A.M., Maillart was in the offices of the Zurich building department for a meeting with two city councilmen, the secretary, three city engineers, and the chief city architect, Hermann Herter.[11] It proved to be an acrimonious encounter at which Ritter's report was argued in his absence. Killer, clearly proud of his discovery, was Maillart's principal adversary. He was upset that Maillart had not made the extensive calculations prior to construction just carried out by Ritter. Killer insisted that strengthening was necessary because the stress calculated by Ritter was 100 percent higher than the code allowed. He rejected Maillart's proposed reinforcement unless it could be checked, presumably by Ritter.

Maillart's defense was simple. At the outset, he stated: "I confess that there has been a serious error and I shall not shirk from the consequences." But he objected to Ritter's calculations because they did not reflect the actual structure and were therefore pointless for him. Maillart again insisted that the only way to overcome doubts about the structure was by a load test. In response to the city's new argument that Maillart should pay not only for the repairs but also for Ritter, he responded that the consulting bill (for 2,850 francs) was much too high.

Alone among the city officials, Herter was partial to Maillart and raised a completely different set of issues. He pointed out that in fact the project had first been studied by a Dr. Suter, on whose death the design was given to Maillart, whose design allowed the city to save between 40,000 and 50,000 francs. Because the previous engineer had been paid, "in order not to have costs rise too much, Maillart had made do with a modest design fee." These arguments fell on deaf ears and the discussion ended by throwing the question of Ritter's payment to Killer, Maillart, and Ritter to settle. Maillart was angered by the meeting because it was clear to him that none of his adversaries had made any attempt to understand his structure. Maillart could not fight bureaucrats, but at the very least he wanted the incident to demonstrate that his theories worked.

The next week, the city requested drawings from Maillart for his proposed reinforcement, to be completed by July 4 for Ritter's review.[12] Maillart chose to ignore the deadline.[13] Ritter continued to avoid a public confrontation with Maillart: He never attended a meeting on the gym controversy, and he never answered Maillart's criticism of his calculations. In spite of Killer's objections and without Ritter's opinion, the city went ahead to approve Maillart's August 14 scheme for strengthening the gym roof.

PRESTRESSING THE GYM

In late August, Maillart received a brief note from the city reminding him that the issue of Ritter's payment had still not been settled. The engineer responded angrily that he had begun the reinforcement work under the assumption that he would not be responsible for paying the consultant. He felt cheated and told the city so in no uncertain terms.[14] A few days later the city called a meeting without Maillart and reconfirmed its view, over Herter's objection, that Maillart should pay Ritter.[15] This outcome was galling to Maillart. He decided not to comply and to let the affair proceed into court.

Maillart did not stay angry for long. In fact, he even turned this distasteful episode to his own advantage by using it to develop a new building technique. He reacted to the mistake much as he had earlier to those at Zuoz and Aarburg – making an example of it to improve the profession's understanding of structure. Additionally, he decided to write an article to explain his design, to clarify the mistakes in Ritter's thinking, and to show the lack of grounds for the city officials' worries. Throughout September, Maillart supervised three workers in the construction of sixteen horizontal struts in the gym roof, using an early application of prestressing (Figs. 152 and 153).[16]

In late November, Maillart received word of Zu-

rich's ruling against him: He was ordered to pay Ritter. Dating from before World War I, the feud between Maillart and Ritter climaxed over the disputed payment. But Maillart obstinately refused to capitulate; the final decision came in the new year.

THE ATTACK ON ZURICH

On February 15, 1935, Maillart signed an agreement with the city of Zurich whereby Ritter's 2,850-franc fee was divided equally between himself and the city.[17] One month later, the usual Saturday edition of the *Bauzeitung* carried Maillart's article on the gym. Jegher sent a copy to city councilman J. Hefti, a stout opponent of Maillart's. In a covering letter, Jegher sarcastically congratulated the city for getting without any additional cost a building that could now carry much more load thanks to "the humane qualities of my colleague Maillart; for whose ability this matter speaks with the utmost clarity. . . ."[18] Jegher went on to observe that "you have me to thank, since I published the drawing showing the reinforcing steel . . . and thus the error first came to light!" Jegher implied that not only was the city incapable of finding the error on its own, but that the final result was a testimony to Maillart's generosity of spirit rather than to their action.

Infuriated, Dr. Hefti and his engineers began to prepare a detailed response to Maillart's article. What had Maillart written? After repeating the arguments he had made to the city, Maillart reasoned that before the defective hanger could fail, much of its load would have to be taken by the sloping eave beams.[19] In Maillart's simplified analysis, on which the design was based, the middle (mistakenly underreinforced) hangar carried all the load of the central half of the span and the other hangars carried nothing. For that large load, he had designed the reinforcement, but unfortunately only about half that amount had been used. To the city, he had argued that such a simplified analysis was very conservative because the adjacent hangars must take some of the load, and hence relieve the defective

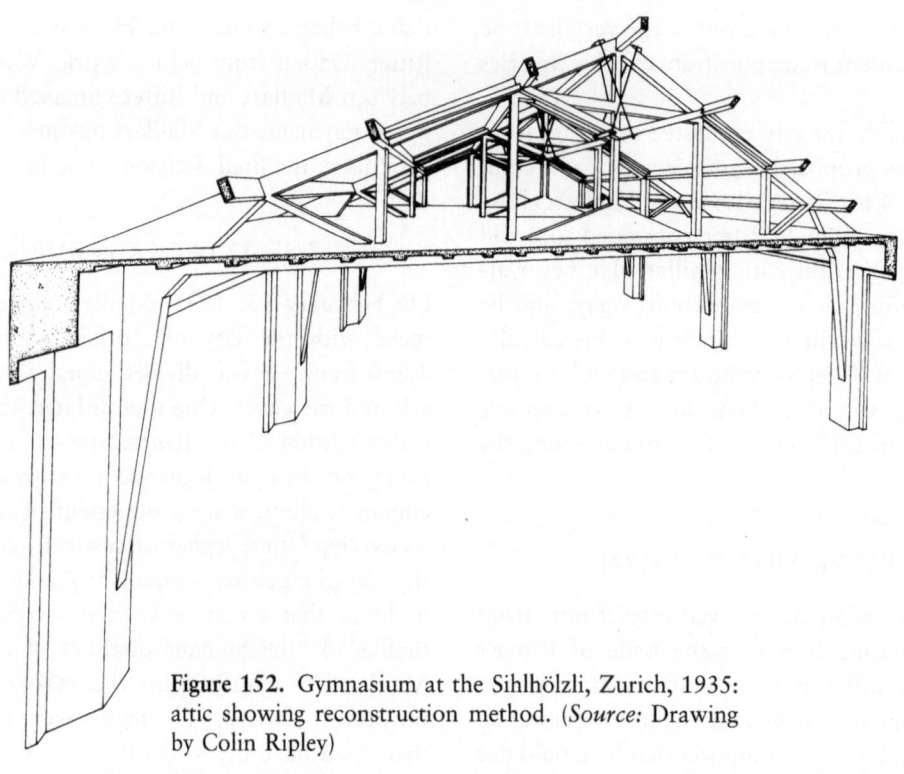

Figure 152. Gymnasium at the Sihlhölzli, Zurich, 1935: attic showing reconstruction method. (*Source:* Drawing by Colin Ripley)

hangar of much of its assumed load. Maillart calculated by simple means that the eaves would relieve the hanger of 45 percent of its load; his full-scale measurements showed that the eaves actually took more, and, hence, the hangers were much less in danger than even he had thought.

He could estimate such reductions because, as usual, he saw the structure as a single entity, visualizing the forces as a designer and not as an analyst. Like a concert pianist, Maillart sometimes missed a few notes here and there, but he never missed the meaning of the music and he never allowed a discontinuity in the performance. The pedants who complained of the wrong notes and missed the beautiful whole were like the authorities who insisted on useless numbers. Maillart's original theme had been changed slightly by his wrong note; but he proceeded to incorporate that misstep into a variation on his theme and thereby created something slightly different.

Maillart argued that a direct means of strength-

ening (adding a new hanger) would have been difficult to construct. Therefore, "I proposed ... a more elegant and more easily constructed solution ... a system made of [horizontal] struts in between the two middle compression posts." The construction difficulties stimulated him to seek another solution that would be both easier to build *and* more elegant.

THE CITY FIGHTS BACK

On May 15, Dr. E. Ammann, secretary of the building department, sent a letter to Carl Jegher stating four major objections to Maillart's article.[20] Then Maillart composed an acerbic reply counterattacking each of Ammann's arguments before the latter's article was published.[21] Maillart took a huge risk by writing the article in the first place and then by defending himself against the city's attack on it. It is almost unheard of that a professional engineer would publish his own mistake and then proceed

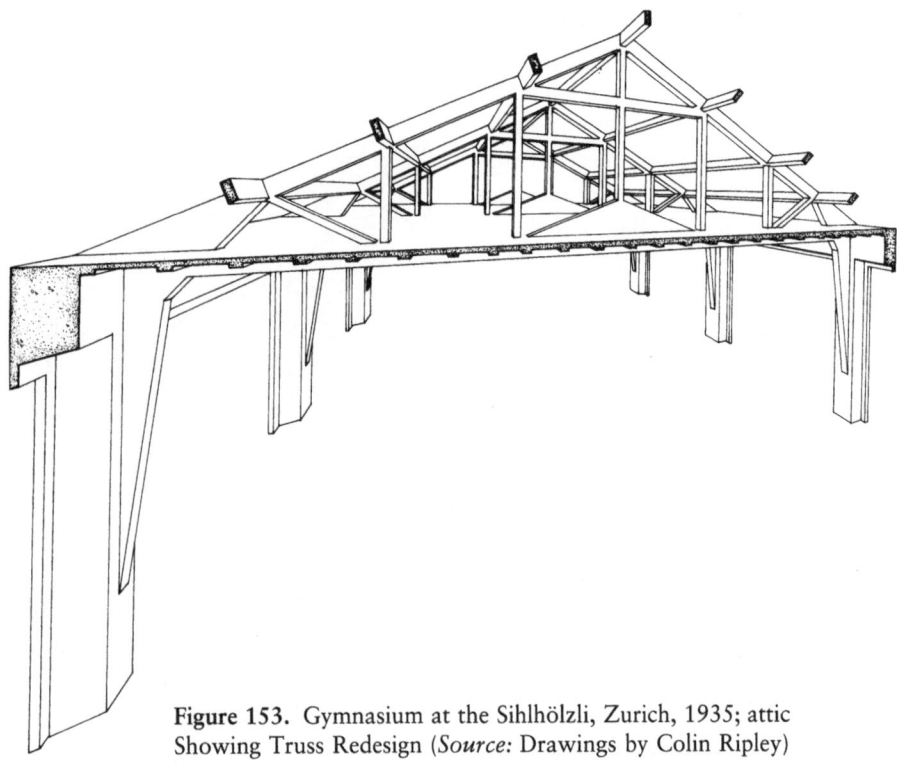

Figure 153. Gymnasium at the Sihlhölzli, Zurich, 1935; attic Showing Truss Redesign (*Source:* Drawings by Colin Ripley)

to denounce a public authority who had insisted that he correct it. Yet Maillart's reply was so pointed and so unanswerable that, after several months of self-examination, the city had no option other than to withdraw its letter in order to avert publication of Maillart's reply.

Maillart showed in his reply that the city failed to understand how the structure was loaded and that it objected to a concrete stress that was well within the limits of safe practice.[22] Finally, as to the neglect of certain loadings, Maillart warmed to his subject by observing that even the city's expert (Ritter) found no problems when one put "100 men per frame [in the unused attic] with a full snow load [on the roof] and a hurricane during which ten men enjoy gymnastics!" [on the equipment hanging from the frame]. The image of intense activity in the face of a hurricane, which, incidentally, also allowed the full snow load to remain on the pitched roof, reduced the debate to a joke. When abstract numbers turn into imagined realities, then the logic

of loads can be assessed. Maillart's method may have been unorthodox, but its effect was telling.

He concluded with a clear expression of his basic ideas. "The claim for refined calculations cannot be upheld. The simplest method is always enough, when one neglects no harmful but rather only helpful conditions. The great calculation effort," he went on, "is lost when it does not lead to tangible savings." Indeed, he claimed that "the more rigorous but thoroughly voluminous calculations of the expert . . . could have been the basis for saving something like 6 francs per frame, thus for the 16 frames about one hundred francs on the entire structure." When we recall that Ritter had charged 2,850 francs just to make those "voluminous calculations," Maillart's intent is obviously to ridicule the city's reliance on the professor's results.

Maillart raised in his unpublished reply to the city the fundamental issues of how to evaluate good designs and how to reward designers for them. The bureaucratic mind could not escape the trap of

complex mathematical analysis as the way to evaluation and it could not admit that a large savings
in construction should result in a financial reward
for the designer. In addition to being forced to pay
half of Ritter's fee, Maillart had not received full
payment for his design. These issues remain as
unresolved a half century after Maillart's reply as
they did in the 1930s. When collected together,
Maillart's ideas show how they form a basic part
of a new theory of structural engineering that has
fundamental implications for the future of building
as well as for its political, scientific, and aesthetic
context.[23]

Figure 154. Huttwil Railway Bridge over the Huttwil-
Wolhusen Railroad. (*Source:* Madame M.-C. Blumer-
Maillart)

New Forms and Lost Competitions
1935–1936

THE CHALLENGE IN MINUSCULE BRIDGES

Although he built later in Zurich, Maillart's contentious gym article may not have helped him secure commissions from the city of Zurich,[24] but it
did publicize his talent for solving difficult problems
in a simple, inexpensive way. He was very much in
demand in 1935 as a consultant and as a designer
for small, complex structures. Whenever central authorities like Ritter were absent, Maillart had opportunities such as those small overpasses in small
towns where he explored beam forms as he had 30
years ago explored arch forms in the Graubünden.

While under *Hausarrest,* as Maillart called some
recent head cold and dental problems, he had
sketched a new bridge at Huttwil.[25] Along with the
equally small Liesberg Bridge, whose design was
completed in 1935, this bridge represented a new
departure for Maillart (Fig. 154). Until 1933, he
had seen small bridges (up to 30 meters in span) as
thin arches stiffened by straight deck beams used as

parapets. These visually strong, wall-like beams
made possible the thin arches below, but in 1932,
Maillart began to question their heavy appearance.
At Schwandbach, and Töss the following year, he
had reduced their depth and achieved a lighter overall effect.

Then in late 1934, he began to see a new possibility: to use the beam alone, without any arch at
all, thus eliminating the expensive arch scaffolding.
But how was he to avoid the appearance of heavi-
ness in the straight beam? Maillart's solution at
Liesberg and at Huttwil was to form haunches
(smoothly curved increases in depth of beam) over
the two columns supporting the main span. These
columns in turn were shaped in profile and made
horizontally thin in contrast to the vertically deep
horizontal beams (Fig. 155). The results did not
completely satisfy him, but they marked the beginning of his search for an elegant form in small
bridges whose spans and clearances were too slight
to make arches logical and where economies could
be achieved with straight elements only.

Of all his small-bridge projects, the one that intrigued him the most was the railway bridge over
the Birs River near Liesberg. It was Maillart's first
successful attempt to solve the aesthetic problem of
a continuous-beam bridge.[26] Several months earlier
– on his first working day after seeing off Marie-

Figure 155. Birs River Bridge at Liesberg, 1935: full view. (*Source:* FBM Studio Ltd.)

Claire in Genoa – he had sketched roughly a continuous-beam solution in which the top line of the beam rose up over the two column supports in the suspension bridge cable form.[27] He had worked over this strange form [Fig. 156(a)] from October 29 until November 6, 1934, when he settled on a horizontal top line with a haunch below at the very thin columns [Fig. 156(b)]. By the end of the year, he had completed the design, construction began in February 1935, and on May 16, Maillart was called to the site for the removal of the scaffold.[28]

He had devised the novel construction procedure of leaving the two ends of the bridge temporarily unsupported when the scaffold came down. Thus, the bridge ends drooped and forced the main-span girder to move upwards at midspan, counteracting the main-span dead load. By this means, Maillart could keep the girder depth relatively small and still carry a very heavy locomotive loading.[29] He was also pleased with himself for embedding the rails directly in the concrete deck rather than placing them on transverse ties on a gravel ballast. "The Swiss railway people are sick of such audacity," he reported, but he was delighted to have gotten away with this latest "impertinence." Not only did it save money, but Maillart claimed that it also reduced the

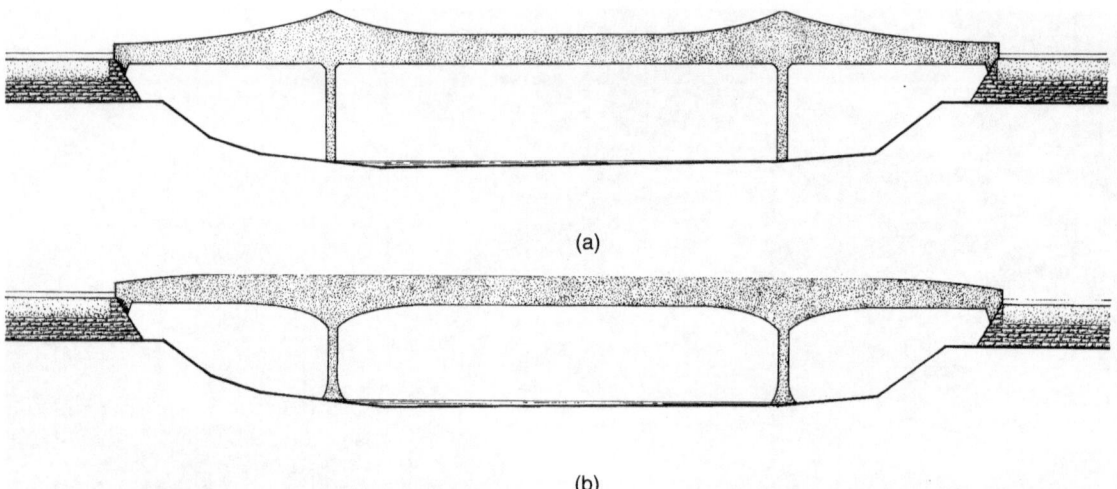

(a)

(b)

Figure 156. Liesberg Bridge: (a) preliminary design, Maillart drawing, October 29, 1934; (b) final design, Maillart drawing, November 6, 1934. (*Source:* Drawings by Mark Reed)

vibrations of the structure under a moving locomotive.

On the last day of May, Maillart went back to Liesberg for the load test directed by Mirko Ros (Fig. 157). As the 112-ton locomotive moved out over the bridge on a wet day, Maillart sat smiling under his umbrella on one of the main girders, with his legs swung over the outside face while the *Bauzeitung* photographer recorded the scene (Fig. 158).[30] Maillart enjoyed his little bridge, which in profile he found resembled a "slug" (*limaçon*). In spite of this strange similarity, the form did not displease him except for the exposed outside (abutment) supports, which were to have been covered with earthwork: "note the unpleasant form (*unschönen Gestaltung*) of the end abutments," he wrote.[31] These supports were left exposed at the insistence of the owners in order to allow a wider open waterway below the bridge for flooding (far left in Fig. 159).

On a more subtle level, Maillart made his design visually dramatic by the use of the haunches at the columns and the curved top edge near the abutments. The bridge is on a 45° skew, meaning that it crosses the Birs River not at right angles to the

Figure 157. Maillart and Ros discussing the load test for the Liesberg Bridge, 1935. (*Source:* Madame M.-C. Blumer-Maillart)

Figure 158. Maillart sitting with his umbrella on the Liesberg Bridge. (*Source:* Madame M.-C. Blumer-Maillart)

river, but at an angle of 45°. Therefore, when looking at the bridge from the river, the haunches of the two beams appear to form arches whose shapes vary as one walks along the river banks (Fig. 160).

In June, Morton Shand, in Zurich to give a lecture, came upon recent photos of the Liesberg Bridge in Maillart's office and begged him for permission to publish them, which he did later that year.[32] What excited Shand was that Maillart could pack so many new ideas into one little bridge built in so inconspicuous a location and for so little money. In the same way, Maillart designed the Huttwil Bridge, which crossed the railway at even a sharper angle (skew of 25°) than the Liesberg Bridge.[33] Likewise a straight, three-span, beam bridge with slender columns, this little work was

Figure 159. Birs River Bridge at Liesberg, 1935. (*Source:* Madame M.-C. Blumer-Maillart)

also of sufficient technical interest to merit publication.[34]

Maillart was moving away from curved elements toward more angular structures. The discipline of economy was motivating a search for play where concrete was still to be thin and light, but now it had also to be simplified to make the scaffold and form work less expensive. After 1934, nearly all of his work would reflect this shift away from smooth curves. Meanwhile, throughout Europe and America, smoothly curved arches remained the dominant arch form, while designers regularly regarded beam bridges as heavy, decorated forms without structural expression.[35] Stimulated by the search for more economy, Maillart was exploring the aesthetic possibilities of straight-line forms.

REFLECTIONS ON DESIGN

By the spring of 1935, the Zurich office had been reduced to three people: his chief engineer Keller, his architect Kruck, and one draftsman.[36] From the beginning of the year, Maillart had responded to the lack of design work, as he had 10 years earlier, with a series of articles that addressed current engineering concepts. This time he focused on aesthetics. He first reacted to an article in France about two three-hinged, hollow-box, arch bridges recently built at Laifour and d'Anchamps over the Meuse River in the Ardennes. Designed by Charles Rabut, principal teacher of Freyssinet, these bridges sought "to keep a general profile to which people are accustomed [even if that leads] to a form which is a little heavy. . . ."[37]

Maillart criticized this attitude in a brief article in which he traced his own development from Billwil and Tavanasa to Salgina and Felsegg.[38] He tried to show that the French designs led to unnecessary construction complication and extra materials. Although Maillart implied, thereby, that the French bridges were uneconomical, they were in fact remarkably inexpensive.[39] Maillart then asked rhetorically, "Is there a marked inferiority, from the aesthetic point of view, in the more honest forms of the Salgina bridge?" Surprisingly, he answered

with a criticism of the Salgina, where, instead of an ogival (broken arch) curve, "in order to respect tradition, the intrados [underside of the arch] was made round near the crown." Only with the Felsegg Bridge, Maillart noted, did he arrive at the more logical form. He had reached a better or more "honest" solution, in his own view, by keeping the "purely structural" conditions foremost in his mind.

Not only did the French article promote the heavier arch as preferable aesthetically, but it also failed to mention Maillart's earlier designs of hollow-box, three-hinged, arch bridges upon which Rabut must have drawn. Realizing that his work was not well publicized among French engineers, Maillart proceeded to publish a more complete discussion of his bridges in the February 1935 issue of the French journal *Travaux*.[40]

Maillart's work continued to excite responses throughout Europe. The French, German, and English technical press publicized his small bridges with numerous pictures. He was surprised that in Germany he received more notice than in France, because, unlike France, he had never succeeded in building there. He wrote the Blumers that "I bought the *Hütte,* a German engineering handbook (for the first time since 1909) and even though the chapter on 'masonry bridges' [concrete bridges] is sparse, I found therein the Rossgraben and Bohlbach bridges. Thus, I am on the way to becoming 'classic'."[41] The most significant of Maillart's own writing to appear internationally was an English translation (by Shand) of his article from *Le Génie Civil* for the May–June issue of *The Concrete Way,* the little British journal devoted to new forms in concrete.[42] As far away as Japan, engineers had heard of his unusual designs.[43]

As August gave way to September, business worsened.[44] Then on September 14 came the official results of a Schaffhausen competition in which neither of Maillart's bridge entries were mentioned.[45] Disregard for the quality of these remarkable designs would elicit strong criticism of the jury 2 months later by the *Bauzeitung.* For the high bridge, the jury chose a steel-beam bridge resting

Figure 160. Birs River Bridge at Liesberg, 1935: arch view. (*Source:* FBM Studio Ltd.)

on "stilts" rather than Maillart's thin, long-span arch, and for the low bridge in the city, they chose a bascule (each half rotates upward to permit boat passage) or a continuous beam in arch forms, where, as Maillart explained, "the architect 'cheated' by making an elegant 'child's sketch' which does not at all conform to the engineer's plan." In describing this to the Blumers, he used the Swiss-German words for cheat (*bechisse*) and a childish-looking sketch (*Helgeli*) to stress his contempt for the entire proceeding. His only support now came from Bern and Geneva, but unfortunately these municipalities had little money for building, although he appeared to have won the large quai Turrettini commission because his design cost 100,000 to 200,000 francs less than the original one made by another engineer.[46] Even with the Turrettini commission, he was forced to cut his staff's salaries.

In mid-October, amid business woes, he began to think about a talk on bridges he had reluctantly promised to give in Zurich to a group of modern architects there. The talk was announced in the popular press as well as in the *Bauzeitung*.[47] To his amazement, by 8:15 P.M., the large auditorium in the ETH Machinery Laboratory Building was overflowing (it holds about 300). His old adversary Arthur Rohn was there as was Professor Jenny-Dürst, who introduced Maillart, followed by Sigfried Giedion, who talked about Maillart's "aesthetic" in such a way that by the time he arose to speak, "I was persuaded myself that decidedly I was an ace!" He showed seventy slides of his bridges, beginning with Stauffacher and ending with his current designs.

The response was enthusiastic, and after the lecture, a group of admiring young architects escorted Maillart back to his little apartment at the Bärengasse.[48] One, Alfred Roth, recalled years later the strange decor and surprisingly conventional interior. These architects were amazed by Maillart's lack of interest in having his apartment reflect his extraordinary originality in form making.[49]

In his talk, Maillart had criticized the Bern building director, Bösiger, for preserving a wooden bridge in Wangen instead of building his own reinforced-concrete design that the town itself preferred. Consequently, Bösiger saw to it that Maillart was excluded from any bridge under his control.[50] Thus, even the trickle of small jobs from this canton dried up. While the *Neue Zürcher Zeitung* wrote about his exploits as "an honored old man," his Zurich office work ceased and reluctantly Maillart had to release Alois Keller, his last engineer there.[51] At the same time, another major, but unremunerative work was slowly taking shape in Geneva.

THE VESSY BRIDGE

On December 11, 1934, Maillart sent to the city of Geneva his initial report on a new bridge over the Arve River at its 180° bend near a recreation area called "le Bout du monde" (the end of the world), close to the suburb of Vessy.[52] He had been rethinking the Salgina design since 1930 and this new design represented his most advanced ideas on angular shapes and on arches made up partly of hollow boxes and partly of channels (the lower half of the boxes as they approach the hinged supports). At Vessy, Maillart moved those hinges out onto the span (Fig. 161), as in the Sirakovo Viaduct design that summer. He also gave a unique x shape to the cross walls supporting the deck, and his calculations show a diagram of forces (a bending moment diagram) of just that shape (Fig. 162). This playful form, although it has some rational basis, is primarily an exhibition of Maillart's personal idea of introducing surprise and amusement into the serious business of engineering. The force diagram shows that the form is not irrational, even if his primary motive was surely aesthetic.

Christian Menn, in describing this type of idea,

states that "one cannot completely deny that even decoration by an engineer endowed with artistic talent may, by all means, have an appealing aesthetic effect." He illustrates this idea with the cross walls of the Vessy Bridge. Stimulated by this work of his predecessor, Menn goes even further:

Even dalliance and static nonsense, like a properly dosed pinch of salt, are capable of reviving the aesthetic appeal of the structure of a secondary supporting element [the x-shaped cross walls]. Being extremely delicate, such variations must be carried out by real artists only; the normal engineer is advised to desist from it.[53]

Although Maillart would not have labeled his structure "dalliance," he would have agreed with Menn's basic idea of the real artist and he was therefore pleased when the city told him in mid-July of 1935 that it might go forward; and so for a ridiculously small sum "he reworked the design carefully."[54] The simple bridge, like Schwandbach, presented him with the opportunity for new forms, this time within his basic three-hinged arch type. He had received the design contract thanks to his friend Braillard, an architect and city councilman who strongly supported Maillart's work.

On August 26, 1935, Maillart submitted a cost estimate for the bridge, and early in the fall, the city received eight bids all substantially below Maillart's estimate.[55] This was an indication of the economic Depression as well as the fact that the design seemed relatively easy to build. Construction did not start at Vessy until early in May 1936 (Fig. 163), even though Maillart had urged city officials to begin in the autumn to avoid the high waters of late spring. He worried additionally about the inexperienced builder, who because of his low bid had been given the contract against Maillart's advice.[56]

Completed over 1 year later amid considerable bickering over design costs,[57] the canton of Geneva also questioned Maillart's load-test results for Vessy and he was obliged to write a defense of the structure.[58] Once again, officials worried that the calculations were too simplified and, in this case, that they did not take into account the abutment movements measured during the test. Maillart's defense demonstrated his pragmatism and an inconsistency: The small abutment movements had minimum influence because the three hinges allowed for adjustment without stresses; these hinges were not perfect (they act like partly rusted hinges), but that was all right because the bridge vibrations were thereby reduced; and the calculations did not need to include the deck because its influence was favorable to the arch (it reduced the arch stresses).[59]

These arguments would never stand up logically against the academic objections then prevalent at Maillart's alma mater. First, if the hinges allowed abutment movements with little stress, then they also permitted relatively large vibrations; these two effects were in direct conflict, that is, the more flexible the bridge (the more the hinges are free to rotate), the more the vibrations and the less the effect of abutment movement. One cannot have it both ways, but that is how Maillart defended his bridge.

The second questionable argument was his justification for a simplified calculation that did not include interaction between the deck (over the cross walls) and the arch below. Maillart had always argued that the designer should see the structure as a whole and take advantage of the monolithic nature of reinforced concrete. This is just what he did in his deck-stiffened arches, but in defending the Vessy load test, he seemed to be saying that such interactions were to be neglected.

Maillart was right in each of his arguments taken separately, but taken together, each effect was irrelevant to the design. The measured foundation movement was slight and not dangerous and hinges rendered it meaningless; it would probably have been insignificant for an arch without hinges had the designer provided the proper reinforcement as Freyssinet did. The hinges were not perfect because they were concrete hinges, not lubricated metal ones. Maillart made his three-hinged arches very stiff by fusing them in the center half with the deck; thus, it was irrelevant whether the hinges were perfect or not, vibrations here were not a problem. Finally, unlike the case for the very stiff deck and

very thin arch in the deck-stiffened designs, the deck over the outside quarter spans had little influence on the overall behavior.

Maillart's logic was one of priorities; he was happy to give up consistency in analysis as long as he could see that the neglected effects had little importance. What outraged the academics was that this neglect had the effect of rendering their sophisticated analytic research meaningless. Maillart's calculations and Ros's test proved that the Vessy Bridge was sound and that the simplified calculations provided a sufficiently accurate basis for design. More sophisticated analyses have the tendency, as Maillart argued, to make the analyst less sure of a design than the reverse and hence less willing either to accept new forms or to reduce materials in standard forms. The reason that rigorous analysis goes together with restricted design vision is that the analysis, however complex, is always far from reality and the exponent of rigor always sees the errors. The reason that great rigor does not aid design is that reality cannot be controlled beyond a certain point and only people with deep experience *both* in construction (literally the making of forms) and in calculations (in the verification of performance) can judge how simplified to make both the form and the formula.

Maillart's seemingly illogical defense was a tongue-in-cheek attack on the academics, the codes, and the authorities who followed them unthinkingly. The mind focused on the logic of calculations seemed destined to neglect the facts of observed behavior. Bridge calculations are essential, but they always remain abstract and hence susceptible to misuse.

An example of this strange result of extra rigor (already noted in Chapter Five) comes from research carried out in the United States during the 1930s on arch bridges and culminating in the article by Nathan Newmark, perhaps the most brilliant structural analyst of his day, in which he commented on the difficulty of performing a rigorous solution of arch and deck interaction. Although he was capable of such analysis, he considered it impractical and hence advised designers to neglect the

Figure 161. View of the completed bridge over the Arve River at Champel-Vessy. (*Source:* FBM Studio Ltd.)

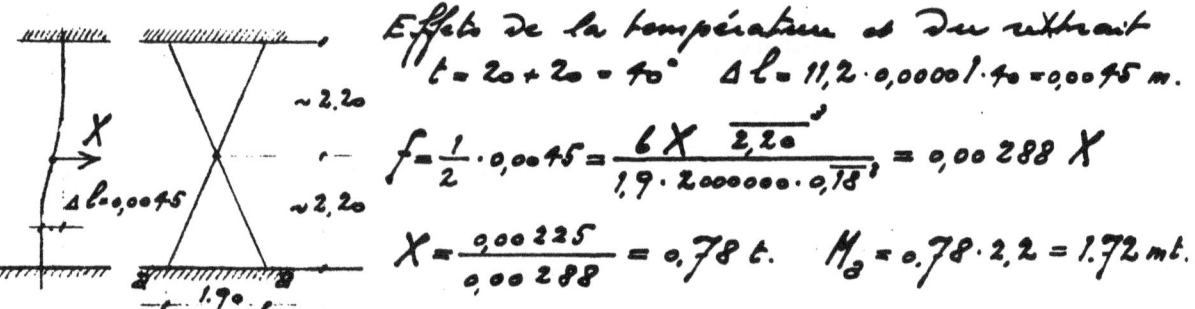

Figure 162. Maillart Calculations, Bridge over the Arve River at Champel-Vessy. (*Source:* Maillart and Company, Drawing No. 319/21, p. 9, October 26, 1936)

Figure 163. Construction of the Bridge over the Arve River at Champel-Vessy. (*Source: Schweizerische Bauzeitung*, Vol. 112, No. 24, December 10, 1938, p. 290)

influence of the deck stiffness on the arch performance. His concern for analysis led him away from the elegant designs of Maillart and helped prevent the deck-stiffened arches from entering American practice until the 1970s.[60]

"SAGESSE SANS DIPLOME"

The masterpiece at Vessy (Fig. 164) and the adulation of architects in Zurich did little for Maillart's business, which by 1936 had declined to 55 percent of what it had been in 1932. Even when the federal government asked Maillart to prepare a design for a new Zurich telephone office in reinforced concrete to compete with a steel design, he was dubious about his ability to win the commission because "the stupid code gives advantage to steel."[61] In spite of his pessimism, his design was judged to be superior and he did eventually get the project. But not in time to prevent the Depression from forcing him to reduce salaries by 10 percent and further cut costs by having his staff work one less hour per day.

On Thursday, January 23, this bad news was compounded by Paul Maillart's death. Though no surprise, Maillart was deeply upset: Paul was the first of his generation to go. The loss made Maillart more acutely aware of his own ill health and especially of his legs, "which are always too heavy so that I am less agile."[62] Sleep had become a problem, and he often wrote letters at night now instead of on the train as before. In nearly every one to the

Blumers during this time, he would include some personal introspections.[63] He mused about his tendency to take on the academics,[64] remarking how glad he was that Marie-Claire had from him some of this inner conceit as well as what he referred to as his "unaccredited wisdom" ("*sagesse sans diplôme*"). And he readily admitted to his daughter that her natural talent "for often being automatically at the center of a party is not something which you inherited from me."[65]

Ros had just released his report on the Thur River Bridge at Felsegg, whose load test of 1934 was finally documented on February 12, 1936. Ros confirmed what he and Maillart had predicted with such self-assurance when the bridge was completed: "a very interesting structure, an original and economical solution, whose remarkable style allows no reservations from a technical standpoint."[66] Unfortunately, Maillart's 1936 competition design for a similar form in Bern would not receive similar praise, thanks to a prejudiced jury dominated by Ritter and Bühler.

THE BROKEN INSPECTOR

In early 1936, Europe had more serious problems to contend with than biased bridge juries. The previous October Mussolini had begun his attack on Ethiopia. By the time Maillart vacationed in Venice after the New Year, Italy was at war.[67] Pierre Laval, the French Prime Minister, resigned over his failure to resolve the Ethiopian crisis and King George V died the next day (January 20). On March 4, Hitler announced his movement of German troops into the Rhineland.[68] Even the technical *Bauzeitung* departed from its usual subject matter to publish on January 7 Carl Jegher's review of Dutch cultural historian Johann Huizinga's new book, *In the Shadow of Tomorrow*. The editor urged his audience to take seriously the current European crises and to read Huizinga's analysis of it.[69] Maillart wrote the Blumers that both in Bern and in Zurich he was devoting most of his time to studying the construction of bomb shelters.[70]

His conflict with Ritter and Bühler did not stop clients all over Switzerland from consulting him on

Figure 164. Rendering of the Arve River Bridge at Vessy. (*Source:* Maillart and Co.)

difficult problems. Especially where cracks appeared unexpectedly in concrete structures, they called on Maillart's diagnostic skill. "I am the inspector of failed works," he wrote.[71] The strain of work and travel and the worry about family, business, and politics contributed to Maillart's debilitation after the car accident. On Friday, July 3, he dined in Zurich with Alexandre, who in remembrance of their strange, unforgettable Russian life had brought him some fresh caviar. In spite of that generosity, Maillart beat his friend in Honeymoon Bridge; he then returned alone after midnight to his little apartment in the Bärengasse. As he started to undress, the bedside rug slipped on the well polished linoleum and he fell sideways. In great pain, he realized immediately that his left leg was useless, so he crawled to the telephone and called the city ambulance.[72]

Maillart was quickly transported to the cantonal hospital where x-rays revealed that he had broken his thighbone. He was put in traction and moved into a general ward because no private rooms were available. The next morning, the engineer rather enjoyed the company of the young people with whom he found himself sharing the ward; he was particularly amused by their conviction that his injury was the result of a motorcycle accident, prevalent among his fellow warders.

That afternoon the engineer was moved to a private room, where he was able to receive numerous visitors and where he immediately began to telephone his offices. On July 8, he felt well enough to write his regular letter to Indonesia in which he revealed his accident in a typically bemused way to the Blumers:

You may ask why I write you from Zurich since I usually am still in Geneva on Wednesdays. It is because there is a change of program which obliges me to prolong my stay in Zurich. It is a rest cure which I spend not in the somber interior of the Bärengasse, but in a beautiful room full of flowers on the Zurichberg with a large balcony and view of the woods.

He then described the events of those past few days and begged them not to worry for "it could have been much worse."

Toward the end of the month, his recovery was seriously set back by a thrombosis in the injured leg. Pain and high fever lasted for several weeks.[73] As painful for the engineer as his leg was, the enforced absence from work hurt even more. This was especially true when the torrential spring swelling of the Arve River washed away part of the Vessy Bridge scaffold just as Maillart had feared it might.[74] His frustration was compounded by the fact that not a single new project had come in since his accident a month before.[75]

He was pleased, however, to see the publication of his article on the Vevey quai and to learn of final arrangements for construction of the new quai in Geneva.[76] In spite of his poor health, Maillart insisted on accompanying a group of engineers and architects from Geneva the day after his return (October 14) to the Vessy site to explain the design to them. By mid-November, Maillart was still weak, and the leg still painful, but at least he could walk with a single cane. He immediately returned to work and visited several sites with Stettler.[77]

LOST COMPETITIONS

On September 26, as he had expected, Maillart learned of his failure to receive a prize in the Sittertobel (Kräzernbrücke) competition in St. Gallen. Announced in the *Bauzeitung* on June 6, Maillart had been excited by this new challenge at a low point in his work. But again, with Bühler and Ritter sitting on the jury, the odds were against him. Furthermore, his hospitalization had prevented him

from completing the design, which he gave to his former chief engineer, Alois Keller, to submit.[78]

Much more important to Maillart was the major bridge design competition sponsored by the Swiss Federal railways for a prominent crossing of the Aare River next to Maillart's Lorraine Bridge of 1930.[79] Maillart's entry had continued his recent development of three-hinged arches that had begun with the Salgina and included the Felsegg, the Vessy bridges, and especially the studies for the Sirakovo Bridge (Fig. 146). The competition provided a showcase for Maillart's work, but it also reflected his continuing problems with the engineering establishment. Prejudiced by Bühler's championing of his own official design and Ritter's complicity with him, the jury aroused the ire of Carl Jegher, who, himself, published a criticism of this bias. For Maillart, it was impossible to take seriously such an obviously rigged contest.[80]

Maillart found intolerable the way the competition proceeded. There was an official design (by Bühler) against which the competitors proposed alternatives [Fig. 165(a)]. The railways had worked on it for 6 years, whereas the competitors had only a few months and meager funds.[81] "Bühler and Ritter – my two fervent 'friends' – dissected my designs and calculations without me or the other members of the commission being able to contradict them."[82] Had the contest been fairer, that railroad bridge might have become Maillart's best-known work because of its size (150-meter span) and its prominent location in the Swiss capital. As it was, Maillart's design represented only his "unaccredited wisdom."

Within a month, Maillart had received Bühler's report of the jury, or "consulting commission" as it was officially called, and which he described bitterly as "a tissue of malevolent and petty criticisms not only about my designs but also about the others. Since no one else on the commission could undertake a check of his [Bühler's] words, they swallowed them and the conclusion is naturally that the official design [by Bühler] is the best."

The railways had allocated 100,000 francs for prizes, but the commission "had the courage" to distribute only 78,500 francs, "in spite of the enormous amount of work required of the contestants."

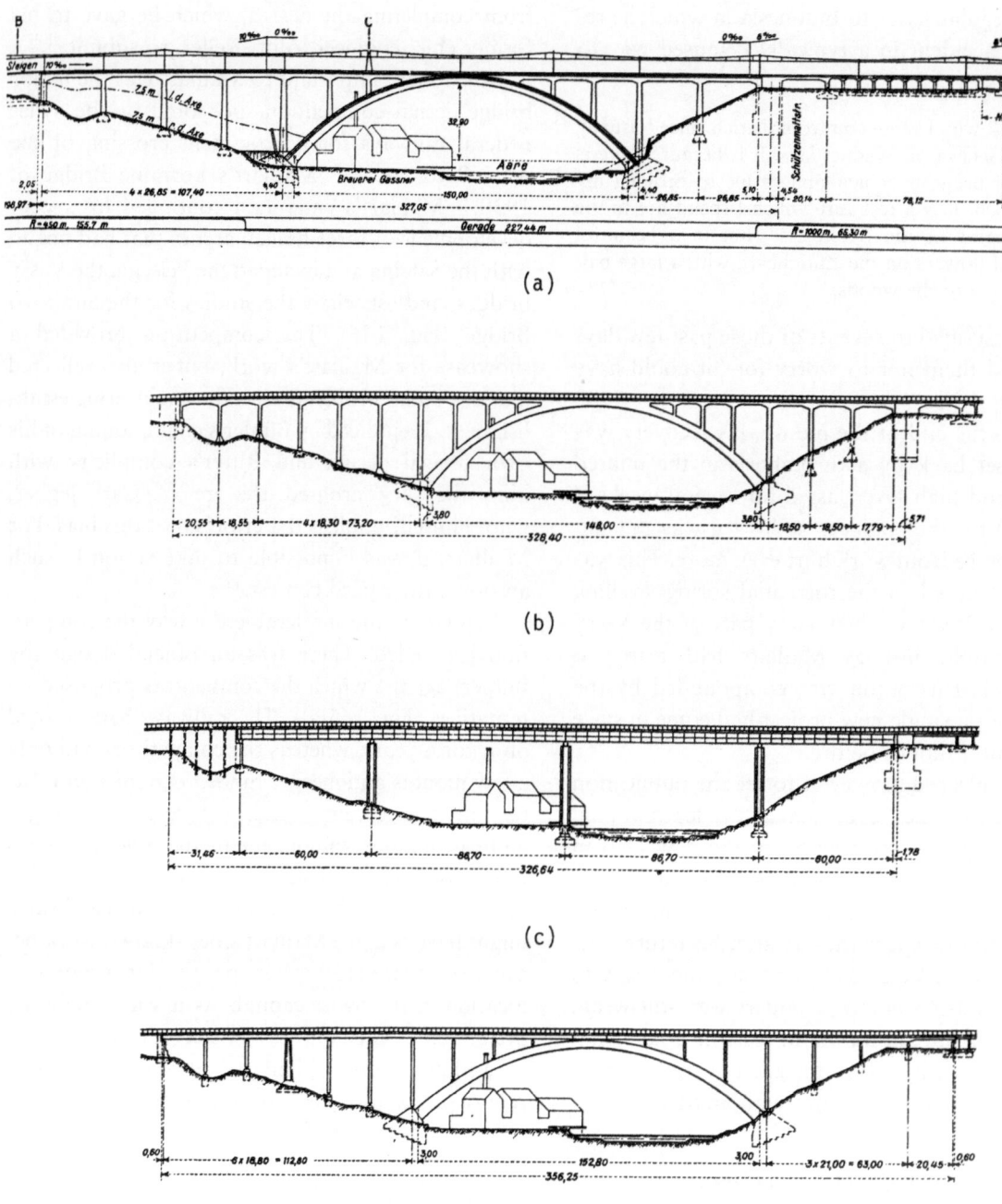

(a)

(b)

(c)

(d)

Figure 165. Bern Railway bridge competition, 1936: (a) official design (by Bühler), (b) first-place design (by Rothgeb and Weiss), (c) second-place design (by Salvisberg), (d) third-place design (by Salvisberg). (*Source:* Schweizerische Bauzeitung)

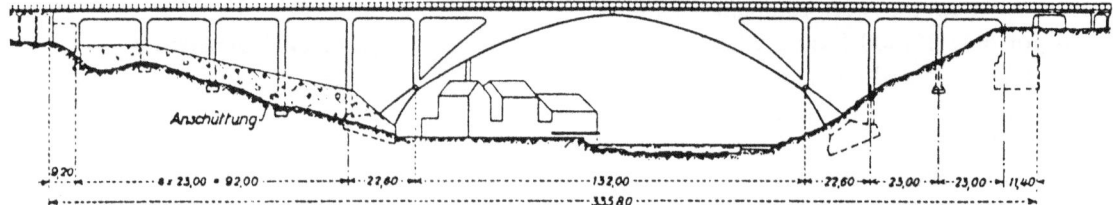

(a)

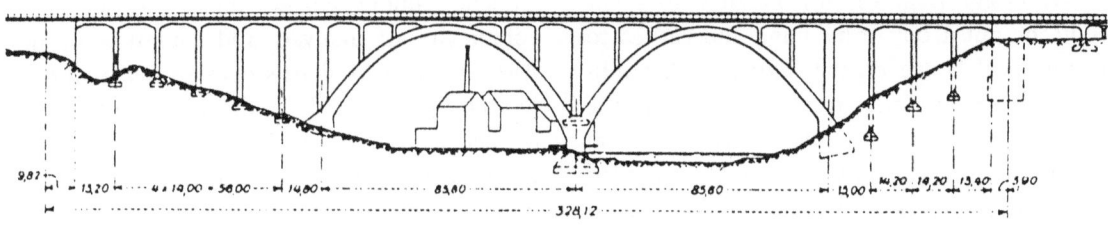

(b)

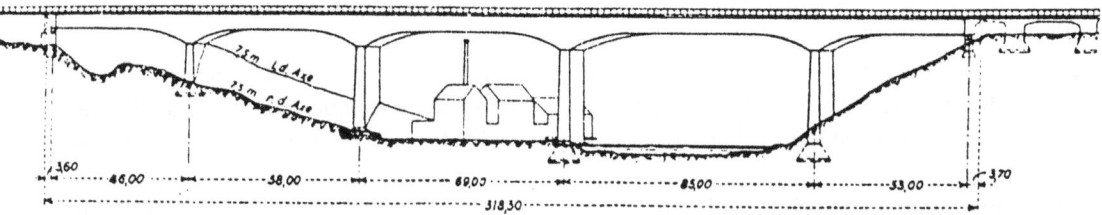

(c)

Figure 166. Bern Railway bridge competition, 1936: (a) fourth-place design (by Maillart), (b) fifth-place design (by Pulfer), (c) sixth-place design (by Bollinger and Vonmoos). (*Source:* Schweizerische Bauzeitung)

Maillart noted that the steel companies suffered most and got only 18,000 francs, whereas he got the second highest award of 12,000 francs, which would just cover his expenses. Moreover, the commission placed Maillart's entry fourth only [Fig. 166(a)], a low ranking Maillart was sure that Bühler instigated because "he guessed that my design was the most threatening to him! Thus they gave me a little more money to keep me quiet. As one might have expected, all that was a cabal mounted in advance."[83]

On Tuesday, November 3, Maillart traveled to Bern to see an exhibition of the competition designs. His proposal would have cost 250,000 francs less than Bühler's and also less than the first-place alternative design [Fig. 165(b)]. Faced with a *fait accompli,* he put it out of his mind and turned to other matters.[84] However, the *Bauzeitung's* publication of the competition results, as well as the announcement of an evening's discussion on the Schaffhausen competition of 1935, reminded Maillart that in his 34-year career, he had yet to win first prize in a juried bridge competition in Switzerland.[85]

1936 had been a bad year for Maillart. It had begun with the death of his brother Paul and had ended with 6 months of painful, slow recuperation from his broken thigh bone. The big bridge competitions of Schaffhausen, St. Gallen, and Bern were all lost and the economic Depression seemed to have no end. Germany's power under Hitler increased and was made clear to engineers in Berlin and Munich in October at the Second International Congress of Rohn's new Association. Mussolini's conquest of Ethiopia proved the League of Nations to be powerless.[86]

Against this dramatic background, the little Vessy Bridge moved slowly to completion, while Maillart's battles with the authorities reached a new level of animosity over the newly published Swiss code for reinforced concrete, over bridge competitions, and over the ideals of engineering. Occupied almost exclusively with the inspection of small cracked structures, Maillart stood almost alone against Ritter in Zurich and Bühler in Bern.

Career Summary
Codes, Research, and Design,
1932–1937

MAILLART AND RITTER

In spite of financial and family worries, Maillart kept up his battles with the authorities. With little design work in the early 1920s, Maillart had released his restless intellectual energy by instigating a debate with the Swiss engineering establishment focused on Rohn, but, more deeply, on a growing tendency among engineering academics to honor complexity of analysis and imitation of design. Also, it suited Maillart's puckish personality to set himself up against the authorities and to make fun of their committee-oriented strategy. Rohn left the technical scene with his assumption of higher administrative duties, but his protegé, Max Ritter, replaced him as Maillart's *bête noir.* Ritter was humorless, enthralled by mathematical analysis, and had a natural bureaucratic personality – directing the International Association as its secretary and chairing the National Swiss Committee dedicated in the early 1930s to the writing of a new code for reinforced concrete.

Ritter brought to the committee considerable technical competence and organizational talent. He had worked on many designs, he directed Ph.D. students in research, and he was versed in the current literature on concrete research throughout the world. His goal was to create a code based on the latest research that would be as general (hence complex) as possible and that would perpetuate the prevalent ideal of engineering as a scientific discipline. Maillart's prestige was so great that he could not be excluded from the code committee. Ritter's problem was to isolate his adversary and to convince the other members of the need for his (Ritter's) vision of codification. By late 1931, the committee had produced a draft to replace the official 1909 code.

Maillart's serious accident in October 1931 prevented him from attending meetings of the reinforced-concrete code commission, but he had kept abreast of its activities and had strongly objected to the draft that the commission debated on November 14. The resulting new draft failed to heed Maillart's objections, and Jegher asked Maillart if he would write a critique of it for the *Bauzeitung*. In late November, Maillart began to do this. Especially in the evenings, when he was completely alone in the three-story office building, he read and wrote. He could see that many leading authorities in Switzerland were lining up behind a document that he found unacceptable. When Maillart's discussion appeared early in 1932, it would estrange him still further from the Swiss engineering establishment.

MAILLART DEBATES THE CODE COMMITTEE

Maillart's critique appeared in the *Bauzeitung* on January 30, 1932. His was the only criticism published and the code committee began to think of how to respond.[87] His essential objections also appeared in Emperger's Journal *Beton und Eisen*, where he presented his own far simpler approach:

This simplest of calculation methods suffices for all safety requirements. It offers a great saving in time and paper. Anyone who knows how voluminous structural calculations of the ordinary type can get will know [that] . . . it is enough to have proof that the allowable limits are not surpassed.

The Ph.D. question of the correct value [for the ratio of steel stiffness to concrete stiffness] would be eliminated . . . which certainly strips away a certain mystery that may offend various high priests of reinforced concrete. The practitioner, on the other hand, would welcome the elimination of this pseudo-scientific rubbish.[88]

Once again Maillart was using language that was highly unusual in technical publications. It illustrates how bitter Maillart's feelings about academic values had become. Lack of work and the physical pain from the automobile accident did little to improve his disposition or to restrain his outspoken advocacy for radical simplicity.[89] The first

public response to Maillart's attack came from Professor A. Paris in Lausanne, who in relatively mild language defended the compromises written into the new draft. While recognizing Maillart's experience and achievements, he stressed the need for diplomacy and for abandoning pure rationality in the interest of achieving committee agreements. Paris's arguments were not unreasonable and would strike a chord of sympathy with anyone in the twentieth century trying to ride a committee document through the rocky terrain of volunteer professionals.

Despite its moderation, Paris's discussion irritated Maillart, who quickly fired off a harsh response. For him, committee ideas were something "thrown into a pot" and he asked rhetorically, "is this something 'scientific' that gives the engineer what Professor Paris strives for . . . ?"[90] Such sharpness was hardly calculated to win anyone over. On reading the *Bauzeitung* debate, even Maillart's daughter Marie-Claire in Indonesia observed to her father that he lacked tact with his adversaries. "That is possible," Maillart responded, but "those people worked in an underhanded way. They got together to develop a response and designated a professor to do it." In fact, no such formal response was ever published. Paris was not that professor: His response did not answer Maillart's technical arguments.[91] It was almost certainly Max Ritter whom the committee designated, but Ritter carefully avoided a public confrontation on these issues. Maillart's adversaries continued to wage a kind of guerilla war against him.

On Saturday, April 29, Maillart spent the afternoon with the code committee in Zurich "where they put me on the spot."[92] Maillart's defiance was a major embarrassment to the engineering establishment, members of which kept trying to get him, as Switzerland's best known engineer of concrete, to agree with their new code. In spite of Maillart's opposition to the new code, the code committee had scheduled him for a lecture at their next meeting to take place in Basel on December 3, 1932. Maillart began to think again about his simplified calculations for this "somewhat revolutionary lecture." This would probably be his last chance to convince the committee of the superiority of his methods.[93]

The central issue of the lecture, "Practical Dimensioning of Reinforced Concrete," was the same for calculations as it was for structural design: how to work with minimum effort to produce the most reliable result. In both cases, the conventional argument held that structurally extraneous formulations must be used and the work then fitted to those alien ideals. The scientific method that Maillart decried came from a branch of classical physics called the theory of elasticity. Engineers were being taught this theory and derivatives from it in schools and now the codes were being made to fit that theory. It was a torturous process that eventually would die 30 years later when Maillart's position would be finally recognized as appropriate.[94] But in 1932, Ritter and the others carried the day: Maillart lost his battle over the Swiss code. His adversaries never responded publicly to the formulas that he had published earlier that year and that he presented to the assembly of engineers in December. The code was formally approved on June 10, 1933, at a meeting in Neuchâtel.[95]

THE ATTACK ON "APPLIED-SCIENCE" ENGINEERING

In the summer of 1934, Maillart published an article for the E. G. Portland Cement Association in their journal, *Bulletin du Ciment,* in which he began to reflect on ideas about arch bridges. This six-page article was more critical and more reflective than his earlier papers published in the Liège conference proceedings and in the *Bauingenieur.* He directly criticized two of the most famous Swiss concrete arch bridges for failing to reflect the possibilities of lightness in reinforced concrete. One of these was designed by Max Ritter.[96] He also condemned the growing practice of encasing a steel structure in concrete; this was just the idea Hennebique in France had started with over 50 years ago and had, as Maillart noted, abandoned in favor

of reinforced concrete, "where steel is used exclusively to carry tension."

But Maillart did more than criticize present practice; he also used his own recent examples, especially Schwandbach, to reflect on design. "The image of these bridges at first baffles the eye used to arches of stone." He admitted the problem many people were having with his new forms, but he stressed that "a new material, to be used rationally, cannot and must not be poured into ancient moulds, even if the new forms do not show that degree of maturity which only a much longer development can bring to them." Maillart recognized that his designs represented the beginning of a long development that would someday be as fully mature as, for example, the great stone bridges of the engineer Jean Rodolphe Perronet were in late eighteenth-century France.

In this article, Maillart also spoke of the great arches such as Plougastel by Freyssinet. He recognized that they could have been made much lighter, and he predicted that concrete bridges already competitive for medium spans would soon compete with steel for much greater spans as long as there was a "rational development," that is, a full utilization of materials so that lighter and safer structures resulted, and an absence of nonessential bureaucratic code restrictions.

After his October 1935 Zurich talk to the modern architects, Maillart created an even greater furor by publishing another attack on the Swiss code in the widely read German journal *Der Bauingenieur* (which had invited him to write the article). That would get their attention, he thought, and also make his position clear to his friends abroad. It did indeed create the desired effect.[97]

On November 30, 1935, the Zurich alumni association scheduled a meeting to protest his *Bauingenieur* critique. Maillart welcomed the chance to defend himself. He listened patiently while the president of the Swiss Society of Engineers and Architects led a discussion on details of the new code, adjourning the meeting without calling on Maillart. Before anyone could leave, Maillart stood and challenged the group to face him publicly.[98] On the technical issues, they raised no serious arguments; the force of their objections was much deeper and more closely linked to ethics and the role of the engineer in Swiss society. They had three complaints. Their opening salvo was aimed at Maillart's discussion of how the code led engineers to design heavier, costlier structures, thus permitting them to ask higher fees. Maillart responded that "I do not accuse anyone and that those who are honest, i.e. those present here, can not feel touched and that if others are upset (*betupft*) that would not be so bad!"

They went on to object on nationalistic grounds to Maillart's writing a critique of the Swiss code in a German journal. He answered this by observing that the *Bauingenieur* editor had requested his critique. More important, he stated that "Swiss engineers of reinforced concrete are directed entirely by that which is written in Germany. Proof of this is that my articles in the *Bauzeitung* have never found any response from [you Swiss engineers]. . . ."

In conclusion, the Zurich engineers claimed that Maillart's article put them in a bad light with the Swiss architects, who were often their main clients. He dismissed their argument by noting that Swiss architects did read the *Bauzeitung* but did not read the *Bauingenieur*. In short, Swiss engineers would be judged in Switzerland by what appeared in Zurich not in Berlin. There the matter rested. The profession's leaders failed to get Maillart to retract anything and none of them was able to put into writing any arguments against him. The code was formally accepted in 1935 and stood unchanged for 20 years, and Maillart's objections stood unanswered in Switzerland for as long.

Maillart argued almost alone for a broader type of structural engineer. As he had noted in the *Bauingenieur* article, the engineers who first succeeded with reinforced concrete were largely designers and builders and therefore saw themselves responsible to their client – the owner – for the reliability and the economy of their designs. The 1909 code for Maillart was "simple and generous," thus allowing for progress in design, whereas the 1935 one was complex and restrictive; it had "armored paragraphs" (*paragraphenpanzer*) that would "ossify" progress in reinforced concrete. (Maillart failed to recall his earlier objection to the 1909 code on much the same grounds as he was arguing in the 1930s.) By the 1930s, designers had become separated from builders and, by this new code, were attempting to relieve themselves of responsibility for thinking about performance. The question in the new code was much more "are the stresses below the allowable" rather than "will the structure perform safely and reliably?"

The urge to define quality by abstract measurable quantities was overcoming the vision of quality as full-scale concrete performance. Maillart put the Sihlhölzli gym controversy into general terms by his critique of the code. The debate might be called one between the two Ritters: Wilhelm, the proponent of performance, and Max, the champion of prescription. Not until the 1960s would Maillart's basic objection to stresses as the measure of performance become the foundation for a new and far simpler code.[99]

Following a 3-day visit to René in Marseilles, in January 1936, Maillart returned to grim Swiss realities, which included a reprimand from the Central Committee of the Swiss Society of Engineers and Architects for his German article. He expressed this predicament in a letter to the Blumers:

I have gotten your letter sent by airmail on December 20 and thank you for your good wishes. They include in effect what I wish you also: health and work. As for wealth and honors, I believe that they burden you as much as they delight you and the envy that they provoke among your peers is not always pleasant to feel!

Yes, my lecture in Zurich did not please everyone and my article in the "Bauingenieur" even less. It seems that M. Boesiger has not yet calmed down and the high priests of reinforced concrete and certain heads of offices are angry with me for revealing certain facts. But I believe that it is better to point the finger at mistakes rather than to give an air of perfection which foreigners will not accept anyway. We live too much on our famous technique of earlier times and we [Swiss] are asleep. At the Poly, in spite of costly laboratories and institutes, there is the risk of developing a spirit at once asleep and authoritarian which refuses to accept that which is not sanctioned by the professors.[100]

It was against this spirit that he wrote because the professors "do not want to admit that in fact progress in technology comes always from practice and they only end up consecrating it after having battled against it." Maillart saw the new code as the symbol of a decline in Swiss engineering. The Swiss, he believed, suffered from a loss of private initiative and they "live in a depression under the yoke of a bureaucracy . . . and where everyone thinks only of his own interest."[101] But on reflection, as he wrote to the Blumers, he realized his own limitations: "in this imperfect world it is not only what one produces which is appreciated but also good relationships with people. I am not very strong in that department and at times I suffer from it."[102] Such thoughts continuously surfaced in his letters of early 1936. This had been a problem for him since the 1902 founding of his firm; even then, Maria had urged him to improve his public relations.

In spite of rejection in Switzerland, Maillart's reputation for research reached England at this time. The editor of the London Journal, *Concrete and Constructional Engineering,* had asked him to write his ideas on theory, which he did on the 1937 Easter weekend.[103] Maillart's simplified approach to calculations anticipated closely the practice accepted throughout Europe and America in the 1960s, a quarter of a century later. But in the 1930s, such simple ideas were anathema to the "authorities."

CURRENT ISSUES: INTEGRATION OF FORM AND FORMULA

In a lecture before the SIA (Swiss Society of Engineers and Architects) at the ETH on November 13, 1937, Maillart gave a concise mathematical basis for his vastly simplified method of calculating beams. These elementary formulas embodied the observations he had been making since his 1907

tests on beams and continued through to his September 1937 article in *Concrete and Constructional Engineering.* Maillart was at war on two fronts, one against the Maginot-line aesthetic derived from the past and one against the *Paragraphenpanzer* (armored codes) that threatened to choke creative design with their presumption of accuracy and safety for the future. In both 1909 and 1937, Maillart criticized practice in Germany, where he had been unable to build anything because of these twin problems.[104]

Following Maillart's lecture, Ros spoke about his own 14 years of testing full-scale structures, many of which were designed by Maillart.[105] In effect, the entire program centered on Maillart, his works, and his ideas. This was Ros's testimony to his old friend, and Maillart's other great supporter, Carl Jegher, made sure that the ideas would be immediately available to the profession. He sent Maillart's lecture to press in December and it appeared on the opening pages of the 1938 New Year's edition of the *Bauzeitung.*[106] The article addressed the use of imitative forms and the codification of complex calculations. Maillart criticized the use of needlessly low allowable stresses that prevent the design of thin arches and the copying of wooden or steel-beam floors that rule out the smoothly shaped monolithic forms that are natural to reinforced concrete. He emphasized the need to see structure as a whole rather than as a collection of disparate parts:

Not only does the feeling for beauty awaken the desire to see the structure as a whole rather than as single parts, but also this overall viewpoint results nearly always in an economical advantage. The mushroom slab would be one example. Its emergence led to the establishment of a number of elegant theories about slabs on point supports, which were used as a basis for most calculations. Nevertheless, treating the slab alone does not lead to an economical result. That will be achieved only if one reviews the entire building: slab, columns, and column capitals. Of course that presents a nearly impossible problem to solve by mathematical calculation, so that only model studies and measurements on completed structures can lead to a somewhat more certain result, thereby giving much smaller dimensions than would be the case where the ordinary calculation methods are followed. But still there will be sufficient safety. Such is the essential difference between my methods of calculation for the flat slab and those expounded in most regulations.

Maillart returned here to the old debate between Wilhelm Ritter and Franz Engesser from the 1880s: Swiss load testing versus German mathematical calculations. Maillart was summarizing his ideas and preparing to focus them on the twin impediments of monumental, imitative form and complex, piecemeal computations. He continued in the article:

The engineer should free himself from forms handed down by the tradition of older materials in order to reach the goal of fully using the materials in complete freedom and with a view of the whole. Perhaps then we can arrive, as with modern airplanes and automobiles, at a similar beauty with a new style derived from the properties of the material.

He characterized the calculation problem by the single term n, representing the ratio of stiffness of the steel and the concrete, and explained how it could be eliminated from the calculation. This term stood for the attempt by analysts to pattern the calculations after the theory of elasticity. But like the ether hypotheses of nineteenth-century physicists, attempts to explain actual full-scale concrete performance by using n were getting more and more elusive, so that the most sensible approach was, as Maillart recognized, simply to drop the idea altogether and to start again by reevaluating test results. Maillart argued the same basic case for form and formula; do not imitate stone, wood, or metal, and do not copy classical mechanics.

Following World War II, the engineering profession began seriously to reevaluate tests and carry out new ones to justify the simpler approach advocated by Maillart. They did not, however, take seriously the other half of Maillart's program to create a new style derived from the properties of reinforced concrete. Only a few engineers understood Maillart's ideas fully, and they have created new images by following the precepts on form and formula announced in Zurich in late 1937.[107]

Maillart in Germany
1937–1938

BATTLE IN BASEL

On February 10, 1937, Maillart entrained for Basel to give a talk on his own work[108] and that of Paul Bonatz, the German architect well known for his work on autobahn bridges.[109] Maillart had published in the *Bauzeitung* an article in April of the previous year, entitled "Some New Reinforced Concrete Bridges," in which he explained how his work differed in meaning as well as appearance from the recent bridges of Paul Bonatz's in Nazi Germany.[110] He was only beginning to formulate his revolutionary ideas about bridge design that he had put into practice 6 years earlier with the Salginatobel.

Bonatz's influence in Switzerland had begun almost directly after the World War I, when he served as a juror for several major bridge competitions. In 1937, Julius Hoffman in Stuttgart, Germany, published a book celebrating Bonatz's 30-year career.[111] An exhibition in Düsseldorf devoted to the "creative people" of the Third Reich led to another book in 1937 picturing the "classic bridges" of Bonatz designed for "eternity," as Hitler had described his Third Reich. Also included were idealized portraits of German workmen in the new social realist style borrowed from Stalin's Russia. The book ends with an illustration of a Roman arch bridge.[112]

The previous year, on April 1, just as the Basel Art Museum designed by Bonatz was being completed, the Basel section of the Swiss Society of Engineers and Architects had invited him to lecture – not about that large, stone-faced structure, but rather about the bridges of the Reichsautobahn, Hilter's new superhighways.[113] According to Bonatz, architecture was needed to beautify the engineering of Hitler's new highway bridges.[114] In his April 1936 article, Maillart questioned this concept:

The engineer decides only with difficulty to depart from the forms that conform to tradition and even if he might wish to, owner and public often follow him unwillingly. Small wonder then, if new types of structural forms, coming honestly and relentlessly (without looking to the past) from the nature of reinforced concrete, are more readily brought to completion in rather out-of-the-way places; the cities remain "immune," because there weight will be given to a certain "monumental" appearance. Even with rural reinforced concrete bridges an "architectonic" minimum are the abutments required to frame the structure and isolate it from its surroundings. To do away with this seems as revolutionary as to build a house without a base. In contrast to the oldest of the following examples, the Spital Bridge over the Engstligen, the others [Töss, Liesberg] all show the tendency to let the most integrated possible structural form grow directly from the ground. I feel that this "disregard" [for the monumental] is a gain and even those who cannot overcome their instinctive, i.e., atavistic, antipathy to the thin polygonal arch must venture to admit that the deck-stiffened footbridge over the Töss which directly connects to the rural stream bank gives a pleasing effect.[115]

Formal design of engineering structures, characterized by the works of Bonatz, required monumentality of a type, pleasing to many political leaders, that expressed permanence and grandeur. Bonatz saw this expression in massive, simple forms. He had been working at utilitarian objects since the mid-1920s with the dam barriers of the Neckar.[116] His approach appealed to many nonpolitical engineers both in Germany and in Switzerland who gave the weight of professional opinion and practice to underpin taste in Germany at that time.

For Maillart "architectonic," "traditional," or "monumental" forms were inconsistent with the very nature of reinforced concrete, with its potential for new, thin, forms, and for the intimate connection between structure and landscape. Even a half century later, the forms Maillart published in April 1936 look new. Except for the Spital Bridge, they all connect smoothly and without interruption to the river banks. They are not "natural" forms but rather provide a smooth transition between the natural and the man-made. Having already worked out his new structural forms, he explored ways to make the bridges appear more integral as well as more integrated with their surroundings. This was

just the reverse of the ideals put forward by Bonatz. The forms he was designing for the German autobahn were not integral: columns and beams appeared disconnected, with massive abutments rigidly separating structure from embankment.

It was this theme that Maillart expanded in the lecture he gave early in 1937. Over the next year, he would return to the theme of the monumental architecture-inspired bridges of Germany, which he saw as they were intended to be seen, as direct reflections of the new German state's consolidation of its power and monumentalization of its dominance. To Maillart, this amounted to forcing "atavistic" forms on materials incompatible with their properties.[117] Maillart faulted the very core of Bonatz's thinking – that of collaboration between architect and engineer on bridge design. He urged engineers on their own to think deeply about aesthetics as something arising from the structure and materials themselves.[118]

In early 1937, the design that would best characterize Bonatz's ideas was being completed for a bridge over the Danube at Leipheim. Maillart, however, did not mention that bridge or any other specific bridge by other designers in his lecture. Rather, he concluded the evening with a last slide showing his rejected proposal for the railway bridge in Bern. He criticized the official design, noted that his proposal achieved both visual and technical unity, and observed it to be 250,000 francs less expensive than the design that would be built. "Future generations may judge which solution was more beautiful and better!"

REST AND REWARD

Although the talk left him tired, Maillart's trip to Basel proved his ability after the broken thigh to travel alone, which he celebrated in Zurich the next day at dinner with Nissen and Mlle. Oechslin. On February 18, his old friend, the city architect, Herter, handed Maillart some sketches of a covered swimming pool that the city wanted built there in time for the 1939 National Exposition (Landesausstellung). It was to be one of the largest such works in Europe, and Maillart speculated that its size and complexity had led the city to give the design to "old Maillart." But rather than pleasure over a badly needed job, his first reaction was relief to have a valid excuse to keep his Zurich apartment.[119]

In July 1937, Maillart was diagnosed as being in a weakened condition and given strict orders to rest. After an initial revulsion at the idea of again being exiled from his offices, he gave in, and in late July, checked into the Hotel Waldhaus at Vulpera near Bad Tarasp. The spa was located on a beautifully wooded valley slope in the lower Engadine.[120] There he received word that in 1937, for the first time, the Royal Institute of British Architects had made an engineer an honorary corresponding member. In fact, two had been named: Maillart and Freyssinet. He was pleased, but his personal reticence prompted him to ask his family not to discuss the matter with others.[121]

At the end of 3 weeks (on August 14), Maillart returned to his routine.[122] Thanks to the Quai Turrettini and the restoration of a church in Geneva and the two projects in Zurich, his business limped along. But throughout the rest of the summer, he had few new designs and he spent his time mainly writing consulting reports on which he complained: "one loses one's good humor and one's money."[123] Early in September, the Swiss Society of Engineers and Architects celebrated their centennial in Bern with many honored speakers and participants, but Maillart was not among them. Once again the professional leaders chose to ignore him even if clients and foreign professionals did not.[124]

Even Bösiger could not prevent Maillart from getting commissions such as the Gündlischwand Bridge near Grindelwald – the kind of secondary road job Maillart was given because his designs cost less to build than those of his competitors. Started on September 8, it would be completed by December 20 at a cost of about 180 francs per square meter of roadway surface (length times width) as compared to the wooden bridge over the Schwarzwasser ordered that year by Bösiger that cost about 320 francs per square meter.[125] As Shand put it, to get "self-conscious stylistic titillations," Bösiger was always willing to spend much more money.[126] In a two-part article, the Englishman noted that

neither this country [Great Britain] nor America has had any share in all the recent pioneering work in concrete bridge design. That has been left entirely to continental engineers, and their chief stimulus, their main chance, has been the prosaic one of reducing cost. Of no country is this truer than Switzerland, where Robert Maillart, the most original bridge-designer living, has only been kept busy thanks to the world crisis – which has not prevented the almost gossamer delicacy of his supremely elegant bridges being fiercely attacked as shameful degradations of lonely Alpine valleys.[127]

The little Gündlischwand Bridge engaged the engineer because of its unusual form: skewed (crossing the river at a very sharp angle) and flat (the road crosses the river with very little space between structure and water).[128] In addition, the soil was unstable, further challenging Maillart's design talents. His solution, following similar designs at Liesberg and Huttwil, was a straight beam on either side of the roadway that itself was connected near the midheights of the beams. In this way, the beams extended partly below the roadway and partly above, where they served also as parapets. Unlike his two earlier bridges, Maillart widened the top of the parapet at midspan to provide more stiffness and to break up visually the otherwise heavy looking vertical outside surface of the beams. He referred to this unusual form as his *soucoupe* (saucer); it amused him to find a solution that was at once technically correct and visually curious (Fig. 167).

On October 24, the people of Zurich voted the funds for a new Congress Hall, and by the end of the year, Maillart was under considerable pressure to complete designs for it because the contractor was already driving foundation piles. His part of the big project was the most structurally complicated, and with almost no help in the Zurich office, he was constantly called away from his desk to inspect field work in progress.[129] As the year ended, Maillart's office in Geneva was almost out of work. As Christmas approached, the family began to drift away on its own holidays, and by December 24, Maillart found himself once more alone. Even

Rosa's usual gala Christmas dinner the next day seemed different from past celebrations, with only Robert, Rosa, and Marguerite. The following week Maillart was alone on New Year's Eve.[130]

He again received bad news from his accountant: Business had continued to decline. A bad year generally in the midst of the Depression, 1937 had been the worst year for Maillart of the past 15. In 1936, he had begun twenty-eight new projects in Geneva, whereas in 1937, there were only nine.[131] He was back at the low point that had followed his penniless return to Switzerland in 1919. Six months later, he would again refer to 1937, when in June 1938, an urgent letter came from Edmond. As Maillart had feared, his unemployed son had run out of money. He telegraphed him $500 the next day and then in Zurich on Friday he began to compose a long letter. In sending Edmond this money, Maillart did not want his son to "conclude that you have a rich father." Maillart stressed how badly 1937 had been and that he could only exist "thanks to my moderate way of living." Then he revealed something that he had never spoken of before. Following the broken leg, his insurance had provided him with a significant sum of money, "therefore," he wrote to Edmond, "my contribution [to your finances] comes not from my good head but rather from my bad leg." He then cautioned Edmond that "proceeds of this sort are of course out of the question for the future. . . ." The insurance company had refused to sell him any further disability insurance, so that, Maillart warned, Edmond should not imagine that future accidents could lead to future income. Even within the seriousness of the situation, Maillart could not resist the ironic idea that his accident allowed him temporarily to support his eldest son.[132]

Despite slow business, Maillart's reputation continued to extend in three different directions. As a designer in Switzerland in 1937, he was still the most respected engineer working in reinforced concrete. As a researcher, he had attracted serious attention for fundamental ideas. Finally, as an engineer striving for elegant designs, he had received more attention than anyone else in avant

(a)

(b)

Figure 167. Lütshine River Bridge at Gündlischwand, 1937. (*Source:* Madame M.-C. Blumer-Maillart)

garde circles, especially through the writings of Shand and Giedion.[133]

THE NATIONAL EXPOSITION

Early in 1938, when he had almost given up hope of keeping his three offices, the city of Zurich gave Maillart several important projects. A National Exposition, for which the projects were planned, was being organized for the following year. On Sunday, February 20, a city referendum approved the swimming hall mentioned earlier.[134]

Of far greater significance to Maillart and to the future of structures was a Cement Hall for E. G. Portland, the same cement research group for which

Maillart had done a roadway study in the late 1920s. It was a "small thing which even so required much work," he wrote to Edmond in February.[135] Its smallness was similar to that of the Schwandbach, and like the bridge, Maillart studied the form with great care. That summer he made a cardboard model of the hall that was pictured with elevations and plans in a long *Bauzeitung* issue (September 3) devoted to the 1939 National Exposition (*Landesaustellung*). It demonstrated how Maillart could turn a "small thing" into an original form of thin-shell concrete structure.[136] Nothing like this form had appeared before and it showed dramatically the extraordinary thinness (6 cm) possible with reinforced concrete when given an appropriate shape. Yet the small building was almost totally neglected in reports on the National Exposition abroad.[137]

BRIDGE FORM AND THE GERMAN REICH

In spite of Maillart's inability to get work in Germany, his ideas were crossing the Rhine and being taken up by a few perceptive German engineers. One example was Professor Ulrich Fischer in Breslau, whose academic study of concrete-arch bridge forms was in large measure stimulated by Maillart's recent publications, especially the 1936 *Bauzeitung* article that featured the Thur Bridge at Felsegg. In early October, the former Austrian journal, *Beton und Eisen,* published an article by Fischer on arch bridges that spoke at some length about Maillart's pioneering designs, asserting that "the extraordinarily thin slabs and ribs that Maillart uses are wonderful. . . ." Fischer also noted that Maillart's deck-stiffened arches were very complicated to calculate correctly and that "in reinforced concrete these have only been built in other countries [not Germany], where the calculations have been carried out in a simplified way. Load tests revealed substantially less deformation than these calculations predicted. . . ."[138] The German professor thus acknowledged that Maillart's insistence on simpler calculations not only helped promote new design ideas, but also led to structures that were safer even

than those simple computations predicted. Here was a clear statement of the validity of Maillart's ideas about seeing a structure as an integrated whole and about the German engineers' late discovery of these design potentials.

Not only could other engineers see the uniqueness of Maillart's newer forms, but at this time, his prewar ideas were being rediscovered, in France as well as in Germany. In March, Maillart's prewar chief engineer, Arnold Moser, since the end of World War I a consulting engineer in Paris, published a detailed article on flat slabs in which he gave Maillart's forms, his calculation ideas, and his experimental work of 1912. Moser showed further how widely used the Maillart flat slabs were now in France.[139] By contrast, Swiss academics maintained their indifference to Maillart's ideas.[140] In April, the Swiss journal *Strasse & Verkehr* published an article on the Vessy Bridge written by Maillart's engineer, M. A. Huber, who described the form, gave engineering drawings and photos, discussed construction and costs, and ended with a favorable review of the load-test results.[141] In the same issue, the journal editors wrote an article on the quai Turrettini then being completed. The unusual design of the quai as a hollow box once again showed Maillart's originality with form and his drive toward lighter but stiffer structures.[142]

A second major article in a Swiss technical journal appeared in May on the Vessy Bridge, emphasizing that this bridge was the first in the canton of Geneva "to be built according to the modern vision of reinforced concrete bridge structures" (Figs. 168 and 169). More importantly, it noted also that the "large Danube bridge at Leipheim for the Reichsautobahn had been built according to the Maillart system for hollow boxes." For the first time, these two works of very similar design, Vessy and Leipheim, were connected in a published article.[143]

In September, a German article on the Leipheim Bridge openly acknowledged its debt to Maillart. Considering the symbolic centrality of the Reichsautobahn for Hitler's Germany, the Swiss derivation of one of its principal bridges was remarkable. The *Bautechnik* article described how the Reichsautobahn administration had designed a concrete bridge and a steel bridge and then opened a competition with builders free to submit bids on either of these designs or on designs of their own. The most impressive submission was from the pioneering firm of Wayss and Freitag.

In the endeavor to find the lightest possible structure, this firm proceeded with the ideas taken from the Swiss bridge engineer Maillart, that in an arch wall without outrigger the upper spandrel over the abutment hinge is nearly free of stress and takes just about none of the load; one cuts this out and so gets without any loss of carrying capacity a not insignificant savings in weight, there results an extraordinarily light arch rib in which the building materials are used at their best in all parts.[144]

This is a clear description of Maillart's central design idea; but the German article went on to add that the form had been reworked with professor Emil Mörsch as consultant and with "all questions of visual form studied exhaustively through models in collaboration with Professor Bonatz." In short, Maillart's basic idea was reworked technically by Mörsch and aesthetically by Bonatz, the two best-known German experts in their respective fields. Never able to build anything in Germany, Maillart was now used as an exemplar for a major bridge in one of Hitler's pet projects. But Maillart's form needed reworking by Bonatz to make it an appropriate monument (Fig. 170).

Shortly after the German article appeared, Carl Jegher asked Maillart to publish his ideas on the Vessy Bridge design and the Leipheim's resemblance to it. Although never eager to write articles, Maillart found Jegher's proposal attractive for two reasons. First, he was pleased to see his ideas understood abroad and therefore welcomed the chance to explain his designs in the light of the German example. Second, he recognized in the Leipheim Bridge the hand of Paul Bonatz in the making of forms as monumental and nationalistic symbols that were far removed from his motives at Vessy (and indeed at Zurich for his Cement Hall as well). In both bridges and roofs, Maillart found strength in geometric form rather than in solid mass; his goal was to find forms that reduced forces

Figure 168. Arve River Bridge at Vessy by Maillart, 1936. (*Source:* E. Oberweiler)

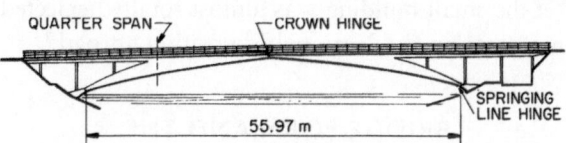

Figure 169. Arve River Bridge at Vessy by Maillart, 1936. (*Source:* Drawing by A. Evans and T. Agans)

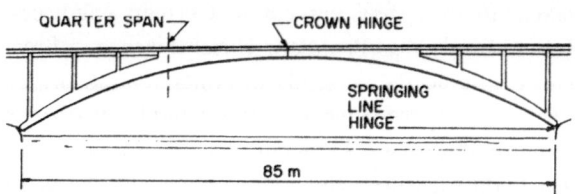

Figure 170. Danube River Bridge at Leipheim by E. Morsch and P. Bonatz, 1937. (*Source:* Drawing by A. Evans, T. Agans)

rather than shapes to express monumentality. He often said, "I build for eternity and if it falls before, I am awfully sorry."[145] It was a very different attitude from the posture of infallibility adopted by those associated with the Third Reich.

Maillart divided the article into three parts: a description of the Leipheim Bridge, one of the Vessy Bridge, and finally a comparison of the two.[146] For Leipheim, he was somewhat critical of its extra material but pleased that "through this work there has been a departure from worn-out paths" so that the designers made "better use of the nature of rein-

Figure 171. Bridge over the Arve River at Vessy. (*Source: Schweizerische Bauzeitung*)

forced concrete by an appropriate fusion of deck and arch." For the Vessy Bridge, Maillart emphasized its lightness, its exceptionally low cost, and its excellent performance under Ros's load test.

Maillart's writing takes on more vigor with the comparison, for here he culminated his yearlong worries about Germany and his feelings against monumentality. Whereas he considered his Arve River Bridge to be "a useful object of the purest type, the Danube Bridge by contrast has the character of a monument to the German *Reichsautobahn;* it aims not to be a pure useful object, but rather a representative landmark of the German art of building." By "pure useful object," Maillart meant one that had no extraneous material and no excess of cost, but in which a new beauty was still found through the aesthetic choice of the design. By contrast, at Leipheim, he wrote: "The beauty has been chosen to conform to the type of structure advocated by the collaborating architect, Professor P. Bonatz; the resulting structure has thus been made to fit with [his] aesthetic criterion." Particularly

striking was the smoothly curved underside compared to the broken arch form that he used at Vessy to express the hinge and to permit a deeper profile at the quarterspan (Fig. 171).

GERMAN THREATS AND MORAL REARMAMENT

Maillart's severely critical attitude toward German building reflected a widely held distaste in Switzerland for Hitler's Reich. Indeed, the neutral nation was now the only German-speaking country not under the iron boot of the Führer. Engineers were particularly upset by the fate of Austria. On October 1, 1938, the *Bauzeitung* announced that the venerable 90-year-old Austrian Society of Engineers and Architects had ceased to exist, its journal disbanded; "it had been quenched," along with about fifty other Austrian technical societies. Their activities had been taken over by the National Socialist Bund for German Technology with headquarters in Munich. The editors of the *Bauzeitung* barely disguised their consternation.[147] Austria henceforth would be a provincial extension of the Reich, with Munich, not Vienna, at the center of technical action.

Having watched with horror Hitler's invasion of Austria in March, the suppression of Austrian engineering in October, and aware now of Hitler's designs on Poland,[148] the Bauzeitung published a "call for the moral rearmament of Switzerland," which appeared in the October 29 issue. It proclaimed that service to the fatherland included service to humanity and that "thy will be done is not only a request for leadership but rather a call to action. Then thy will is our peace."[149] The reference to leadership (*Führung*) was directed toward the leader (*Der Führer = Hitler*) and the call to action was to the tradition of democracy and freedom in opposition to the direction in which central Europe was heading. The Nazis already were putting pressure on the Swiss-Germans to join the Third Reich. As the *Bauzeitung* call emphasized, "the world is watching Switzerland as the oldest existing democracy and expects us to be an example of true democracy. . . ." Switzerland would not stand for an Anschluss, nor would it let other democracies permit its dismantlement; it was rearming and without breaking its neutrality, it was communicating its resolve to the Germans.

CHAPTER NINE

Honored at Last

1938–1940

The Last Visit

1938–1939

REUNION AND CURE

Maillart was at the station in Geneva to meet Marie-Claire and her 8-year-old daughter on Monday morning, July 4, 1938. The two young Maillarts were returning after an absence of 4 years. To Marie-Claire, her father looked pale, thin, and slightly bent; his eyes had lost none of their sparkle, but she was struck by his frailty. They went to his office-apartment and spent the afternoon together with other family members. For supper, as Maillart had done also for Marie-Claire's first visit, with the help of Amélie, he prepared a feast, highlighted by caviar and well-chilled vodka. They drank to those no longer with them and to Eduard still in Tarakan.[1]

After a visit with Nissen, Mlle. Oechslin, and Alexandre in Zurich, the three left on July 10 for the Waldhaus Hotel in Vulpera for Maillart's 3-week cure.[2] True to character, on his first day, Maillart established a routine. Early each morning he walked with Marie-Claire and his granddaughter down to the valley where he was one of many who strolled around (Fig. 172), sipping the foul-tasting mineral drink to the tunes of a little orchestra.[3] They returned by bus to the hotel where Maillart then rested while Marie-Claire went off to play tennis or golf. Maillart enjoyed the company of his first grandchild (Fig. 173). In the evenings, dressed formally, they attended balls and concerts. While

Marie-Claire danced, Maillart sat in an easy chair, observing, smoking his pipe, and smiling in his slightly bemused way. Others would come up and talk to him, but he made no effort to retain their

Figure 172. Maillart and his daughter, Marie-Claire, in front of the Mineral Water Hall, Vulpera, July 1938. (*Source:* Madame M.-C. Blumer-Maillart)

company. Marie-Claire was surprised by how well known he had become.

THE BRIDGE CHAMPION

Mainly English, Americans, and Swiss patronized the fashionable Waldhaus and many of them were addicted to contract bridge. The big annual tournament, called the Challenge-Cup of Vulpera, took place toward the end of their stay and Maillart off-handedly asked Marie-Claire if she would like to enter as his partner. She had the impertinence, not unusual in their relationship, to ask, "are you really up on the latest rules? Are you not still used to the old way of playing?" Cut to the quick, Maillart shot back at her, "all right, this evening you will not go dancing, rather you will come to my room and teach me all the latest rules."

Marie-Claire had been playing bridge continuously in Indonesia and had become something of an expert, winning many competitions there. By 1938, the game had become immensely complex compared to the simple version of auction bridge that she had played so often as a child with her father in Russia. Maillart's bridge since then had been mostly with the ladies in Geneva, where the competition was slight. He had no experience with duplicate bridge let alone in a tournament.

That evening, she explained the most recent bridge conventions to her father, who sat attentively, saying little, his eyes half closed in concentration. They sat in his narrow room facing each other across a table. Without cards or writing, Marie-Claire explained the new conventions. From time to time, Maillart would ask a question. The whole session lasted barely 2 hours. Although she did her best to bring him up to her level, she had little real hope of succeeding.

It was, therefore, with meager expectations that they sat down to their bridge table the next day with sixty-two other contestants, as the all-day grand tournament began. It was a great surprise, both to them and to everyone else, when the announcement came that first place had been won by M. Maillart and Madame Blumer.[4] "Picci and I

were astonished, because we were thoroughly unknown outsiders," Maillart later wrote to Edmond.[5] Shortly afterwards, the vacation ended. Marie-Claire stayed on a few days to compete in a golf tournament, so Maillart left alone on July 31 for Geneva.[6]

FINAL FAREWELL

On Monday, August 15, an American architect visited Maillart at his Geneva office to discuss a New York project (which never materialized). Maillart asked the architect who had recommended that he visit him. The American replied, as Maillart wrote Edmond, "that had been unnecessary since my name is known and appreciated in every technical

Figure 173. Maillart and his granddaughter, Marie-Claire Blumer, at Vulpera, July 1938. (*Source:* Madame M.-C. Blumer-Maillart)

Figure 174. Marie-Claire, ·Robert, Eduard Blumer, and Elspeth Blumer, fall 1938. (*Source:* Madame M.-C. Blumer-Maillart)

school in America!" That architect's strong belief in Maillart's American fame was, in fact, misplaced; in almost no American engineering school was he heard of in the 1930s. Undoubtedly, Maillart was familiar to him through the writings of people like Giedion and Shand.[7]

In September, with Eduard returned, the Blumers accompanied Maillart to an engineering meeting at Chillon. One event there was the remembrance of the late Karl Hilgard (1858–1938). He had been the last man alive to serve as an assistant to Carl Culmann, and had gone on to a distinguished 15-year career of railroad building in the United States. Returning to Switzerland in 1897, Hilgard became a

professor at the ETH, but unlike Maillart, after his return to Switzerland from Russia, Hilgard became a deeply unhappy man when he found it impossible to adjust to the narrow possibilities of his native land after the excitement of huge building projects in an immense, rapidly developing nation. Maillart, too, had completed large works abroad, and he also missed the large-scale buildings of his early career, but after a few difficult years in Geneva, he had adjusted to the limitations of his life there, finding within it new directions for his creative energies. Hilgard was never able to do this and, in the words of Carl Jegher, the "shadow" that hung over his last years made him seem "tragic."[8]

In early October, the Cement Hall scaffold went up and Maillart took the Blumers to the lakeside to show them this most recent design of his.[9] The occasion fit neatly into his fall routine, which consisted of 3 days spent nearly every week in Zurich with the Blumers. Apart from one small bridge project, he had no work in Bern and his Geneva office limped along. Still the big Zurich designs, now under construction, promised to make 1938 almost profitable.

Later that month, Maillart proposed that he again conduct a small excursion to the Graubünden, this time to see his Salginatobel Bridge. Neither Marie-Claire nor Eduard had ever seen it and so they planned a 2-day trip that included Eduard's sister, Marieli Vögeli-Blumer, and her daughter Elsbeth.[10] They drove by the Walensee, to Chur, and then over the Julier pass to St. Moritz.

The next morning they followed the Inn River to Zuoz, where they stopped to see Maillart's 1901 bridge before going over the Flüela pass into Davos, down to Klosters by his railroad bridge of 1930, and finally to Schiers. From there they took the narrow little unpaved road up the Schraubach valley to where it crosses the ravine (Tobel) of the Salgina brook. There they dismounted and gazed at the bridge, its white concrete form beautiful and powerful when seen from the meadow to the south of the bridge.

Eduard's sister took a picture of the group (Fig. 174) and then Maillart stepped away, toward the

Figure 175. Construction of the Cement Hall, October 29, 1938. (*Source:* Prader and Company)

bridge, while the others hung back. For a while he stood, hands behind his back, slightly stooped since his accident, staring intently at the 8-year old-bridge. It had been a long climb from Zuoz to the Salgina, 30 years during which time his life had changed radically. What went through his mind as he gazed at that bridge? To Marie-Claire, her father seemed subdued, pensive, even somewhat sad. She sensed that he knew himself to be seriously ill. The excursion had not been joyous, not full of the usual banter and wit typical of so much of their previous time together. When he turned back to them, walking with the slight limp, they quietly returned to the car, and drove back to Zurich.

In November workers sprayed concrete onto the wooden forms for the Cement Hall by a process known as "gunite" (Figs. 175 and 176). Maillart frequented the construction, but was decidedly annoyed that the instability of his left leg made him feel uneasy at a building site.[11] By December 1, the concrete shell stood free of its wooden forms and Maillart took his daughter to see it.[12] As the time for her departure approached, Maillart arranged such technical visits and also discussed a much more ambitious trip, stimulated by the largest design project in his Geneva office. Maillart had been working for the Swiss firm, Baur and Company, on a large apartment house to be built in Colombo, Ceylon. He had begun in late 1938 to talk to the Blumers about the possibility of his visiting Ceylon to see it and thence of course completing the trip via Indonesia.[13] This idea seemed to ease his mind and help compensate for the impending isolation.

On Monday, February 6, 1939, Maillart celebrated his sixty-seventh birthday in Geneva. Early the next week, the Blumers, accompanied by Nissen, took the train from Zurich to Geneva, where they picked up Maillart, and continued on to Genoa, where their Dutch ship, the Mornix van St. Aldegonde, awaited them.[14] It was a painful time for everyone. Marie-Claire found this separation harder than any previous one because of Maillart's age, his failing health, and the political uncertainty.[15] Late in the day, the ship pulled away; they waved to each other until distance dissolved their signals.

Figure 176. Construction of the Cement Hall, November 11, 1938. (*Source:* Prader and Company)

1939

PENEY

During the Blumers' last month in Switzerland, Maillart's most important project was a new bridge across the Rhône between Aire la Ville and Peney, west of Geneva. The canton had decided in 1938 to build a dam downstream from those towns to be connected by a new bridge across the higher and wider waterway behind the dam. The canton gave Maillart the design commission, but decided to hold a special competition for the construction of the 187-meter-long crossing. Contractors would be permitted to bid on Maillart's design, or on one of their own, or on both. A jury would then decide on prizes and give a recommendation for the construction contract award.[16]

Maillart turned to this design early in the new year and produced final drawings by January 17.[17] Ever since the Liesberg design in 1934, he had been thinking about beam bridges of three spans – one longer span in the middle and two shorter symmetrical spans on either side (Fig. 177). At Peney, this meant a central span of about 75 meters and side spans of 53 meters, about double the dimensions of his previously built longest-spanning, beam bridge at Gündlischwand. The Peney design is startlingly original (Fig. 178). Maillart took elements devised for earlier works and, as at Salginatobel, combined them into a unique new work. He made the beams and roadway into a hollow box by adding a lower slab but only near the two river supports. In the center half of the midspan and over the outer parts of the side spans, the section is an inverted U shape (Fig. 179). He thus opened the box in a manner similar to the Salgina (and Tavanasa before), except now the box openings were in the bottom slab. He also curved the walls of the beam in the vertical plane to make them much deeper over the river supports and far shallower elsewhere. Finally, he designed the two river sup-

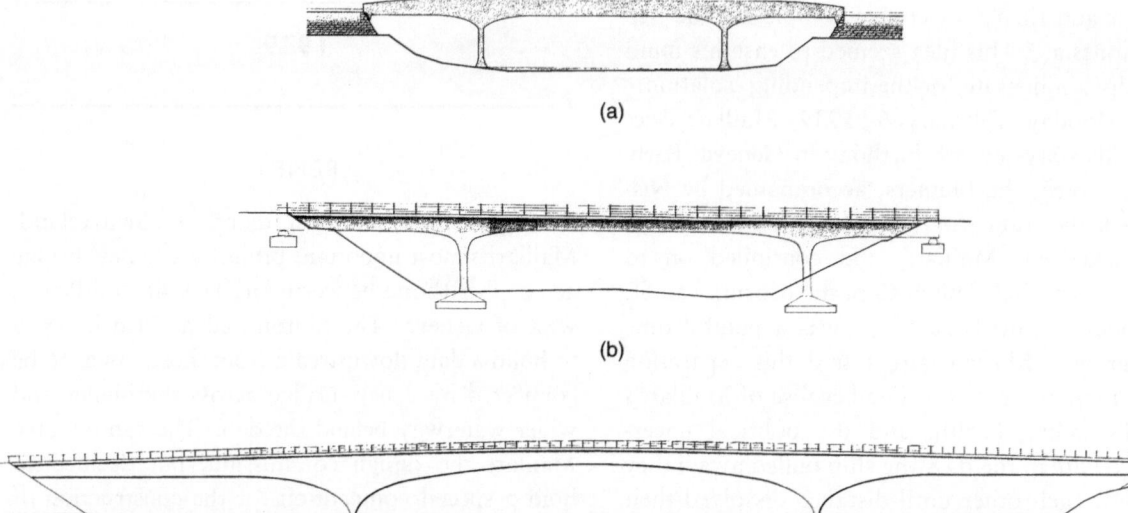

(a)

(b)

(c)

Figure 177. (a) Liesberg Bridge, 1935. (*Source:* Drawing by Mark Reed). (b) Weissenstein-strasse Bridge, 1938. (*Source:* Drawing by Mark Reed). (c) Peney Bridge, unbuilt, 1939. (*Source:* Maillart and Company)

ports to almost vanish in a thin profile (Fig. 180). This profile is a haunched beam with an exceptionally light center region and with thin extensions toward each bank.

The design contrasted starkly with Maillart's other bridge of January 1939, the Unterwasser near Gadmen on the new Susten pass road in the Bernese Oberland.[18] There he had designed a three-hinged arch only to have his nemesis Bösiger order that it be faced in stone.

Bids on Maillart's Peney design and on alternate designs in both concrete and steel were submitted on May 12 to the jury, which included Braillard, recently defeated public works commissioner and Maillart's good friend. It also included Professor Paris of Lausanne, and architect F. Gampert, president of the new Architects group on international matters, to which Maillart had been appointed. On June 26, after posing questions to Maillart, largely based on criticism of his design made by those submitting alternates, the jury proceeded to rank the entries. They awarded first place to a five-span continuous-beam bridge in reinforced concrete for which the bid price was 324,000 francs. The lowest bid for Maillart's three-span bridge was 396,000 francs. Both bids were by the same contractor, Bertelletti. While awarding the project to Bertelletti for the five-span design, the jury felt constrained to apologize for rejecting Maillart's three-span design. "If the commission envisages therefore the construction of a variation in reinforced concrete, this does not mean that it fails to recognize the technical and aesthetic advantages of the official design of M. Maillart and of the variations in steel."[19]

Maillart's reaction was perhaps best summed up in the *Bauzeitung* later that year, when it noted that

Figure 178. Rendering of the Peney Bridge. (*Source:* Maillart and Company)

the cost differential was essentially due to the span differences, spans of 38.5 meters being less costly than those of 75 and 53 meters. The public authorities either should not have permitted five spans or they should have expected such a cost difference and judged the bids on the basis of a budget for the official design. Maillart's design, for a 75-meter-span bridge, would have cost about 212 francs per square meter of bridge deck surface, a reasonable figure for such a span length. To get some idea of how reasonable the low bid of 174 francs per square meter (for the five-span design) was, we can take Maillart's Weissensteinstrasse Bridge of 1938, which has about the same main span length and which cost 132 francs per square meter. There is little doubt that Maillart could have made a design of five spans that would have been less costly, and far more elegant, than the winner.

In addition to the disparity in span, Maillart criticized the aesthetics of the shorter-span bridge. The *Bauzeitung* expressed this by stating that the winning design appealed to the jury because of the "present view with its predilection for straight beams, which is an appropriate form for steel bridges . . . [but which] appears very oppressive [in reinforced concrete] when compared to the elegant official design [by Maillart]."[20] The *Bauzeitung* implied that in allowing a shorter span, the Peney jury had been guilty of changing its rules during the judging, that it had rejected a truly outstanding design, and that it had followed the predilection for straight beams then characteristic of so many

Reichsautobahn overpasses of the late 1930s.[21] Here was a case of a cheaper concept with no aesthetic distinction taking precedent over a more generous idea with a unique elegance. Unlike Vessy, where a column in midstream might have been dangerous and thus the span was determined by the waterway, at Peney, the calm water behind the new dam allowed numerous supports limited only by the need for a small navigable channel.

The jury's justification for the winning design demonstrates the vacuity of a vague aesthetic. The five-span design has, the jury wrote, a "simple and tranquil aspect in harmony with the character of the countryside [*aspect simple et tranquille en harmonie avec le caractère du paysage*]."[22] What can such language mean? The countryside consists of green banks gently rolling down to the river. How is a somewhat heavy white concrete beam completely straight for 186 meters in harmony with such a scene? Straight beams had become popular for the Reichsautobahn overpasses and they reflect a formal, nonstructural, approach to bridge design. Compare this formal approach to Maillart's Weisensteinstrasse Bridge where, as at Peney, there is play in the underside curves and the lightly tapered extra-thin supports.[23]

The stiff, straight form marching dutifully across the dammed Rhône characterized for Maillart the playless form of a political system committed to order without freedom. It may seem tenuous to emphasize Maillart's technical-political view and to apply it to a harmless little bridge destined for well-

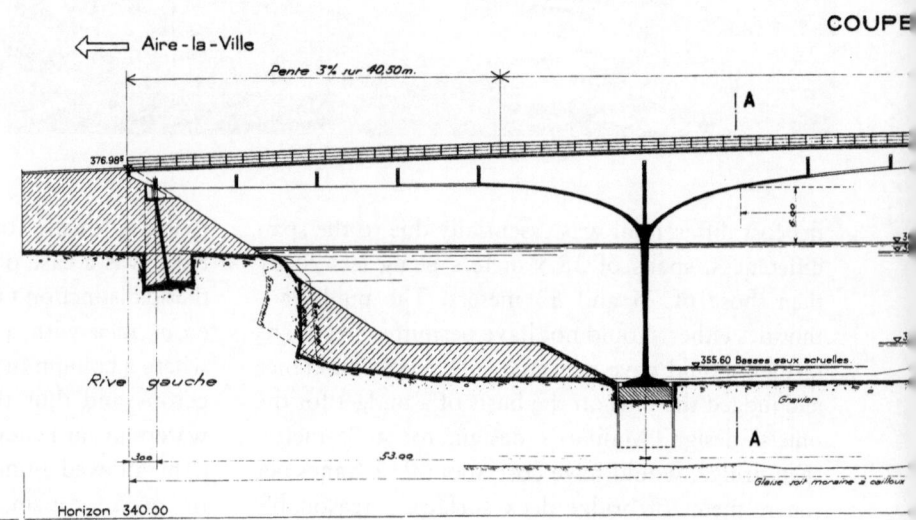

COUPE

Aire - la - Ville

Pente 3% sur 40,50m.

A

376.98⁸

355.60 Basses eaux actuelles

Rive gauche

Gravier

Horizon 340.00

Glaise soit moraine à cailloux

SITUATION
1:2000

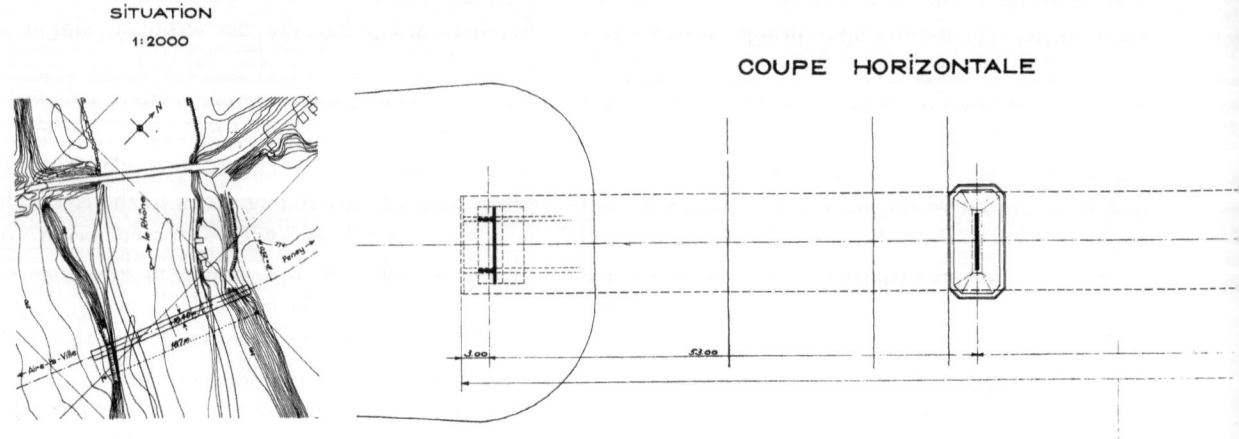

COUPE HORIZONTALE

Aire-la-Ville

Figure 179. Peney longitudinal section. (*Source:* Maillart and Co., Drawing No. 386/19, January 17, 1939)

LONGITUDINALE

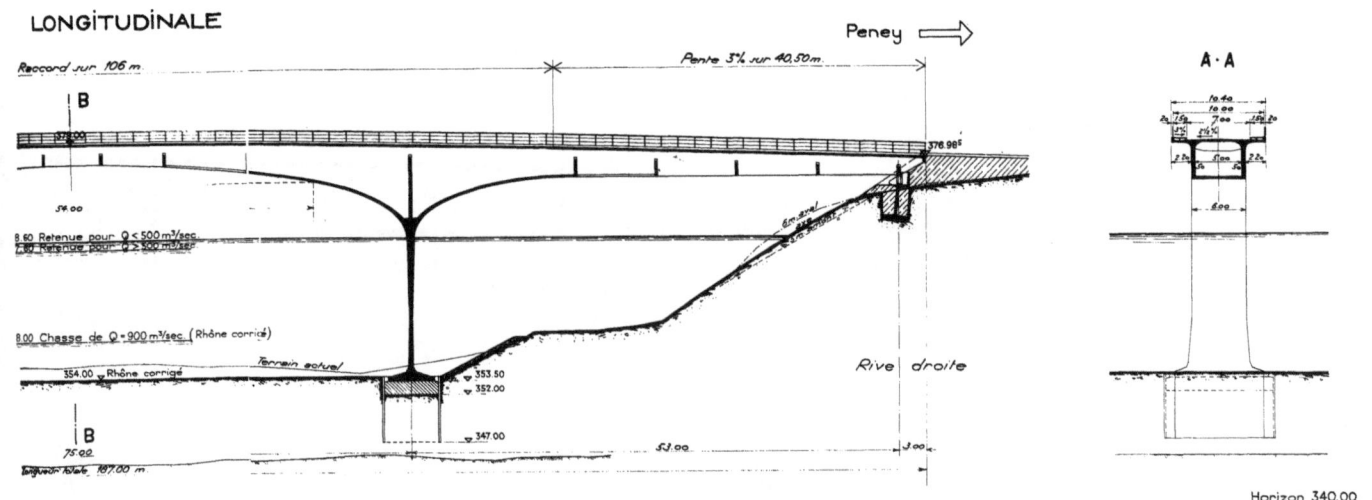

VUE PAR DESSUS

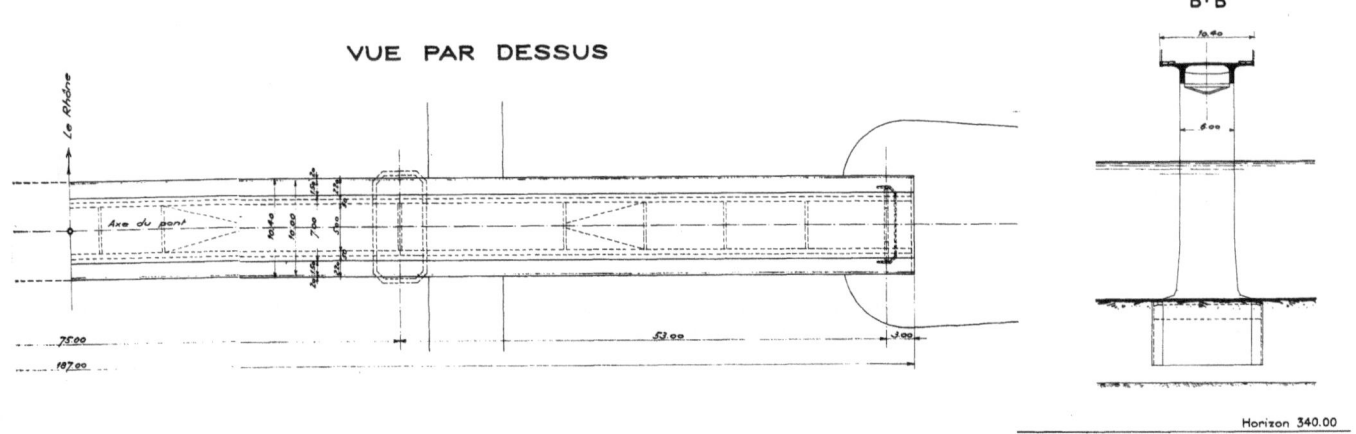

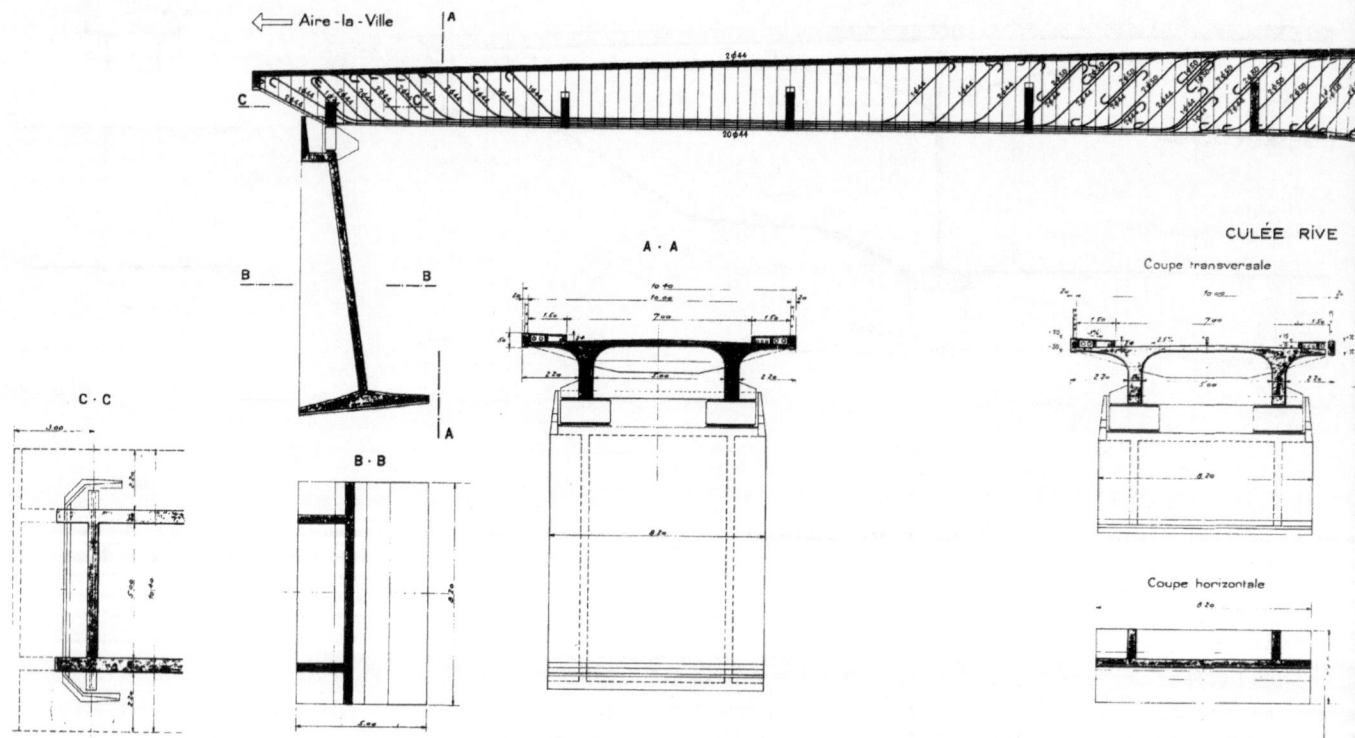

Figure 180. Peney Bridge: reinforcing details of girder and main column. (*Source:* Maillart and Company, Drawing No. 386/20, January 17, 1939)

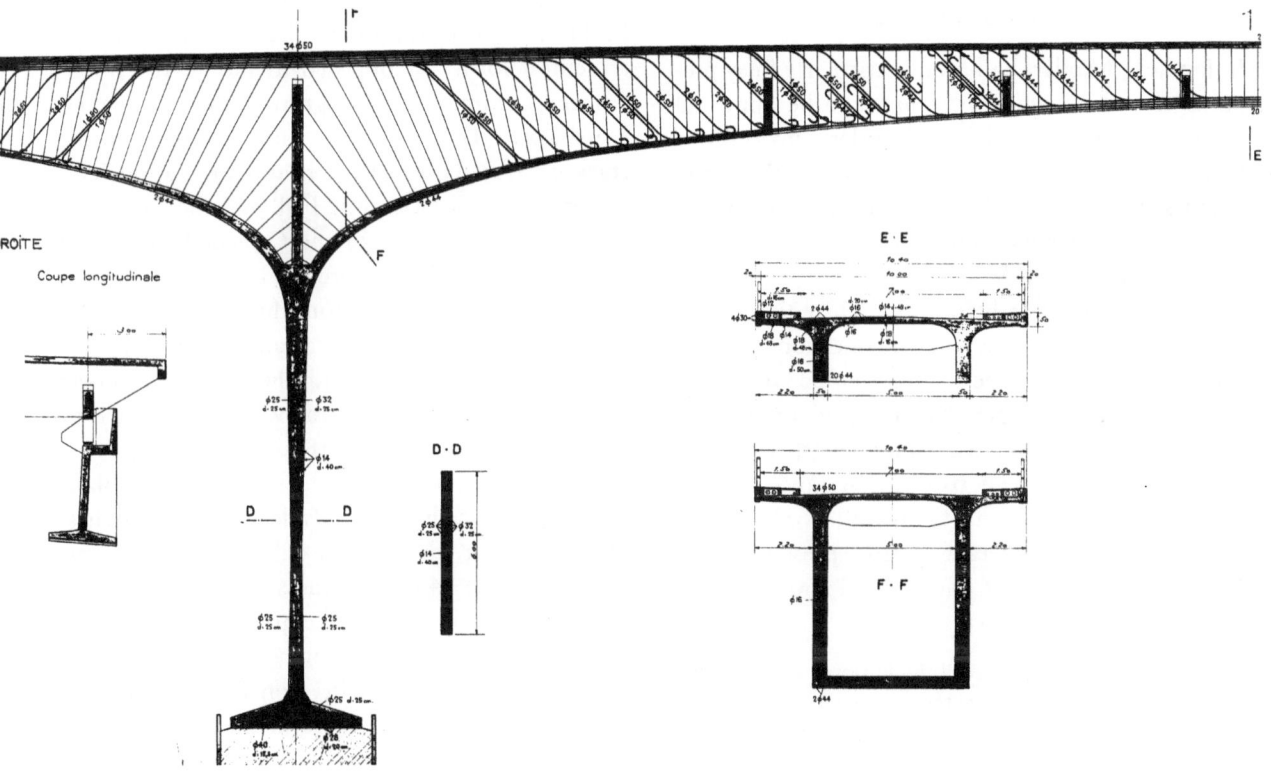

DROITE

Coupe longitudinale

E·E

D·D

F·F

deserved oblivion in Geneva; but such a view was part of his temperament and, especially, part of that independence of vision found in Switzerland. In Germany, there was no journal like the *Bauzeitung,* no set of tests like those recorded in *Bericht 99,* and no writing like Maillart's New Year's Day article of 1938 or his essay comparing Vessy to Leipheim.

AUTHORITARIAN THREATS VERSUS A MASTERPIECE

In January, Maillart had been particularly busy preparing for the *Landesaustellung,* and he appeared weary to those around him.[24] Hitler's invasion of Czechoslovakia in March and Mussolini's subsequent seizure of Albania gave him cause for worry about Alfred Maillart and his Italian wife, who were still in Rapallo.[25]

As it had the year before, the *Bauzeitung* again opened its April 8 issue with the question of "Moral Rearmament." The editors noted the wide response to their September 1938 moral proclamation and they quoted at some length from Dr. Carrard, a well-known psychologist and head of the Institute for Practical [Industrial] Psychology at the ETH. For the Jeghers, Carrard represented "Ein Psychotechniker" [their underlining] and they concluded with his summary:

What the world and above all the business world needs, are pioneers, men, who have the courage to take the way God wills. We do not need a "Führer" unless we mean by that God the Almighty rather [we need] several leaders who show us his way and go courageously along it themselves. It is a question therefore of building up new relationships with transformed men. Will you, reader, help with this, or do you know a better way?[26]

This peculiar Swiss rejection of central authority explains in part how it was that Maillart could succeed so well in building pioneering structures. The distinct political regions felt themselves relatively independent; they resisted central authority and authoritarian personalities. This attitude had allowed Maillart to build in the Graubünden, perhaps the most independent-minded of the cantons; and it accounted for his success with the secondary road authority in Canton Bern, which could not be controlled by the canton building director, Bösiger. It was this road authority that had allowed the Schwandbach Bridge to be built in spite of its unauthorized calculation methods.[27]

On March 11, the *Bauzeitung* had published a long description of the Zurich Exposition, including a two-page spread on the Cement Hall with illustrations of construction and of the completed structure: "designed by R. Maillart and built finely by Fl. Prader." The editors extolled its "fully exposed structural form" as opposed to the nearby brick-covered Ceramic Hall, designed by the architect H. Leuzinger.[28] They especially called attention to the stark contrast between the heavy foundation reinforcement for the Ceramic Hall and the light reinforcement at the base of the Cement Hall, which sat on four slender, tapered columns. Maillart in private referred to these "four thin legs" and the "6 cm thick vault" as having "a rather daring appearance."[29] He was pleased with the look of this novel structure in which all parts worked together to make an integrated whole (Fig. 181).

The profile view shows the shell's extreme thinness, which recalls Maillart's deck-stiffened arches. Here, however, the shell is only 6 cm thick compared with the 14-cm thickness of the Töss Bridge, his thinnest arch. This shell expresses visually what Maillart had expressed verbally in his 1931 article on "Mass or Quality in Concrete Structures": where "lighter, less costly, and thereby more durable construction, [means] in other words . . . high quality work."[30]

But having merely lighter structures can lead to disaster if the designer does not correctly visualize how the form will carry the loads. The deceptively simple Cement Hall is a highly sophisticated structure whose performance could not easily have been predicted even by experienced engineers. The way Maillart approached this design provides a summary of all his ideas on structure developed since the Zuoz design 38 years before.

He envisioned the shell in two parts: the upper curved shell as a thin arch and the lower slanting walls as cantilever beams (Fig. 182). The thin arch

Figure 181. Cement Hall, completed. (*Source:* Prader and Company)

Figure 182. Cement Hall, drawing. (*Source:* Drawing by Mark Reed)

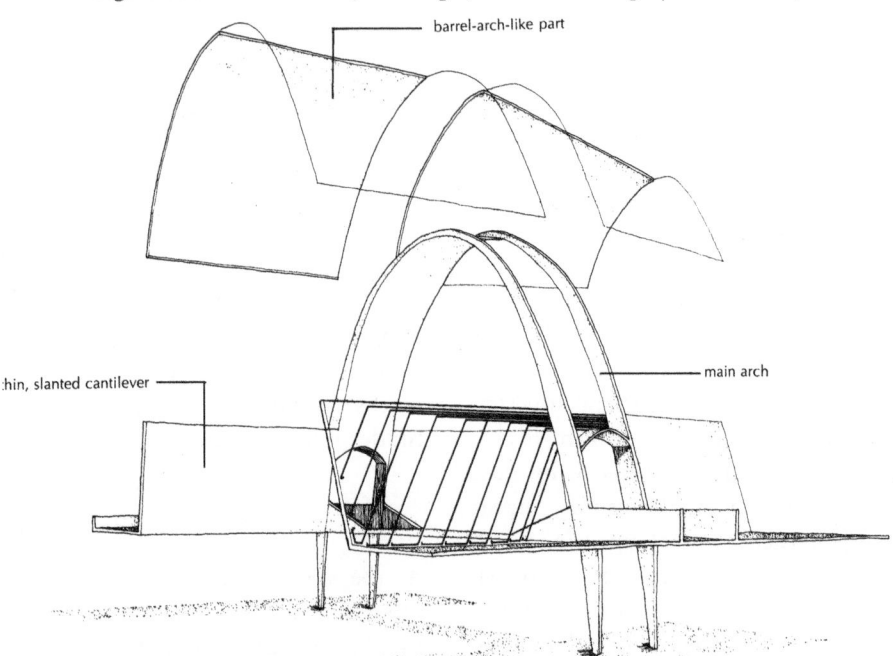

easily carried its own weight plus any snow load, just as does the thin arch at Töss, but unlike Töss, there was no deck above to stiffen the even thinner arch below. The lack of bridge loading permitted less material, but mainly it was the deep arches midway back from the shell edge that stiffened the slender roof.

Maillart marked off the lower parts of the 11.7-meter-high shell – almost straight but slanted walls each about 4 meters high – and provided heavy reinforcement to permit them to carry the entire thin-arch load as beams cantilevered from the stiff arches. The beam reinforcement illustrates to the engineer just how Maillart made these lower parts carry the entire roof by the use of tension-resisting steel bars extending over the walkway opening and then dropping diagonally to transfer the load properly (Fig. 182). The concrete beneath the walkway carried the cantilevered-beam compression.[31]

On Wednesday, May 3, completion of the Congress Hall was marked by a special ceremony that Maillart attended.[32] (The Swimming Hall unfortunately was nowhere near completion and would not be finished until 1940.) The Cement Hall opened with the rest of the National Exposition 3 days later. Maillart's spirits were high: Not only was the Exposition a success in general, but particularly so was his Cement Hall. Maillart reported to Edmond: "Both technical people and lay persons gazed at it in astonished admiration!"[33]

It was, in fact, the most striking example of reinforced-concrete construction at that time. Its curved roof contrasted dramatically with the other chief contender for this distinction: the curved roof (with an 825-foot diameter) for Hitler's new Reichstag for which a model was made in 1939. The latter was clearly an aesthetic shape superimposed on structural form. Hitler conceived it as a steel skeleton covered outside and inside by false façades, which made the external masonry appear to be an ancient lanterned dome at gigantic scale.[34] Unlike Wren's ingenious three-tier dome of St. Paul's in London, Hitler's dome idea arose at a time when such immensity could be achieved in rein-

forced concrete and new forms were being developed by German engineers in this material.[35]

Because of the *Austellung,* the June 1939 central committee meeting of the International Association of Bridge and Structural Engineering was held in Zurich, where the principal item on the agenda was planning for the Third Congress of the Association to be held in Warsaw on September 3–8, 1940. As a member of the permanent central committee, Maillart attended the 4-day meeting in Zurich.[36]

The committee noted that the largest memberships were German, followed by the French, then Swiss. The association was thus fulfilling the mission envisioned by Rohn in 1922 in bringing together the belligerents from the Great War and thereby he hoped it would contribute to a peaceful future between France and Germany through Swiss mediation. The major symbol of this idea in Switzerland was the recently completed headquarters of the League of Nations in Geneva. Indeed, on the inside cover of the May 6, 1939, issue of the *Bauzeitung* celebrating the opening of the National Exposition, there appeared a full-page photo of the "Palais des Nations, Genève." By then, it had proven fruitless as an instrument for peace.

Less surprising but disappointing to many technical people, the engineering association, which met in Zurich, had likewise had no mediating influence at all; one might almost say that it had served negatively by keeping controversial issues submerged and by trying to emphasize personal relationships in the face of real political tensions. This conciliatory attitude typical of twentieth-century technical groups did not appeal to Maillart whose New Year's article on the Leipheim Bridge had raised the critical, but pertinent questions of how public works should be designed and of how political attitudes influence their appearance.

THE LAST BRIDGES

At the same time, Maillart designed a startlingly original bridge at Garstatt in the Bernese Oberland. In May, he completed the final drawings for the

Figure 183. Railroad overpass at Lachen, 1940: bridge profile. (*Source:* FBM Studio Ltd.)

new bridge (begun in 1935), for which he designed a hollow-box, three-hinged arch bridge, but this time with no curvature to the arch, which is made up of two triangles meeting at the crown in a "broken-arch" junction as at Vessy. The bottom line of each half is now no longer curved (or arched), but rather perfectly straight. The bridge is also skewed, further complicating the design.[37] Once again Maillart was striking out in a new direction, as he had done with the Cement Hall in 1938 and the Peney Bridge in 1939.

In late June, Maillart began what was to be his last bridge design. That month, the Swiss railways had solicited bids for two bridges over their rail lines along the south shore of the Zurich lake between the towns of Altendorf and Lachen. The railways had made a design and contractors were permitted to make bids on the basis of alternate designs. Having finished the Cement Hall, Florian Prader put together in July a bid based on new designs by Maillart.[38] In spite of his poor health, Maillart's design for the larger of the two bridges showed a freshness of vision.

For this 40-meter-span, skewed crossing at Lachen, Maillart projected a hollow-box, three-hinged arch similar to Vessy (Fig. 183). He wanted a flat, angular shape for the arch halves (Fig. 184), but he retreated from the uncompromising straightness of Garstatt. For the shorter bridge at Altendorf, he designed a three-span continuous beam because the spans are far less than at Lachen and the roadway is only one lane whereas Lachen is two. Altendorf continued the ideas Maillart had worked on since 1934 with Liesberg (then Huttwil and Gündlischwand): a relatively deep continuous beam over three spans with its total depth unchanged except at the two intermediate supports where small haunches make a transition between the beam underside and the thin columns. At Altendorf, the side spans are shorter than in any of the earlier beam bridges and the beam depth slightly greater, so the visual effect is of a powerful, even heavy, beam. Maillart sketches (a rare instance where they were preserved) show a downward curve in the end of the beam, but the bridge was built without it [Figs. 185(a) and 185(b)]. The result is a design that appears stiff, although relieved by the slender columns and gentle haunches below.

Figure 184. Railroad overpass at Lachen, 1940: bridge showing the skew, box arches, the hinge and buttress Support, and the expression of form through the pattern of form boards on the arch profile (*Source:* FBM Studio Ltd.)

By the end of the month, Maillart's physical disabilities once again took the upper hand: His doctor ordered a complete rest at Bad Gurnigel, forbidding even walks. Despite his strict regime, he was able to send off a detailed critique for a new bridge near Château-d'Oex with sketches to his chief engineer in Bern.[39] Maillart left Bad Gurnigel in mid-August,[40] at about the time the railways recommended that Prader and Company be awarded the contract to build the two bridges at Lachen and Altendorf on the basis of Maillart's designs. In addition to low cost, the railway report considered "both designs to be functional and outstanding in their beautiful appearance. The problem with the foundation for the arch bridge [Lachen] was very well solved." They added that "in spite of small dimensions the stability and stiffness are fine."[41]

On September 3, following Hitler's invasion of Poland, Britain and France declared war on Germany.[42] Swiss mobilization suddenly stopped almost all business and the National Exposition closed (it reopened a short while later). Maillart's entire office force disappeared into the military except for Stettler, who worked into the nights to complete what projects Maillart had in hand.[43] To complicate his life further, Maillart had to relocate just then in Zurich because the Bärengasse was at last torn down.

Maillart's life had been shaped by political events. Now, as if his physical condition mirrored Europe's political disorder, on Wednesday, October 11, he suffered a massive hemorrhage. Alone he called a taxi and, clutching a small valise, managed to get to the hospital, where tests revealed an infected kidney. The condition required immediate surgery, but there was a question about the ability of Maillart's heart to withstand such a procedure. With no other option available, on October 24, Maillart submitted to what seemed to be a successful operation. However, 2 days later, he suffered a

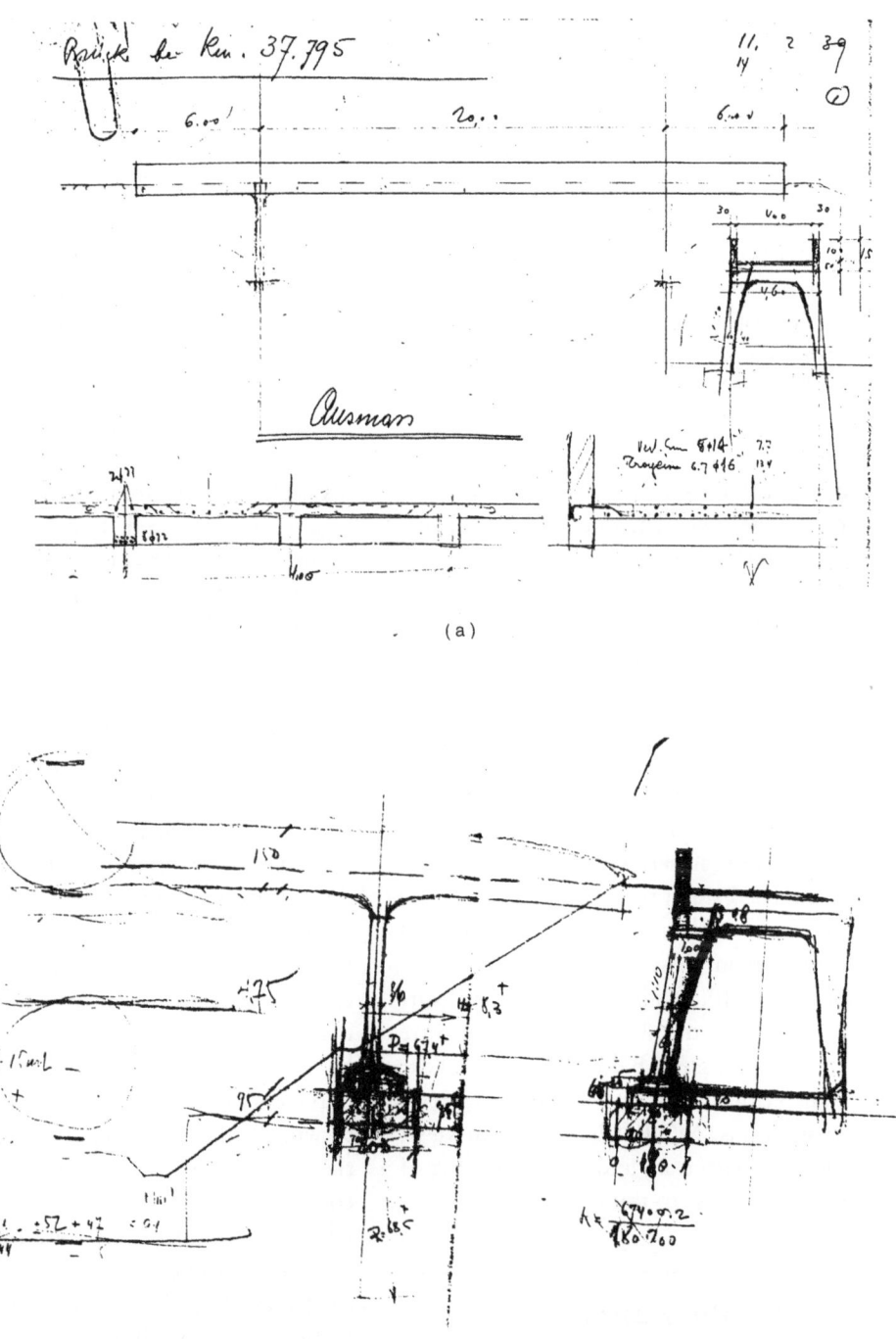

Figure 185. Maillart's studies for the Altendorf Overpass: (a) beam longitudinal profile with transverse view of supporting frame; (b) longitudinal haunching for beam profile and transverse haunching for supporting frame. (*Source:* Zurich Maillart Archive)

massive heart attack followed by another heart attack within 2 weeks. For a month, he hovered between slow recovery and severe setbacks.[44] At the end of that time, using all the strength he could muster, he wrote Marie-Claire his first letter in over 6 weeks; it was difficult, but he succeeded in describing his ordeal.[45] Finally, on December 5, he was able to leave the hospital: A family friend drove Maillart home in "pomp," as he wrote his daughter, to make the distinction with his lonely hospital entrance 8 weeks earlier.[46]

Maillart's friends and family were amazed at his recovery, the sharpness of his mind, and his almost healthy appearance. "It is a true miracle," wrote Rosa, who, with her usual devotion, was by his bedside most of the time. The doctor gave permission for him to return to his normal Bern–Zurich rounds on December 20 provided that Amélie accompany him.[47] They arrived at Zurich in good spirits and Amélie quickly put his new apartment in order. Maillart was pleased to be back at the center of the Swiss engineering world with Ros, Jegher, and of course Nissen. He marveled at the apartment's modern amenities that the colorful Bärengasse rooms had lacked. Maillart cared little for the new luxuries in themselves, but he saw clearly that in his weakened state, they had become necessities as was the fact that this apartment formed part of his office as in Geneva.[48]

On January 1, 1940, Maillart found himself surrounded by ten family members in a small private dining room of the Brasserie Bâloise. The New Year had begun auspiciously in Geneva and, in spite of wartime constrictions, Maillart had work in view, a new article in preparation, and most engaging of all, the Cement Hall test to think about.[49]

To the Master

THE AUTHORITIES

The article that engaged Maillart's attention at the beginning of 1940 addressed issues that he had been grappling with for 40 years. Reflecting on his persistent independence, he began with the question, "What are the constraints on the engineer?" He gave two: the laws of structure and the prescriptions of society. The first comes from "the professors, then the books, the journals and is completed by the engineer's experience born from observations of work done by the engineer or by others." The second comes from codes or rules aimed at "keeping him from being too rash or endangering public safety."[50] These are the disciplines of engineering. Now a lifetime of accomplishments provided Maillart with solid evidence for the validity of his contributions: his flat slabs, his hollow boxes, his deck-stiffened arches, and above all *economy of means*. He stressed that the typical twentieth-century process of codes, teaching, and public approvals had impeded the development of new forms. And he repeated his oft-cited example of Germany's handicaps in this regard.

Even near the end of his life, the very word "authorities" stimulated a vigorous reaction in Maillart. It brought out his ire, sarcasm, and humor. Despite his typical, Swiss bourgeois life, he had always relished being a rebel, professionally, against that tradition, whose essential image was the self-assured, humorless authoritarian whether Robert Moser, François Schüle, or, later on, Arthur Rohn, Max Ritter, or Adolph Bühler. For Maillart, each represented authority without imagination. It is almost as if he needed such people to bring out his most innovative ideas. They were his essential opponents in the game of design, which he set out to win. Without opponents, there can be no victories. These opponents always defeated him in closed competitions, on juries and in committees. But where the competitions were open – in wilderness communities like Schiers and Schwarzenburg or in the pages of the *Bauzeitung* – Maillart could score, often with such telling effect that his victory was complete.

If Maillart's career showed nothing more, it would say to the future, beware of all pronouncements from those who use their position of authority as their shield. Beware of all Führers, of all those around whom others fawn. But in so doing, beware also of the rebel without a positive record. Maillart could rebel because he knew how to build. His distaste for the authorities grew out of his successful structures.

He was playing a game with strict rules, consistent with his favorite parlor games and his general mode of thinking. But there is more to Maillart than that. During the 46 years of his career, he had had the rare chance to see an entirely new and commanding technology begin, develop, and mature. Only Emperger and Freyssinet were his peers in that sweep of events and neither of those pioneers ever ran his own construction company and design office. It does no disservice to the Austrian or the Frenchman to say that by 1940, only Maillart had integrated all aspects of design into a controlling vision.

By contrast to the authorities, Maillart's ideas came directly from built structures or from the practical need to win a construction or a design contract. He was less interested in what he considered secondary, such as a consistent general theory, a calculation for stresses, a detailed codification, or international congresses.

Intellectually, Maillart was something like those great natural athletes whose improvisations on the field can never be explained by rules, traditions, or training even though all are essential. Confronted with a solid wall arch at Zuoz, Maillart opened it up at Tavanasa; confronted by unexplained channel-beam test results, Maillart made a fundamental (shear-center) discovery in mechanics; confronted by cracks in the deck of the Aarburg Bridge, he saw a new deck-stiffened form; confronted by impossibly small budgets, he devised the Schwandbach and Vessy bridges. All of these creations baffled and sometimes antagonized the "authorities." Maillart derived part of his joie de vivre from confounding these people and he was sometimes quite undiplomatic in doing so. Indeed, Maillart could be frighteningly sharp and brusque.

There was a deep passion beneath Maillart's deceptively quiet bourgeois exterior. In the end, it was the passion that frightened the authorities because the bureaucratic mind cannot deal with it. When Maillart concluded his article on "the Engineer and the Authorities" with the italicized exclamation *L'economie des moyens!* he was in fact expressing a proven theory. Research is only valuable if it leads to that economy, teaching must center on it, codes must allow for it, and every design is to be judged by it. That was Maillart's integrated vision, the final results of which are those unforgettable forms that have outlasted all authorities whether they were professional in Switzerland or political in Germany.

THE THIRD ZURICH VISIT: REST IN PEACE

Maillart's most powerful propaganda tool in his criticism of Germany's massive and supposedly permanent works was the light, temporary Cement Hall. Scheduled for destruction in late January 1940, a surprising fate made it over the next half century one of the most enduring images in modern structural history of the potential for reinforced concrete. In spite of its physical disappearance, its photographic permanence is secure and its record of performance fully documented in a way that no German work of that period even approached. It would even become the stimulus for one of the most ubiquitous commercial symbols of the late twentieth-century America, the McDonald's arches (Fig. 186).[51]

Ros began on January 26 to carry out systematic and detailed tests, recording deflections and stresses under a wide variety of loadings. The tests continued on January 27 and were supposed to conclude 2 days later with a loading up to full collapse; but,

Figure 186. Diagram for the Cement Hall as the inspiration for the McDonald's Logo (*Source:* Maillart and Co.)

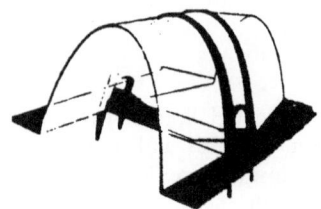

"due to the unexpectedly greater buckling strength of the two-rib stiffened shell," the structure could not be destroyed and Ros had to postpone the collapse test in order to get the explosives needed to blow up the recalcitrant roof.[52]

Ros was happy to have an excuse to postpone final proof of Maillart's talent until the weakened engineer could be at the site for the climax. This happened on Friday, February 2: Maillart stood by Ros as the latter gave the signal to detonate the walkway connecting the two supports for each arch [Fig. 187a]. The shock broke windows 100 meters away, but destruction of the structurally vital tension tie still did not bring down the thin shell. In defiance of professional calculations, it stood cracked but whole for 24 hours more [Fig. 187b]. Then, finally, on the next day, with Maillart and Ros again standing together, the demolition crew finally succeeded in pulling down the structure by attaching a grappling hook to the columns and tugging sideways until the ribs gave way.

Ros, dressed in his usual bizarre field-test garb, and Maillart in his long winter coat stood watching intently. Once all was quiet, Maillart moved slowly forward. He went alone straight for the rubble, picked his way through it, and peered silently at the mangled pieces of his design [Fig. 187(c)]. Some workers began climbing over the debris, joining the designer, who later commented that they came to "strip the cadaver with pneumatic hammers in order to recover the steel." The war had shot up the price of steel reinforcing bars already by 60 percent and thus the essential skeleton of the Cement Hall was saved and reworked into some other form by another builder. Maillart ended his own personal description of these events with "R.I.P.," rest in peace.[53] But he also observed that "the back of the elephant lived" and, therefore, the delicate eggshell ultimately became the powerful pachederm.

(a)

(b)

(c)

Figure 187. Cement Hall: (a) test with Ros and Maillart, (b) demolition with Ros and Maillart, (c) Maillart inspecting the ruins. (*Source:* Report No. 99 by Ros, 2nd Supplement, 1940)

RETURN TO GENEVA

Once before, Maillart had stood before a ruined design – in 1927 at Tavanasa, where fragments of his first masterpiece had lain strewn along the Rhine riverbank. After that experience had come the great Salginatobel, inaugurating the most creative decade of his designs. The ruin of the Cement Hall ended that decade. There was no trace of bitterness or nostalgia in Maillart's response to this planned demolition. He was still focused on future events and eagerly awaited Ros's report to study and to send to the Blumers.

In Geneva, the family prepared to celebrate the engineer's sixty-eighth birthday on Tuesday, February 6. Rosa provided duck stuffed with chestnuts, followed by bridge until midnight after which they all reassembled at Maillart's apartment for a final grog. As usual, sleep remained difficult for the engineer. In the following weeks, his condition deteriorated. His swollen left side appeared to be caused by a kind of hernia and he was obliged to continue wearing a disagreeable corset prescribed by the doctor. Since his cure at Vulpera in the summer of 1938, Maillart's weight had dropped from 174 to 149 pounds.[54]

By early March, Maillart was unable to leave his bed and was too weak even to write Marie-Claire. Amélie took the unusual step of reporting on his illness to the Blumers. Although he had told them of his general problems, he had left out much. For 3 weeks, he had shingles and furthermore the stitches from his operation had not held, so that another operation was inevitable. Maillart was miserable; he hated confinement and stubbornly resisted treatment.[55]

On the same day that Amélie wrote Marie-Claire, the structural engineering section of the Swiss Society of Engineers and Architects assembled in Zurich, and out of their membership of 200, they elected "Robert Maillart as their first and at present only Honorary Member."[56] This unprecedented honor, clearly created expressly for Maillart, was quite a surprise coming from those against whom Maillart had fought on so many issues. It was an action undoubtedly instigated by Ros and Jegher.

The architects had honored him 5 years earlier and now the more conservative engineers were catching up. Maillart's former rivals were continuing their own work, both presenting papers in Zurich the next day: Ritter on the new idea of prestressing and Bühler on his railway bridge in Bern.[57] These men refused to acknowledge publicly Maillart's greatness, but the profession as a whole had done so.

LIBERATED AND TRANQUIL

On March 12, Maillart was ordered back to the hospital for tests.[58] His weight continued to drop and the doctors now discovered an intestinal ulcer.

Figure 188. Rene, Robert, and Paul Nissen, mid-1930s (*Source:* Madame M.-C. Blumer-Maillart)

Figure 189. Tomb of Robert Maillart. (*Source:* Madame M.-C. Blumer-Maillart)

Figure 190 Paul Nissen and Eduard Blumer, circa 1960. (*Source:* Madame M.-C. Blumer-Maillart)

On Easter Sunday, March 24, René arrived from Aix and kept a vigil by his father, who by Tuesday seemed to have rallied.[59] Rosa now took up the correspondence that was beyond her failing brother.

René stayed close to his father throughout the week, lightening Maillart's otherwise oppressive hospital room. But on Thursday evening, March 28, he had to return to his business. He had buoyed up Maillart's morale, which remained good for Nissen's Sunday visit (Fig. 188). The two old friends talked at length; Maillart was out of bed and standing while his ever-active mind raced ahead, as usual, to future projects.

The next day he was suddenly worse. Fearing uremia, a blood poisoning usually related to kidney disease, the doctors telegraphed René to come back immediately. Amélie stayed with the engineer through the night. Tuesday was even worse, and on Wednesday, when Rosa entered the hospital, Maillart was barely conscious. Then in the afternoon, he suddenly murmured, "tea, coffee, Sumatra": his thoughts were with Marie-Claire, moving toward the long-planned trip that would have taken him to his construction site in Ceylon and thence to Sumatra to see the exotic setting of the Blumer's last 11 years.[60]

René came at noon on Thursday, April 4. his father opened his eyes to recognize his youngest child, then closed them again. He could no longer speak. René stayed as his father slept "like a child and it is thus liberated and tranquil that he departed at 4:00 A.M. exactly on Friday, April 5, 1940."[61]

With Edmond in the United States and Marie-Claire in Sumatra, only René represented Maillart's children. On Monday, there was a service at which the old Pastor Vallette officiated and Lucien Meisser, Maillart's chief in Geneva, gave a brief eulogy. Then they all moved to the Petit Saconnex graveyard, where Maillart was buried (Fig. 189). Paul Nissen stepped forward; Maillart's oldest and dearest friend (Fig. 190) spoke just a few words, ending with "a last adieu to my friend Robert." Condolences came from around the world, among them an immense arrangement of red and white flowers sent by Mirko Ros from the Federal Material Testing Institute. It bore a wide ribbon with the inscription:

TO THE MASTER

Notes

Chapter One

Copies of documents not otherwise located are in the Princeton Maillart Archive. All private family letters are in the Zurich Maillart Archive with copies in the Princeton Maillart Archive. The *Schweizerische Bauzeitung* is abbreviated to *SBZ*.

1. *Dictionnaire Géographique de la Suisse* (Neuchâtel: Attinger Frères, 1902), 1:220.
2. Ed. Sénemaud, *Généalogie de la Famille Maillart* (Mézières, 1868), 9 pages. See also *Biographie Nationale de Belgique* (1895), 13:171–86. The Belgian Maillarts moved in 1852 to Carouge, a suburb of Geneva, where their two sons Edmond and Hector grew up as Protestants.
3. *Dictionnaire Historique et Biographique de la Suisse* (Neuchâtel, 1928), 4:417. For the family details, see *Küpfer, a Genealogy of the Family,* Bern, pp. 13–15, 32–3.
4. *Küpfer,* 33. See also *Extrait de registre des baptêmes, Canton et District de Berne,* May 21, 1872. Further details of the family came from a letter from Gottfried Buchmüller in Bern to Marie-Claire Blumer-Maillart, April 24, 1950.
5. Wilhelm Oechsli, *History of Switzerland 1499–1914* (Cambridge, 1922), 271–2, and *Dictionnaire Historique* (1926), 3:187.
6. Johan Huizinga, *Homo Ludens: A Study of the Play Element in Culture* (Boston, 1955; completed in 1938; published in Switzerland in 1944), 10.
7. *The Chess Friend* (*Der Schachfreund*), publication of the Anderssen Chess Club in Bern. Publisher R. Maillart, Lorrainestrasse 38. There are seven issues of this publication. The dates are November 5, 19, December 3, 17, 1888, January 7, 25, and February 4, 1889.
8. *Dictionnaire Historique* (German edition, 1934), 7:512. Widmann's father, Josef Vicktor Widmann (1842–1911), who came from Austria, was well known as a critic and writer and a friend of many nineteenth-century composers. He wrote a book on Brahms.
9. E. Briner, "Kleines Feuilleton: Fritz Widmann und Robert Maillart," *Bund Morgenblatt,* Bern (April 27, 1940), also "Aus der Jugendzeit eines grossen Schweizers: zum Andenken an Robert Maillart," *Oberlandisches Volksblatt Interlaken* (May 1, 1940). For Widmann's career, see G. Gamper, *Fritz Widmann,* Zürcher Kunstgesellschaft, Neujahrsblatt 1938, Zurich. 32 pages plus 12 plates. After Widmann's death, Maillart came to a memorial exhibition of his paintings in the Zurich Art Museum, where he showed to those present a box filled with Widmann correspondence. The correspondence has not been found.
10. Maillart's school records; Zurich Maillart Archive.
11. For his matriculation exam, he received, on September 12, 1889, the grades of 4 in French, German, and natural history; 5 in history, mathematics, physics, chemistry, and drawing; and 6 in descriptive geometry. The grades went from 1 to 6, with 6 the best. He received, overall, a *good* in his grades and a *very good* in his deportment during the school years. In his one year in Geneva, 1889–90, his grades were 4 in practical work (watchmaking), 6 in design, and 6 in mechanics. He only took exams in mechanics and received $5\frac{1}{2}$. See documents in the Zurich Maillart Archive.
12. For a fuller discussion of the founding of the Federal Polytechnical Institute in Zurich, see David P. Billington, "Wilhelm Ritter: Teacher of Maillart

and Ammann," *Journal of the Structural Division,* ASCE, 106 (ST5) (May 1980): 1103–16. From 1855 to 1911, the school was called die Eidgenössische Polytechnische Schule (Federal Polytechnical Institute). In 1911, the name was officially changed to the Eidgenössische Technische Hochschule, or ETH (Federal Technical Institute). G. Guggenbühl, *Geschichte der Eidgenössischen Technischen Hochschule in Zürich* (Zurich: Buchverlag der Neuen Zürcher Zeitung, 1955), 136–37.

13. W. Ritter, "Karl Culmann," *Allgemeine Deutsche Biographie* (König: Akademie der Wissenschaft, 1903; reprinted by Duncker and Humblot, Berlin, 1971), 47:571–5.

14. K. Culmann, "Der Bau der Holzernen Brücken in den Vereinigten Staaten von Nordamerika" and "Der Bau der eisernen Brücken in England und Amerika," *Allgemeine Bauzeitung,* Vienna, 16 (1851): 69–129, and 17 (1852): 163–222.

15. C. Culmann, *Die Graphische Statik* (Zurich, 1866).

16. Ritter, "Culmann," 572.

17. G. Thurnheer, "Wilhelm Ritter," *Vierteljahrsschrift* der Naturforschenden Gesellschaft in Zürich (Zurich, 1906), 1:510–20, and E. Meister, "Professor Dr. W. Ritter," memorial in *SBZ,* 48(17) (October 27, 1906): 206–8.

18. Ritter and Ludwig Tetmajer, "Der Bericht der Eidgenössischen Experten über die Mönchensteiner Brücken-Katastrophe," *SBZ* 18 (October 31, 1891): 114–15. This introductory report and a series of later parts to the main report led to the Federal Law of August 19, 1892, concerning the calculations and testing for metal bridges and roofs for the Railway; see Thurnheer, "Ritter," 515. *"Professor Dr. W. Ritter,"* Ritter's monumental coauthored reports of 1901 and 1902 on the failure of a concrete building in Basel led directly to the formation of a commission to prepare a Swiss building code for reinforced concrete.

19. Anon., "Ueber den Werth der Belastungsproben eiserner Brücken," *SBZ* 20(2) (July 9, 1892): 12, and W. Ritter, "Ueber den Werth der Belastungsproben eiserner Brücken," *SBZ* 20(3) (July 16, 1892):14–18.

20. W. Ritter, *Der Brückenbau in den Vereinigten Staaten Amerikas* (Zurich: Albert Raustein, 1895).

21. Ibid., 52.

22. W. Ritter, "Statische Berechnung der Versteifungsfachwerke der Hängebrücken," *SBZ* 1 (1, 2, 3, 4, 5, 6) (January 6, 13, 20, 27; February 3, 10, 1883): 6–7, 14, 19–21, 23–5, 31–3, 36–8.

23. Robert Maillart, *Brückenbau: Notes from Lectures by W. Ritter* (Zurich Maillart Archive, 1893). The same form appears in Ritter, *Der Brückenbau,* 12.

24. "Schiefe Strassenbrücke nach System Monier in Wildegg," *SBZ* 17(11) (March 14, 1891): 66–7.

25. Fritz von Emperger, "The Development and Recent Improvement of Concrete-Iron Highway Bridges," *Transactions of the ASCE* (1895).

26. "Situationskizze, Stollen" ("Location Sketch for Tunnels"), Stolzenmühle Hüseren, April 1894. For Maillart's employment, see the letter from Maillart in Zurich to Pümpin and Herzog in Bern, February 20, 1894, and the letter from F. Hittmann of Pümpin and Herzog in Bern to Maillart in Zurich, February 22, 1894. For Maillart's moves, see *Dienstbüchlein für Robert Maillart,* issued September 1891 in Bern, 28–9. This is the official record of Maillart's military service and it includes at each stage in his life the community in which he lived.

27. This complaint is referred to in a letter from A. Jaggi in Zurich to Maillart in Bern, May 17, 1894.

28. For examples of his contacts with classmates, see the letter from A. Jaggi in note 27. See also the letter from E. Lubini in Zurich to Maillart in Bern, May 15, 1894.

29. M. Jotterand, *75ᵉ anniversaire du BAM: 1895–1970* (Apples: Chemin de fer Bière-Apples-Morges et Apples-L'Isle, 1970), 2–7. For descriptions of these towns, see *Dictionnaire Géographique* (1902), 1:250, for Bière, and 1:79, for Apples; also (1903), 2:601, for L'Isle.

30. Documents on the two bridges over LeVeyron brook are dated 28 XII 94. These were saved by Maillart and much of the writing on them seems to be in his hand.

31. *Dictionnaire Géographique* (1905), 3:633.

32. "Inauguration du chemin de fer Apples-L'Isle," *L'Ami de Morges* (September 16, 1896).

33. *Bauwesen der Stadt Zürich,* Auszug vom 15. Dezember 1896. Dick had worked in Ritter's consulting office in 1895 and 1896 and then had taken a job with the Zurich Tiefbauamt.

34. Letter from Pümpin and Herzog, Bern, November 7, 1896.

35. Maillart's home town, in the strict Swiss sense of the word, was Bremgarten, canton of Bern, a northern suburb of the city of Bern. Although the family never lived there, they became officially citizens of Bremgarten in 1886, probably because it was less expensive to purchase citizenship there than in Bern proper. See *Dictionnaire Historique*, 638.

36. R. Moser, "Ueber Brücken," *SBZ* 25(21) (May 25, 1895): 146-9. As Moser showed, the Kornhaus Bridge of 1898 in Bern, the iron bridge being built when Maillart moved to Zurich, cost 31 percent less than the stone alternative.

37. For a view of Zurich at this time, see V. Wenner, "Brückenbauten," in *Die Bauliche Entwicklung Zürichs in Einzeldarstellungen* (Zurich, 1905), 105-58. Eiffel's attitude toward the Tower is discussed in David P. Billington, *The Tower and the Bridge* (New York: Basic Books, 1983), 86-7. When he discussed the tower himself, he "argued the tower's aesthetics solely on the basis of its pure engineering form."

38. V. Wenner, "Die neue Stauffacher Brücke in Zürich," *SBZ* 33(9) (March 4, 1899): 80-3. Wenner had worked for Pümpin and Herzog from 1888-93, leaving just before Maillart came there. He was made Zurich city engineer in 1898. For a discussion of the Stauffacher design, see David P. Billington, *Robert Maillart's Bridges* (Princeton, Princeton University Press, 1979), 15-18.

39. W. Ritter, "Consultation over the Design of a Stauffacher Bridge in Zurich," Zurich Tiefbauamt, August 9, 1898, p. 1. Ritter based part of his report on the precedent of an earlier bridge. See W. Ritter, "On the New Construction of the Coulouvrenière Bridge in Geneva," *SBZ* 27(14) (April 11, 1896): 100-1.

40. Wenner credited Maillart with the design ("Die neue Stauffacher Brücke," 80-3), and in a speech of his own years later, Maillart recounted how the design came to be accepted; see R. Maillart, "Lichtbilder Vortrag Basel," unpublished lecture, 1937, Zurich Maillart Archive.

41. "Auszug aus dem Protokolle des Stadtrates von Zürich," Vol. 30 (August 1899).

42. Froté came from Bern and graduated from the ETH in 1892. Westermann, from Geneva, graduated from the ETH in 1896 as a mechanical engineer and immediately joined with Froté to found their firm. See *XLV Adressverzeichnis der Mitglieder der Gesellschaft Ehemaliger Studierender der Eidgenössischen Technischen Hochschule* (Zürich, August 1914): 106, 338.

43. W. Noble Twelvetrees, "François Hennebique: A Biographical Memoir," *Ferro-Concrete* (a monthly review of Mouchel-Hennebique construction) 13(5)(November 1921): 119-44. See also Peter Collins, *Concrete* (New York: Horizon Press, 1959), 64-75, and G. Huberti, "Die Erneuerte Bauweise," *Vom Caementum zum Spannbeton, Beiträge zur Geschichte des Betons*, 1B:64, 116-28. Collins, giving a version slightly different from Twelvetrees, indicates that the original design had timber beams and that Hennebique used iron rods (*Concrete*, 64). In a version described by Hennebique himself, the floor beams were iron, but were to be encased in wood. He replaced them by a solid concrete slab in which were embedded iron bars top and bottom. F. Hennebique, "Troisième Congrès du Béton de ciment armé: Opening address," *Le Béton Armé* 1(11 and 12) (March 10 and April 10, 1899).

44. *Relevé des Travaux Executés Année 1902* (Paris: Béton Armés Système Hennebique, 1903).

45. Ibid., 1898.

46. "Table des matières," *Le Béton Armé* 1 (June 1898 to December 1898).

47. Ibid., 1(1) (June 1898): 1-12.

48. R. Maillart, "Das Hennebique-System und seine Anwendungen," *SBZ* 37(21) (May 25, 1901): 225-6.

49. Ibid., p. 226.

50. Maillart never mentioned these bridges in any professional writing, but he did refer to them in private corespondence. In a letter to his future wife from Zuoz on October 13, 1901, he speaks of several small bridges in Pontresina, and on a postcard to her from Rheinegg, he notes his work there on another small bridge.

51. The bridge design is described by H. Schleich, "Die Drahtseilbahn des Rigiviertels in Zürich," *SBZ* 38(17) (October 26, 1901): 179-81. For Ritter's procedures, see Wilhelm Ritter, "Die Bauweise Hennebique," *SBZ* 33(5, 6, and 7) (February 4, 11, and 18, 1899): 41-3, 49-52, and 59-61. The Hadlaub Bridge stood without any significant difficul-

ties for over 75 years before being removed in 1978. Maillart never wrote about this bridge himself, but he did describe it near the end of his life in his 1937 retrospective lecture given in Basel.

52. Anon., "Sanatorium Schatzalp bei Davos," *SBZ* 39(2 and 3) (January 11 and 18, 1902): 13–17, 28–31. Because construction began in May 1899 and the structure was completed by that November, Maillart did not do the original calculations, but he closely followed the work, which later would lead to other substantial contracts in Davos with that same architectural firm.

53. The usual English name for this canton is the French *Grisons,* but because that is the only one of the four official Swiss languages not spoken in the canton, I prefer to use the German name. Froté and Westermann, *Solisbrücke: Notiz zur Statischen Berechnung* (Zurich, May 4, 1901). The design was by Hans Studer. See also Hans Studer, "Steinerne Brücken der Rhätischen Bahnen," in *Schweizerische Ingenieurbauten in Theorie und Praxis* (Zurich; Internationaler Kongress für Brückenbau und Hochbau, 1926).

54. Letter from Froté and Westermann in Zurich to Maillart in Paris, September 25, 1900.

55. Paul Christophe, *Le Béton armé et ses applications* (Paris, 1902), 6.

56. Froté and Westermann [Maillart], *Report for a Bridge over the Inn near Zuoz designed as an arch bridge of Reinforced Concrete* (Zurich, August 13, 1900).

57. For a detailed description of the Zuoz Bridge design and its revolutionary vision, see Billington, *Robert Maillart's Bridges,* 15–29.

58. Maillart's idea was to design the scaffolding, often called falsework, to support only the lower curved slab; once hardened, that slab then supported the vertical walls and horizontal deck that completed the hollow box. Thus, the bridge not only used less concrete than contemporaneous arches of comparable span, but also required less scaffold, a significant construction economy.

59. A full description of the procedure is given by W. Ritter, *Bericht über die Belastungsprobe der neuen Inn-Brücke bei Zuoz,* Zurich, Winter 1901/2, 1–3. This consulting report describes the test, gives measurements of deflections, and comments on the finished work.

60. Letter from Robert in Zuoz to Maria in Zurich, October 10, 1901, 8:00 P.M.

61. Letter from Robert in Zuoz to Maria in Zurich, October 13, 1901, morning.

62. Ritter, "Bericht", 6–7, shows that the bridge under its own dead weight had deflected downward 70 millimeters at the center of the span after 2 days, whereas under full gravel loading, it only deflected 15 millimeters, and under the loaded cart, only 13.6 millimeters. About half of the dead-weight deflection occurred right away and about half developed gradually due to creep.

63. He had written an authoritative section on reinforced concrete; see R. Maillart, "Armierte Betonbauten," in *Schweizerischer Bau- und Ingenieur-Kalender* (Zurich: Casar Schmidt, 1901), Sec. III, 108–10.

64. Letter to Madame Wild from Edouard Elskes, August 12, 1901. "Il est très sérieux et grand travailleur et quoique relativement jeune, a, me disait M. Koller, le caractère *rassis* d'un vieux!" Maillart had worked for Koller in 1895–6.

65. Letter from Robert in Zurich to Maria in Bologna, Wednesday evening, August 21, 1901.

66. The arch casting began on July 24 and ended on July 27. See J. A. Splitzer and A. Nowak, "Bogenbrücken und Uberwolbungen" in *Handbuch für Eisenbetonbau: Brückenbau,* ed. Fritz von Emperger (Berlin: Verlag von W. Ernst and Sohn, 1908), 3:154–7. This material was taken from an article by Edouard Elskes in the *Bulletin Technique de la suisse romande* 29(3) (February 10, 1903): 33–35.

67. G. Bener, *Ehrentafel Bündnerischer Ingenieure und Ingenieurwerke* (Chur, 1927), 73–6. The highest pass tunnel in Europe, the Albula was begun in 1898 and, after difficulties, was proceeding rapidly by August 1901. It would be completely cut through in May 1902 and the rail line to the Inn Valley opened on July 1, 1903. For a brief history of Zuoz, see C. Wieser, *Zuoz* (Bern: Paul Haupt, 1965).

68. Wieser, *Zuoz,* 18. The hotel still existed 100 years later but under the name of the "Engadina."

69. M. C. Blumer-Maillart, *Recollections* (Princeton: Princeton University, Department of Civil Engineering, 1978). Probably the most famous description of the *Swiss Table d'hote* is given by Thomas Mann in

the *Magic Mountain,* a novel that takes place in a Swiss Sanatorium in Davos before World War I.

70. Letter from Maria in Bologna to Robert in Zurich, August 24, 1901.

71. The firm of Wild and Abegg had large cotton mills in the Valle de Susa near Turin. See *Dictionnaire Historique,* 1:31–2.

72. Letter from Maria in Zurich to Bertha in Bern, August 1, 1902.

73. Letter from Robert in Zurich to Maria in Zuoz, August 3, 1901.

74. Maria's father, Benedetto Ronconi, was from Vicenza and her mother came from Genoa. Maria was educated at convents in Italy and Belgium. See letter from Maria in Milan to Robert in Zurich, August 5, 1901.

75. The telegram sent at 9:55 P.M. On August 6 reads: "Maillart Bodmerstrasse 3 Zürich – oui à vous pour la vie – ronconi."

76. Letter from Robert in Zurich to Maria in Bologna, August 7, 1901.

77. Letter from Robert in Zurich to Maria in Bologna, August 9, 1901.

78. Letter from Robert in Zurich to Maria in Bologna, August 11, 1901. Although he always wrote Maria in French, Maillart stated to her that it was not his mother tongue. However, to judge by these letters and his later ones, Maillart was fully bilingual. He always wrote his wife, his daughter, and his younger son in French; and he wrote his mother and his older son in German.

79. Card from Maria in Florence to Robert in Zurich, August 12, 1901. Because Maria had gone to the convent school near Florence, she knew many of the Sisters there.

80. Letter from Robert in Zurich to Maria in Bologna, August 14, 1901 (afternoon).

81. Letter from Maria in Zurich to Bertha in Bern, September 3, 1901, in which Maria tells of traveling on September 2, presumably leaving Bern. The formal announcement printed in Bern and from both Maria's aunts and Robert's mother is dated August 1901.

82. *Zürcher Wochen-Chronik,* III: n. 36 (Saturday, September 7, 1901), 291. Of the fourteen engagement notices listed in the Chronicle, all merely gave the two names except for Maillart's, which was in larger print and with a box around it.

83. Letter from Maria in Zurich to Bertha in Bern, September 18, 1901, in which she describes her plans for furnishing the apartment.

84. Letter from Robert in Zurich (Bodmerstrasse) to Maria in Zurich (Seidenfädeli), September 7, 1901 (evening).

85. Note from Robert in Zurich to Maria in Zurich, no date, but headed by "Under your window at 5: 30 o'clock in the morning."

86. Letter from Robert in Zurich to Maria in Zurich, October 18, 1901; see also letter from Robert in Zurich to Bertha in Bern, October 27, 1901.

87. Letter from Robert in Zurich to Bertha in Bern, November 12, 1901, with some opening words by Maria.

88. Letter from Robert in Zurich to Maria in Bologna, August 22, 1901 (morning).

Chapter Two

1. Letters from Maria in Zurich to Bertha in Geneva, November 21 and 28, 1901. Also a card from Robert in Geneva to Maria in Zurich, November 23, 1901.

2. Letters of Maria in Bern to Robert in Zurich, December 9 and 10, 1901 (dated December 9 also, but most likely written the following day), as well as the letters from Robert in Zurich to Maria in Bern, December 9, 1901 (7:00 P.M.). Letter from Maria in Zurich to Bertha in Bern, December 29, 1901.

3. This trip is recorded in a letter from Robert in Turin to Bertha in Bern, January 12, 1901, and one from Maria in Bologna to Bertha in Bern, January 20, 1902.

4. Letter from Maria in Bologna to Bertha in Bern, January 21, 1902.

5. Letter from Robert in Bologna to Bertha in Bern, January 22, 1902. Maria added several paragraphs to this letter. Maria's letter of January 11 states that they will return on Sunday at the latest.

6. Letter from Maria in Zurich to Bertha in Bern, undated but probably February 27, 1902.

7. Letter from Maria in Zurich to Bertha in Bern, March 10, 1902.

8. Letter from Robert in Zurich to Maria in Bern, April 4, 1902.

9. Letter from Robert in Zurich to Maria in Bern, April 23, 1902.

10. Letter from Maria in Zurich to Bertha in Bern, April 25, 1902.

11. Robert Maillart, "Bogenträger aus armiertem Beton." Patentschrift, No. 25712 Eidgen. Amt für geistiges Eigentum, Bern, February 18, 1902.

12. The origins of "Kultur" engineering are described in Wilhelm Oechsli, *Geschichte der Gründung des Eidg. Polytechnikums mit einer Uebersicht seiner Entwickelung. 1855–1905* (Frauenfeld: Huber & Company, 1905), 342, 357.

13. Letter from Maria in Zurich to Bertha in Bern, March 15, 1902.

14. Particularly discouraging was Maillart's loss of the new bridge over the Sihl River in Zurich. It was a copy of the Stauffacher Bridge. *SBZ* 40(3): 34, and also letters from Maria in Zurich to Bertha in Bern, May 23 and June 5, 1902, and letters from Maria in Zurich to Bertha in Geneva, May 7 and 9, 1902.

15. David P. Billington, "Unknown Contributions of Robert Maillart to Thin Shell Concrete Structures," in *Jahresbericht* 1981 (Zurich: Schweiz. Zement-Kalk-und Gipsfabrikanten), 64–72.

16. "Schlussbericht über den Gaswerk-Neubau der Stadt St. Gallen" (St. Gallen, 1906), 7–8. After graduating from the ETH in 1881, François Schüle had spent 10 years working for Gustave Eiffel, mainly on large metal bridges. Returning to Switzerland in 1891, he worked first for the Swiss Railroad in Bern and then as a professor at the engineering school in Lausanne until 1901, when he was named professor at the Federal Polytechnical Institute in Zurich and Director of the federal materials testing laboratory, also in Zurich. There he had immediately begun a systematic program for testing reinforced-concrete structures and he quickly established himself as an expert in the new field; see F. Schüle, "Résistance et déformations en Béton Armé sollicité à la flexion," *SBZ* 40(22): 237–42; 40(23): 248–54; 40(24): 264–65. Schüle reported on these studies in his 1904 discussion in St. Louis, *Transactions of the ASCE* 54E (1905): 549–51.

17. Letter from Maria in Zurich to Bertha in Bern, probably June 14, 1902.

18. Letter from Maria in Zurich to Bertha in Bern, June 16, 1902. See also letter from Maria in Zurich to Bertha in Bern, June 18, 1902. The reinforcing steel cost 80,000 francs, whereas the Zuoz Bridge cost 27,000 francs.

19. In 1907, Fritz von Emperger (1862–1942) published a detailed survey of such tanks and singled out Maillart's design as exemplary. Without von Emperger's writing, which reproduced Maillart's calculations, the nature of Maillart's major innovation would have been lost, because his office records for this period have not survived. R. Wuczkowski, "Flüssigkeitsbehälter," in *Handbuch für Eisenbetonbau*, ed. F. von Emperger (Berlin, 1907), 3:348–51, 3:407–13. It was this survey that showed Maillart's tanks to be the largest yet built.

20. The exterior wall requires about 2,700 cubic meters in the city design compared to less than 600 cubic meters in Maillart's design, which takes full advantage of reinforced concrete to reduce the total weight by relying on reinforcement to carry the tension.

21. The substantial research program at the University of California at Berkeley is summarized in Alexander C. Scordelis and Per K. Larsen, "Structural Response of Curved RC Box-Girder Bridge," *Journal of the Structural Division,* ASCE, 103 (ST8) (August 1977): 1507–24.

22. When Wuczkowski, who was Fritz von Emperger's chief engineer, surveyed the history of cylindrical tank analysis in the second edition (1910) of von Emperger's *Handbuch,* he reported that "the first known study of this type [which included the ring and the base restraining behavior] was carried out in an empirical [graphical] way by the firm of Maillart and Company for the gas holders in St. Gallen." These extraordinarily original structures did have some later difficulties unrelated to Maillart's wall design. "Schlussbericht," 8.

23. *SBZ* 39(36) (June 28, 1902): 292. The design appears on a drawing dated March 1902 and entitled *Structural Analysis, Bridge Over the Steinach,* signed by L. Kilchmann (1852–1926), Director of the City Building Department for St. Gallen. There also is a detailed scaffold drawing dated February 19, 1902, a location plan dated February 19, 1902, and roadway detail drawing of March 1902. Each was seen and signed by Maillart on July 17, 1902. The earlier competition had begun while Maillart was still with Froté and Westermann, when the

question of a new bridge over the Steinach Brook in St. Gallen had been the subject of a series of proposals from builders.

24. In his design, Maillart relied on the examples of recent railroad bridges near Lausanne and of the Coulouvrenière Bridge in Geneva to show that his ideas had a basis in recent successful structures.

25. Letter from Maillart and Cie. in Zurich to Kilchmann in St. Gallen, July 7, 1902.

26. P. Christophe, *Le Béton Armé* (Paris, 1902), 654–5. Christophe in describing the problems of hingeless arches cites as a major point that the scaffold movement can cause serious cracking. He is arguing here that three-hinged arches are the most rational type in concrete.

27. Maillart and Company, *Preisliste für Erstellung einer gewölbten Brücke über die Steinach* (Zurich, July 7, 1902). This gives Maillart's bid of 48,105.55 francs. In letters of July 15 and 16, 1902, Maillart reduced his price to 38,600 francs. The actual cost was just under 46,000 francs; see Maillart and Company, *Final Accounting* (September 26, 1903), 6. A part of this final difference had to do with the building department's insistence on an arch heavier than the one in Maillart's own redesign. The city building director, L. Kilchmann (1852–1926), succeeded in getting Maillart to reduce his bid because most likely there were competing bids below Maillart's original offer. Maillart's correspondence during that period indicates his criticism of the city design in 1902. In this respect, Maillart's letter of July 7, in which he submitted his bid, is strikingly different from similar letters of the three firms that had submitted bids for designs in the earlier competition.

28. This dialogue reveals Maillart's characteristic desire to eliminate useless material. His calculations showed that the arch thickness could be substantially reduced from that given in the building department design of March, and he therefore based his design on reductions of about 30 percent in the concrete for the main arch. Building director Kilchmann worried about this and eventually decided to object. This correspondence also illustrates the primacy of aesthetics for Maillart. Kilchmann ordered Maillart to lower the arch by 25 cm to make room for gas and water pipes above. Maillart gave a two-part argument against this change: first was the familiar objection that his bid was based on information received before he signed the contract and that changes afterwards would increase his costs. But Maillart's second argument was that the change would lead to a less attractive structure because a flatter arch in that setting would appear heavier. Costs and aesthetics seen together gave Maillart the basis for his argument and Kilchmann never raised that question again. Letter from Maillart and Company to Kilchmann, August 13, 1902.

29. Letter from Kilchmann to Maillart and Company August 18, 1902. The significance of Maillart's response to Schüle lies largely in how he defended the ring method for construction (building the arch in two layers or rings). Schüle had objected because Maillart's calculations did not reflect his method of construction. Schüle was on good ground for raising such a question. But Maillart's response does not at all meet Schüle's concern about stresses. Rather, he responded in a different way by referring to the successful experience with the Solis Bridge, which had been built in three rings. Thus, he argued, if going from one ring to two is considered questionable, going to three is even more so. But because three rings had already been done and proven, then there will be no problem with only two rings.

His major argument, however, is that the heavier single ring of cast concrete will be more prone to cracking than the block arch built by the ring method. Therefore, his response is not to show by new calculations that the stresses are low, but rather to argue from experience that by comparison with other recently built works, his will be fully satisfactory and even less subject to difficulties than works already built and proven. Naturally, this argument has a commercial as well as a structural motive because any delays during construction are costly to a builder and by mid-August he had a work force at the site.

30. Letter from Maillart and Company to Kilchmann, August 18, 1902.

31. This controversy can be traced through the correspondence: letters from Kilchmann in St. Gallen to Maillart and Company in Zurich, August 18, 19, 21, 23, and 27, 1902, and Maillart and Company in Zurich to Kilchmann in St. Gallen, August 18, 20, and 20 (later), 1902. The building department

calculations on the basis of which Kilchmann criticized Maillart's design are found in two drawings made in St. Gallen in August 1902.

32. Letter from Kilchmann in St. Gallen to Maillart and Company in Zurich, August 27, 1902, and letters from Maria in Zurich to Bertha in Bern, August 26 and 28, 1902 (7:00 P.M.).

33. Letter from Maillart and Company, Zurich, to Kilchmann in St. Gallen, December 30, 1902. Actually, Schüle never returned to the project, a substitute consultant was hired, and in December 1902 he ruled, rather arbitrarily, that the extra cost was to be split evenly. Maillart objected, but still agreed.

34. Letter from Kilchmann in St. Gallen to Maillart and Company, in Zurich, September 24, 1903.

35. Letters from Maria in Zurich to Bertha in Bern, July 21 and 25, 1902. Letter from Maria in Zurich to Bertha in Bern, August 1, 1902.

36. Letter from Maria in Zurich to Bertha in Bern, November 6, 1902. See also one of March 20, 1903.

37. *SBZ* 41(20) (May 16, 1903): 230.

38. Letters from Robert in Zurich to Maria in Bern, May 19 and 25, 1903.

39. Letter from Maillart in Zurich to Bersinger in St. Gallen, May 30, 1903.

40. Letters from Maria in Zurich to Bertha in Bern, June 22, 25, and 27, 1903.

41. Letter from Ritter in Zurich to Bersinger in St. Gallen, June 17, 1903.

42. Letter from Maillart in Zurich to Bersinger in St. Gallen, August 7, 1903. Report from Ritter in Zurich to Bersinger in St. Gallen, August 22, 1903. Letter from Maillart in Zurich to Bersinger in St. Gallen, August 28, 1903. For a more detailed discussion of the Billwil design, see Billington, *Robert Maillart's Bridges*, 31–4.

43. *Wyler Bote,* Wil, April 26, 1904. See also an article by Bersinger, "Die neue Strassenbrücke über die Thur bei Billwil-Oberbüren, Kanton St. Gallen," *SBZ* 44(14) (October 1, 1904): 157–9. The bridge was rehabilitated in 1975, but without any change in its structural form. See newspaper account, June 20, 1975.

44. Letter from Robert in Zuoz to Maria in Oleggio, September 8, 1903, and letter report from Maillart in Zurich to Schucan in Chur, September 26, 1903. Other new information from Zuoz came when the well-known local engineer, Giovanni Gilli (1847–1913), showed him for the first time Ritter's 1902 favorable report on the 1901 Zuoz Bridge load test, written before the wall cracks appeared. Giovanni Gilli, *Ehrentafel Bündner. Ingenieure und Ingenieurwerke* (Chur, 1927), 59–61. From 1898 until his death, Gilli was chief engineer of the Rhätischen Bahn.

45. Letter from Marie in Zurich to Bertha in Bern, October 2, 1903.

46. For his economizing, see letters from Robert in Zurich to Maria in Bern, May 18 and 27, 1903, and a letter from Maria in Zurich to Bertha in Bern, June 20, 1903. For his touchiness and then his repentence, see letters from Maria in Zurich to Bertha in Bern, October 5, and 9, 1903. Maria felt so much better after Maillart's repentence that she wrote a comical letter to Bertha written as if from their new cat, Gyp, to her "venerable mistress," October 10, 1903.

47. "Wettbewerb für eine Utobrücke in Zürich," *SBZ* 44(7) (August 13, 1904): 80. The Jury report appears in that number on pp. 76–80 and in 44(8) (August 20, 1904). The Jury report was dated July 16, 1904.

48. R. Maillart, "The Construction and Aesthetic of Bridges," *The Concrete Way* (May–June 1935): 305. The statement reads: "The realization that as a link between the arch and the platform solid spandrels serve no useful purpose, except in the middle of the bridge, and that close to the abutments they exert a useless dead weight which is positively a potential danger, has led to the practice of slotting triangular cavities out of them."

49. Stanford Anderson, "*Sachlichkeit* and Modernity, or Realist Architecture" in *Otto Wagner, Reflections on the Raiment of Modernity,* ed. H. F. Mallgrave (Chicago, 1993), 322–360.

50. Recent advances in steel material research has led to a new type of steel on which a corrosion layer forms early in its life that protects the steel from further damage. Exposed steel has been made fireproof by the unlikely method of building it in tubular form and filling the tubes with circulating water to carry away the fire's heat. In structures such as the Eiffel Tower or the Hancock tower in Chicago, exposed iron or steel can create a powerful aesthetic.

51. Maillart described his trip in cards to Maria in Farnbühl on July 30, from Samedan on July 31; one from St. Moritz and one from Zuoz (with a picture of the town and the bridge), one from the Bernina pass on August 1, and one from Tavanasa on August 2 (with a picture of the wooden bridge). For Maria's description of his trip and her reflections on their anniversary, see her letter to Bertha in Bern on August 1, 1904. By a curious set of circumstances, this bridge remains Maillart's one major work with the least documentation. In addition to the absence of documents in the canton archives, there is nothing in Maillart's 1904 letters to Maria because she returned to Zurich early in September and there are only four more surviving letters in that year. Also, no Swiss publication addressed the bridge until the *Bauzeitung* article of 1914, 10 years later. See the *Amtsbuch* (Chur, 1904), 464, for the competition and also see J. Solca, "Die Rheinbrücken bei Tavanasa und Waltensburg," *SBZ* 63(24) (June 13, 1914): 343–6. The lateness of this article arose because Solca wished to describe the construction of a new road above Ilanz completed in 1913. At one point, the road crossed a new bridge finished in 1911. It was a three-hinged, reinforced-concrete arch of 34.5-meter span described in J. Solca, "Die Glennerstrasse von Ilanz nach Peidnerbad," *SBZ* 43(23) (June 6, 1914): 336–8. Then in the next *SBZ* number, he described Maillart's Tavanasa Bridge and the Waltensburg Bridge, near Tavanasa, designed and built by Westermann (in St. Gallen) in 1912. The Westermann Bridge, still standing, is also a three-hinged, reinforced-concrete arch; it has a span of 50 meters and the arch is a hollow box, but not connected to the deck as a box. Visual comparison of the Waltensburg with the Tavanasa shows clearly the great difference in appearance between two bridges that solve almost the same problem. The cost of Maillart's bridge was 144 francs per square meter in 1905 and Westermann's 225 in 1912. The Tavanasa Bridge also appeared in *Hoch- und Tiefbau* 14(47) (November 26, 1915): 362, 366, in a brief note on Maillart's bridges at Tavanasa, Wattwil, and Ibach.

52. Drawings Nos. 332 and 333, *Brücke über den Rhein bei Tavanasa,* September 30 and October 1, 1904. Both drawings are in color, are signed by Maillart, and are stamped with the date "1 Oct. 1904." The final drawings are nos. 359–62, *Brücke über den Rhein bei Tavanasa,* January 13–14, 1905. All but 360 are colored drawings almost identical to 332 and 333 except more detailed. Drawing 360 is in black and white and is a detailed calculation sheet using the graphic statics of Wilhelm Ritter.

53. Solca, "Tavanasa," 344.

54. For Mörsch's career, see Huberti, *Beiträge,* 110–14. He was a professor in Zurich from 1904 to 1908. From 1901 to 1904 and 1908 to 1916, he was technical director of Wayss and Freytag. He became a professor at Stuttgart in 1916, retiring in 1939. His book was *Der Eisenbetonbau: seine Theorie und Anwendung* (Neustadt/Haardt: Wayss and Freytag, 1902). In a review article late in life, Mörsch referred to his consulting for the Tavanasa design. See *Beton und Eisen* 41(15/16) (August 15, 1942): 153.

55. One card from Maillart in Rorschach and two cards in Ilanz all to Maria in Zurich, March 7, 7, and 8, 1905. A card from Maillart in Tavanasa to Maria in Zurich, March 24, 1905, and a card from him in Sargans to Maria in Zurich, April 10, 1905. He took photos nos. 141, 142, 143 of the scaffold layout on April 12, 1905. For summer construction, see letters from Robert in Zurich to Maria in Morgins, August 2, 3, and 4, 1905, and to Maria in Bern, August 8, 1905. Also a card from Robert in Ilanz to Maria in Bern, August 9, 1905. The scaffold was damaged in a storm on August 6 but quickly repaired. For the load test, see card from Robert in Tavanasa to Maria in Bern, September 28, 1905.

56. Christian Menn, "New Bridges," in *Maillart Papers* (Princeton: Princeton University, Department of Civil Engineering, 1973), 99–118.

57. Felix Candela, "New Architecture," in *Maillart Papers* (Princeton: Princeton University, Department of Civil Engineering, 1973), 119–26; and David P. Billington, "Heinz Isler as Structural Artist," in *Heinz Isler as Structural Artist* (Princeton: Princeton University, The Art Museum, 1980), 9–24.

58. Emil Mörsch, "Die Isarbrücke bei Grünwald," *SBZ* 44(23) (December 3, 1904): 263–7, and 44(24) (December 10, 1904): 279–83.

59. Josef Melan, "Theorie des Gewölbes und des Eisenbetongewöbes im besonderen," in *Handbuch*

für Eisenbetonbau, ed. Fritz von Emperger (Berlin, 1908), 1: 425. For an American perspective, see D. A. Molitor, "Three-Hinged Masonary Arches; Long Spans Especially Considered," *Transactions of the ASCE,* 40, paper no. 834 (1898): 31–85.

60. Alan H. Mattock, "Concrete – Challenge and Opportunity," *Concrete International* 12(5) (May 1990): 54–60.

61. Letters from Robert in Zurich to Maria in Farnbühl, June 28 and 29, 1904.

62. "Der Wettbewerb für eine neue Utobrücke über die Sihl in Zürich," *Beton und Eisen* 3(5) (1904): 264–81, and 4(1) (1905): 6–11. There is no author listed, but von Emperger edited the journal and most certainly wrote the critique of this competition. For a fuller discussion of von Emperger's critique, see David P. Billington, "Robert Maillart and Swiss Bridge Competitions," in *The Development of Long Span Bridge Building* (Zurich: ETH, 1979), 129–33.

63. *SBZ* 38(9) (August 31, 1901): 96. See also A. Geiser, W. Ritter, and F. Schüle, *Expertenbericht [consulting report] betreffend den Gebäude-Einsturz in der Aeschenvorstadt Basel am 28. August 1901* (Zurich, November 1901). For a clear interpretation of the failure, see F. von Emperger, "Der Expertenbericht über den Hauseinsturz in Basel," *Beton und Eisen,* Vienna, 1(1) (1902): pp. 15–19. The article is dated June 5, 1902. It is unsigned but almost certainly by the editor of the new journal, von Emperger.

64. R. Maillart, "Neuere Anwendungen des Eisenbetons" ["New Applications for Reinforced Concrete"], in *Protokoll der ordentlichen Generalversammlung, am 16. und 17. September 1904, in Basel* (Bern: Verein schweizerischer Zement-Kalk-und Gypsfabrikanten, 1904), App. III, 16–23.

65. Much of his information came from the 1902 book by P. Christophe, *Le Béton Armé* (Paris and Liège: Béranger). For tanks, see pp. 372–80; for pipes, pp. 333–34; for chimneys, p. 381; for piles, pp. 180, 325–27. Maillart refers in his 1904 talk to high chimneys in America, to concrete masts with wooden cores in Grenoble, to concrete piles in Hamburg, and to concrete railroad ties under development in Spain, America, Italy, and even Switzerland.

66. "Tuberculosis," *Encyclopedia Britannica* (Chicago, 1963), 22:531–2. "Tuberculosis was the chief cause of death in the U.S. in 1909. . . ." Tuberculosis, the great nineteenth-century killer, seemed almost completely absent in Davos even though housing there was not very modern. The success of the remote Davos climate in curing tuberculosis created a pressing need for better access. The main problem was the arduous 31-kilometer trip from Landquart, the nearest rail stop to Davos. Civic leaders there proposed in 1886 that a narrow-gauge rail line be built from Landquart to Davos. Construction began in March 1888, and on October 9, 1889, the line opened between Landquart and Klosters. It was completed to Davos on July 21, 1890.

Following that lead, the canton decided to build a narrow-gauge network (the Rhätische Bahn) throughout the entire Graubünden. It took 20 years to complete the network begun by Davos. For a detailed story of the development of this network, see W. Catrina, *Die Entstehung der Rhätischen Bahn* (Zurich: Juris Druck and Verlag, 1972).

With the 1890 completion of the line to Davos, visitors increased rapidly, requiring new building and the largest sanatorium of that period to be built was the 1900 Schatzalp, with a cog railway up to the new sanatorium over 1,000 feet above the town.

67. Letters from Robert in Zurich to Maria in Farnbühl, August 11, 15, 1904.

68. B. R. Barber, *The Death of Communal Liberty* (Princeton, 1974).

69. Thomas Mann, *The Magic Mountain* (Harmondsworth, England: Penguin Books, 1960; first published in 1924), 3–4.

70. A. Nowak, "Tunnelbau, Stadt und Untergrund-Bahnen," *Handbuch für Eisenbetonbau,* ed. F. von Emperger (Vienna, 1907), 3:639–41. The originality of the form arises from two design ideas: First, Maillart avoided the most common solution of building columns on the free side of the track and has rather chosen to support the buttressed wall by a cantilever instead of a beam. Second, the buttresses increase in size as they approach the base, giving the sense of increasing strength for increasing heights of wall.

71. Letter from Maria in Zurich to Bertha in Bern, March 21, 1904.

72. Letters from Maria in Arosa to Robert in Zurich, January 23 and February 1, 1906.

73. For these reactions, see Maillart correspondence, 1906.

74. *SBZ* 50(11) (September 14, 1907): 141–4.

75. Letter from Maria in Zurich to Bertha in Bern, November 18, 1902. See also *SBZ* 40(22) (November 29, 1902): 244. The committee was actually first formed in October 1902; see *SBZ* 41(14) (April 11, 1903): 159. Maria wrote Bertha on November 18 that Maillart was chairman, but Otto Pflaghard presented the report in 1903. Nowhere in the *SBZ* is a "chairman" mentioned.

76. Three significant books, in particular, appeared in 1902: C. Berger, and V. Guillerme, *La Construction en ciment armé* (Paris: Dunod, 1902); Paul Christophe, *Le Béton armé et ses applications* (Paris and Liège: Béranger, 1902); and the first edition of Mörsch's classic book. The first two give a large compendia of completed structures along with the theory for analysis; the third was the first edition of perhaps the most widely known German text on reinforced concrete.

77. See F. von Emperger, *Beton und Eisen*, 15–19.

78. Letter from Maria in Weggis to Robert in Zurich, May 1, 1905, and letter from Robert in Zurich to Maria in Weggis, May 4, 1905.

79. Paul Miescher, "Verlegung des Riehenteiches" in *Technischer Bericht über die Ablenkung des Gewerbekanals und die Errichtung einer Turbinenanlage beim Erlenpumpwerk* (Basel: Basel City Archive, 1906).

80. Letter report from director Paul Miescher to the Vorsteher des Sanitäts-Departments, Basel, March 2, 1907.

81. R. Maillart, "Belastungsprobe eines Eisenbetonkanals," *SBZ* 50(10) (September 7, 1907): 125–8.

82. Letter from Robert in Kilchberg to Maria in Geneva, April 27, 1907. Letter from Robert in Zurich to Maria in Geneva, May 30, 1907. Construction was interrupted early in the summer by problems not connected with Maillart, but by August, he could proceed. Letter from Miescher to the Commission, Basel, October 8, 1907.

83. Card from Maillart in Wädenswil to Maria in Weggis (forwarded to Zurich), May 13, 1905.

84. In Maillart's 1913 list of works, there are eight in 1905 of which only four appeared in the 1940 list published by Mirko Ros; and in Maillart's own 1933 list, he printed in boldface his major works of which the Wädenswil factory is the only one in 1905. Tavanasa is the only boldfaced listing for 1906, the year in which it was apparently officially opened.

85. Letter from Robert in Zurich to Maria in Bern, November 3, 1905. The stationery for the first time uses the title of Maillart & Co., Zurich-St. Gallen.

86. Two drawings: one dated Zurich, January, 1904 (crossed out), and approved by the Concert Hall commission on March 10, 1906, and the other dated November 29, 1907. These are in the St. Gallen archives. The 1907 drawing was reprinted in J. Kunkler, "Die Tonhalle in St. Gallen," *SBZ* 58(17) (October 21, 1911): 227–9.

87. R. Maillart, "Die Sicherheit der Eisenbetonbauten," *SBZ* 53(9) (February 27, 1909): 119–20. Translated into English in Billington, *Robert Maillart's Bridges*, 133–4. Maillart gave the lecture on February 3, 1909, to the Zurich branch of the Swiss Society for Engineers and Architects.

88. The rise of general mathematical theories for cylindrical shells begins with H. Reissner, "Ueber die Spannungsverteilung in zylindrischen Behälterwänden," *Beton und Eisen* 7(6) (1908): 150–5. Earlier theoretical work was A. E. H. Love, *Treatise on the Mathematical Theory of Elasticity* (Cambridge, 1892–3). Love's thin-shell theory first appeared in *Transactions of the Royal Society* 179 (1888). It was his second 1906 edition of the *Treatise* that was translated into German by Teubner in 1907. Probably the earliest study of thin-shell interaction (for boilers) was by F. Grashof, *Theorie der Elastizität und Festigkeit* (Berlin, 1878). V. Lewe, "Die Statische Berechung der Flüssigkeitsbehälter," in *Handbuch für Eisenbetonbau*, ed. F. von Emperger, 3rd ed. (Berlin, 1923), 5:71–181. Maillart's tanks are briefly described on pp. 279–81 and a reference made to the calculations in the second edition of the *Handbuch*. Also in that second edition, there appeared a more detailed graphical method (ibid., 2nd ed. (Berlin, 1910), 4:351–5) developed by the Austrian, Karl Federhofer, whose work was in part based on ideas from W.

Ritter and whose single worked example was Maillart's St. Gallen gasholders. This was removed from the third edition. See W. S. Gray, *Reinforced Concrete Reservoirs and Tanks,* 3rd ed. (London, 1954); for a derivation of the Reissner equations, see pp. 164–8.

89. For a discussion of the load tests and their technical significance, see Billington, *Robert Maillart's Bridges,* Chap. 6.

90. "Note on our Illustration," *Hoch- und Tiefbau* 10(12) (March 25, 1911): 89, 93. The note goes as follows: "We bring as an appendix the reproduction [photo] of a new type of floor. The main feature is that in spite of a 2,000 kg/m² live load, it is built without any beams. For that load the slab is strengthened by capitals over the columns which produce advantages for lighting, ventilation, and the efficient use of space as well as giving a handsome visual [architectonic] form. Moreover, the price is modest for this construction system which has been perfected by the firm of Maillart and Co. on the basis of exhaustive research."

91. The American system, so called, was developed in 1905 by C. A. P. Turner, *Concrete Steel Construction* (Minneapolis, 1909).

92. R. Maillart, "Eine schweizerische Ausführungsform der unterzuglosen Decke – Pilzdecke" ("A Swiss-Developed Form of Beamless Slab – The Mushroom Slab"), in *Schweizerische Ingenieurbauten in Theorie und Praxis* (Zurich: Internationaler Kongress für Brückenbau und Hochbau, 1926).

93. Letter from Robert in Zurich to Maria in Karlsbad, June 22, 1911. The city design is shown in a drawing labeled R27/17 and dated St. Gallen, January 20, 1912. Maillart's design appears in a drawing entitled *Filter-Neubau im Riet,* labeled R27/18 and dated St. Gallen, March 1912. Maillart's name does not appear on this last drawing, but it shows his design as clearly seen in Maillart photos numbered 491, 492, 493, 495, 497, 500, 501, 502, and one unnumbered photo, all in the Zurich Maillart Archive.

94. Letter from Robert in Zurich to Maria in Görwihl, August 19, 1909. Similar letters of August 10 and 18 spoke of that work also. Two drawings signed by Arch E. Heene dated July 14, 1909, were signed by the city on July 30, 1909.

95. Letter from Robert in Zurich to Maria in Airolo, August 6, 1908.

96. Von Emperger, *Handbuch,* 3:112–16.

97. The calculations of weight are as follows:
(a) *Maillart Wall* (dimensions taken from Drawing 1435, August 28, 1909)

$$3.7 \times (0.25 + 0.35)/2 = 1.11 \text{ m}^2$$
$$4.3 \times 0.36 \times 1.1 = 1.70 \text{ m}^2$$
$$2.6 \times 0.22 \times 3.95/9.95 = 0.23 \text{ m}^2$$
$$0.55 \times 0.22 = 0.12 \text{ m}^2$$
$$1.20 \times 0.5 = 0.60 \text{ m}^2$$
$$\text{TOTAL} = 3.76 \text{ m}^2$$
$$3.76 \times 2.2 = 8.3 \text{ t/m}$$

(b) *Heene Wall* (weight taken from June 8, 1909, drawing)

$$16.665 \times 2.2 = 36.5 \text{ t/m}$$

(c) Maillart used 8.3/36.5 = 0.23, or 23% of the concrete in the Heene Wall.

98. The main virture of Maillart's design was that it used the heavy building wall and interior concrete structure, already there for large vertical loads, to resist horizontal earth forces as well. It could do this with no increase in materials because the earth forces here are small compared to the loads on the whole building. Maillart's wall is fully of reinforced concrete, whereas Heene's is conceived more along the lines of a masonry wall where dead weight counters the overturning.

99. Documents for the Adler building are in the St. Gallen City Archive, copies in the Princeton Maillart Archive.

100. The canton of Zurich held a competition for the new main university building to be constructed next to Semper's Polytechnikum. In 1908, the architectural firm of Curjel and Moser won the first prize, and by late 1909, they had made design drawings. Bids for construction were solicited with a deadline of June 6, 1910, for earthwork, masonry, reinforced concrete, and cut stone work. We get some idea of the urge for architectural monumentality by comparing the large cost for masonry cladding of 950,000 francs to the relatively modest cost of about 350,000 francs for the structure. There were twelve bidders for the concrete structure.

101. F. Schüle, *Report on the New University Structure in Zurich* (Zurich, September 3, 1910), and the ar-

chitects response is from I. V. K. Wegmann, *Vergebung von Bauarbeiten Universitätsbauten Zürich* (Zurich, November 28, 1910).

102. *Protokoll des Regierungsrates 1910* (Zurich: Canton of Zurich, December 17, 1910). The canton recognized that the designs were not equivalent largely because Jaeger-Favre's and several others needed a hung ceiling, which Maillart's did not. Still that left Maillart's bid of 341,800 francs well above Jaeger-Favre's (now 321,500 francs) and above two others, which the canton ruled out on engineering grounds. The final cost evaluation left Maillart fifth, but the canton decided to split the contract between the fourth lowest bidder, Morel and Bryner, and Maillart as follows:

Maillart	205,795.10
Morel	118,077.70
TOTAL	323,872.80 francs

as less than either bidder alone. This resulted from a closer study by the two contractors and their favorable view that the split would increase their efficiency in this case. The reason for the split was that the Morel ribbed deck seemed to the canton to be better for the Biological Institute because of the need there of so many pipes for laboratories, which could be more easily placed and replaced beneath a hung ceiling than with a hollow slab.

103. Maillart sent his tower bid in on August 15, 1912, along with a bid of 12,850 francs for a concrete cupola. For that smaller structure, the canton decided instead to accept a bid of 9,015 francs from another builder, whose design was in wood. *Protokoll des Regierungsrates 1912* (Zurich: Canton of Zurich, September 19, 1912). Also *Protokoll der Baudirektion des Kantons Zürich 1912* (Zurich: Canton of Zurich, October 14, 1912); all found in the Zurich cantonal archive.

104. R. Maillart, "Die Eisenbeton-Konstruktionen im Neuen Kulissenmagazin des Zürcher Stadttheaters," *SBZ* 57(1) (January 7, 1911): 12–13. Curiously, Maillart did not include this in his list of ninety-nine works done up to 1913 even though he shows a photo of it under construction in his 1920 Geneva brochure.

Chapter Three

1. By 1908, the rules for building competitions had been set out by the Swiss Society of Engineers and Architects and there was already much experience with bridge competition. In late 1910, that society organized two special committees charged with developing set rules for engineering (Ingenieurwesen) competitions and heavy construction (Tiefbauarbeiten) competitions. Only Maillart and Wenner were on both of the seven- and eight-men committees. See *SBZ* 60(21) (November 23, 1912): 288–9. For the Rheinfelden guidelines, see Der Gemeinderat Rheinfelden, "Program für den Wettbewerb zur Erlangung von Projekten und Angeboten für eine neue Rheinbrücke in Rheinfelden," Rheinfelden City Archive, December 28, 1908.

2. For the Rheinfelden competition, see "Neue Rheinbrücke in Rheinfelden," *SBZ* 53(2) (January 9, 1909): 30–1. The general rules gave the names of the jury and stipulated that prize money (7,500 francs) would be distributed to the three or four best designs. These prize designs would become the property of the community of Rheinfelden, whereas all the others would be returned to their authors. Each design identified by a motto was to be submitted anonymously along with a sealed envelope identified on the outside with the motto and on the inside with the names. This envelope was to be opened only after the jury had written and signed its report allocating prizes. The jury's report was to be published in the *Bauzeitung*.

3. The jury consisted of Gustav Gull (1858–1942); C. Habich-Dietschy, a distinguished citizen of Rheinfelden; Karl Moser (1860–1936), an architect from Baden in Aargau, but then a professor in Karlsruhe; F. Schüle; and Alexander Trautweiler (1854–1920), an engineer from Laufenburg in Aargau. The building site rules described the location, the roadway widths, the required waterway openings to prevent disruption of the river flow, and the need for a temporary footbridge during construction. The specifications stated that the final bridge could not be of wood or of steel and they gave the loading that the bridge must carry. The final section specified requirements for the drawings, the written report, the calculations, and the construction cost. Also, those designers wishing to build their designs had to sub-

mit a detailed cost breakdown to which they were bound for 6 months after submission.

4. "Neue Rheinbrücke in Rheinfelden," *SBZ* 53(23) (June 5, 1909): 303. Forty-five separate designs had been submitted by the April 30 deadline, and on May 14, the jury first met and discarded twenty-three projects because of high cost, misalignment, a lack of architectural form (by which they meant visual form), no special engineering value, and two designs were simply "incomprehensible." All of these criticisms appeared in the report but without identifying the designers. After Schüle and Trautweiler had studied the remaining entries, the jury met on May 28 to decide on the content of the report, which consisted of a general statement on the architecture (the appearance), a general review of how the designs satisfied functional, structural, and constructional requirements, and a specific critique on each of the twenty-two first-round finalists.

For the architecture, the report emphasized the need for a beautiful replacement to the distinguished, covered wooden bridge then standing. It also regretted a number of architectually well-designed entries that had to be ranked low because of technical faults. It reviewed the functional questions of alignment, waterway opening, and vertical roadway slope, and judged that all were satisfactory apart from minor faults, that is, some pylons were not aligned with the river flow. Finally, the report reviewed the structural and cost questions, noting that all but one design used arches and that one very good project (this was Maillart's design, but at this stage unknown to the jury) was inexpensive because it did not use stone cladding.

The bulk of the report focused on the specific critiques. In a second round, the jury eliminated nine more, and after a final round, it selected four prize winners from the remaining thirteen and split up 7,500 francs. One typically Swiss feature of this and all competition reports during this period was the unanimity of result. Like the seven presidents of the Swiss Confederation, members of these juries never acted as individuals, but always as a single, unanimous unit. Only after all members had signed the report did the jury open the identification envelopes and connect the numbers and mottos to names:

I. Professor Melan, Engineer in Prague; de Vallière & Simon, Engineer in Lausanne; and Monod & Laverrière, Architects in Lausanne	2,300
II. Maillart & Co., Engineer in Zurich; Joss & Klauser, Architects in Bern	2,000
III. Alb. Buss & Co., A.G., Engineer in Basel; Emil Faesch, Architect in Basel; Franz Habich, Architect in Rheinfelden	1,700
IV. Wilhelm Storz, Engineer in Strassburg; Paul Schmitthenner, Architect in Colmar; Builder Ed. Züblin & Co. in Strassburg	1,500

The jury critique described the first prize design as made of "reinforced concrete arches following the Melan system." It found the calculations to be very carefully and analytically done, that is, in a non-Swiss way (the Swiss way being graphical, following Culmann and Ritter). The construction plans were also clear and the bridge form was "simple and beautiful in a rational, matter-of-fact way." There were no negative judgments on the engineering, but the cost (558,000 francs) was higher than any of the other prize winners.

For the second prize winner, Maillart's design, the jury's comments were very different from those for the first prize. It emphasized that the great advantage of the unclad concrete-block arches proposed by Maillart would be that during construction, the scaffold could settle without inducing stresses in the disconnected blocks. Also it found that "the calculations following elasticity theory are done with the greatest of care and detail." It judged the design to be "fully worked out and to demonstrate an outstanding achievement." For appearance, the jury commended its "simple overall form," but found some details displeasing. It described the very skillful renderings as worthy of special mention.

5. Letters from Messrs Brunner and Simmen in Rheinfelden to Maillart in Zurich, November 22, 1909, and from Maillart in Zurich to the Council in Rheinfelden, November 23, 1909. In Maillart's bid of 367,969 francs, he had admitted neglecting

a temporary wooden footbridge for 9,000 francs. Thus, his bid had become 376,969 francs. There followed a 2-month delay, and in January 1910, the council asked if his October bid was still valid and noted that Alb. Buss & Cie A.G. (who had won third place in the competition, working with a local architect) had the right to make a new bid, based on a 1897 contract for the building of a steel footbridge. The council asked if, in the light of that competition, Maillart would not consider dropping his price. Maillart responded on February 19; he stayed with his basic prices, but increased his general construction costs for such items as the scaffold and the footbridge to 390,969 francs.

6. Letters from Brunner and Simmen in Rheinfelden to Maillart in Zurich, January 24 and February 9, and from Maillart in Zurich to the Council in Rheinfelden, February 15 and 19.

7. Letter from Robert in Zurich to Maria in Orselina, April 25, 1910.

8. H. von Moos, and Guido Hunziker-Habich, *Bericht der Bauleitung über den Bau der neuen Rheinbrücke in Rheinfelden,* (Rheinfelden, January, 1913). Copies in the Rheinfelden Town Archives.

9. Ibid., 14–16. Because Buss had a previous contract with the community, it did later agree to make that firm an award of 25,000 francs, of which 5,000 was for removing their temporary steel footbridge and the rest was a compensation for losing the main construction contract, not an unfair judgment considering the effort Buss had made in the second competition.

10. J. Bolliger, "Brücke über die Sense bei Guggersbach," *SBZ* 51(8) (February 29, 1908): 107–10. The bridge had been completed in late 1906.

11. In a letter from Robert in Zurich to Maria in St. Magherita, March 23, 1908, he wrote, "This is completely our design but Jaeger and Company have passed it off as their invention."

12. R. Maillart, "Korrespondenz (Guggersbach Bridge)," *SBZ* 51(12) (March 21, 1908): 157. He did note one minor change: "only the roadway [slab] part is changed because the wheel load was increased from 1.5 tons to 3 tons."

13. Ibid., 51(13) (March 28, 1908): 169. This correspondence includes statements by Jaeger and Company, by district engineer von Erlach, and by Maillart. The Jaeger Company had responded

right away. As the *Bauzeitung* put it the following week:

We received from the firm of J. Jaeger and Company a long letter giving the early history of the structure of the Guggersbach Bridge. From this presentation, which has no direct bearing on the contested point, the Messers. Jaeger and Company came to a concluding statement, which we give here verbatim together with the clarification to which they refer: "That the constructed Guggersbach Bridge comes from our design and detailed plans without influence from the design by Maillart and Company is proved by the attached clarification from the cantonal building department in Bern." That clarification merely stated, "that the building of the Guggersbach Bridge by the firm of Gribi, Hassler and Company followed the plans of J. Jaeger and Company. From the design by Maillart and company neither J. Jaeger and Company nor Gribi, Hassler and Company received any information from here."

It was signed by von Erlach, district engineer who had represented the canton during construction.

In that same issue, Maillart replied simply that what von Erlach wrote was correct and that he had never claimed that the *district engineer* had given out any information on his design to Jaeger and Company. "If Jaeger and Company thus had the feeling that they needed such a support as that clarification, they were not able to find it in this document." Maillart claimed that Jaeger stole his design and Jaeger claimed that he received no information on Maillart's design from the *district engineer*. Was Maillart justified in his attack on Jaeger? Because Maillart's documents were never published, and his own copies did not survive, the question can only be answered by other evidence, which is worth some further consideration because of the judgment it permits of his character. Two other types of evidence are available: First, the 1904 Guggersbach design clearly derives from the Steinach Bridge of 1902, which was also built up of two rings of concrete blocks, over which concrete was cast; second, Jaeger submitted a design for the 1904 Uto Bridge competition, which more than any other prize design, took ideas of Maillart's from Billwil. Thus, Jaeger had already imitated a Maillart design once before.

Although none of these factors proves conclusively that Maillart's design was stolen, they make a convincing case. There is no further evidence on the other side to suggest that the Jaeger firm designed Guggersbach on its own, and it could not have been in Maillart's best business interest to stir up a controversy that also involved the canton bridge engineer's office unless the case was very strong.

14. Letter report from director Paul Miescher to the Vorsteher des Sanitäts-Departements (Basel, March 2, 1907), 3, in which he considered the Jaeger firm in a joint venture with a Basler firm to be unreliable. Two years later, the canton engineer of Aargau would choose a Maillart bridge design (for the Aarburg Bridge) over one by Jaeger's firm even though the latter was less expensive by 10 percent. See O. Zehnder, *Bericht* (Aarau, August 19, 1910). Zehnder stated that he did not hesitate to prefer Maillart over the ad hoc firm, of which Jaeger was the engineer, put together for this project.

15. "Rheinbrücke in Laufenburg," *SBZ* 55(18) (April 30, 1910): 244.

16. Karl Mommsen, *Drei Generationen Bauingenieure, Das Ingenieurbureau Gruner und die Entwicklung der Technik seit 1860* (Basel: Gebrüder Gruner, 1962), 310–98, is devoted to the Laufenburg power plant and includes some discussion of the bridge. By 1908, work had begun on the power plant and would continue with many difficulties up to the beginning of the War. Ibid., 349–64.

17. Ibid., 364. See also "Rheinbrücke," 231.

18. "Rheinbrücke," 244. The other members were an architect from Karlsruhe; a government representative from the Grand Duchy of Baden; an engineer as a deputy for Professor Engesser; Professor G. Schönleber from Karlsruhe as artist; the chief engineer for the power company; a councilman of Laufenburg; and the mayor of Klein Laufenberg on the German side.

19. *SBZ,* March 1910.

20. Letter from Robert in Zurich to Maria in Venice, July 26, 1910.

21. Letter from Robert in Zurich to Maria in Venice, August 5, 1910.

22. "Wettbewerb für eine neue Rheinbrücke in Laufenburg," *SBZ* 56(13) (September 24, 1910): 163–9.

23. R. Maillart, *Erläuterungsbericht: Neue Rheinbrücke Laufenburg* (Zurich, June 30, 1910): 8. In one design, Maillart made the arch out of reinforced-concrete blocks 80 cm deep at midspan; whereas in the other design, he envisioned a reinforced-concrete arch with a depth of only 40 cm at midspan. As he pointed out in his competition report, the essential virtue of the lighter reinforced-concrete design was a 40-cm increase in the bridge clearance (distance between the water line and the underside of the structure). With Maillart's light bridge, the jury recognized that materials would be saved, but it expressed concern that the very thinness he prized might lead to vibrations (of the scaffold) and risk during construction.

24. Ibid. The trees could well have been the idea of Joss and Klauser. There is no way of knowing just what they contributed to the design. Still, the way Maillart wrote the report shows he liked the idea of the little border park and it is still today a lively and appealing part of the design. The bridge was rehabilitated in 1982; see Ernst Woywod and Branislav Lazic, "Sanierung der Maillart – Bogenbrücke über den Rhein in Laufenburg," *Schweizer Ingenieur und Architekt* 101(29) (July 14, 1983): 763–8.

25. Letter from Robert in Zurich to Maria in Venice, August 1, 1910.

26. These calculations are dated October 14, 19, 22, and November 9, 1910, but they must have been done in June for the competition and simply given a new date and an office identification after he had gotten the contract. Eduard Gruner, son of Heinrich Eduard Gruner, who was chief engineer for the Laufenburg works, recalls that "It is therefore probable that the Maillart bridge was chosen for economical reasons which its designer had offered due to accurate calculations." E. Gruner, *The River Rhine Bridge at Laufenburg* (Basel, February 3, 1975), in the Princeton Maillart Archive.

27. Maillart had other pictures of these bridges and his brochures all had photos of the completed bridges. At Laufenburg, he showed the arches being built of blocks in a crenellated fashion, whereas in Rheinfelden, he showed only the wooden footbridge. Even his brochures of 1914 and later show more photos of the construction of these two works than of the final bridges.

28. Letters from Robert in Zurich to Maria in Venice, July 23 and 27, 1910. For the bridge, see O. Albrecht, "Die Wasserkraftanlage Augst-Wyhlen: II Das Kraftwerk Wyhlen, Die Kabelbrücke," *SBZ*

62(8) (August 23, 1913): 97–9 In a footnote, the editor states that this article comes from the builder, Maillart and Company. Through the general contractor, Conradin Zschokke, Maillart had received other small contracts; see letter from Robert in Zurich to Maria in Orselina, April 6, 1910. See also G. Hunziker-Habich, "Die Wasserkraftanlage Augst-Wyhlen: I Das Stauwehr, Die Wehrbrücke," *SBZ* 61(15) (April 12, 1913): 196–8. See also letter from Robert in Zurich to Maria in Venice, July 11, 1910.

29. Letters from Robert in Zurich to Maria in Venice, July 27 and August 1, 1910. Over the next weekend, Maillart went to Bern to enter another shooting match and, more important to him, to be away from Zurich while his office was being moved to larger quarters. After another tour to Rheinfelden on Friday, July 29, Maillart was met at the Bern station by Otto Tschanz, who showed him a newspaper announcement of his being awarded the Aarburg Bridge. Possibly this good news stimulated his shooting, because over the weekend, he finally won a prize for that, too. For the award to Maillart of the Aarburg Bridge, see "Aarebrücke in Aarburg," *SBZ* 56(6) (August 6, 1910): 83.

30. Ed. Keller and W. Ritter, *Gutachten über den Zustand der Drahtseilbrücke in Aarburg* (Winterthur and Zurich, January 11, 1887). His coauthored report of January 11, 1887, is the earliest cantonal document indicating the need for a new bridge.

31. O. Zehnder, *Bericht* (Aarau, August 19, 1910). From Zehnder's report, it appears that Maillart must have entered the 1908 competition; but there is no available Maillart document that mentions this earlier competition.

32. *Auszug aus dem Protokoll des Regierungsrates des Kantons Aargau* (Aarau, October 15, 1910). In September 1911, the cantons of Aargau and Solothurn agreed on the source for funding the Aarburg Bridge and on October 7 gave Maillart the contract. *Dekret betreffend den Bau einer Brücke über die Aare nebst Zufahrtstrassen in Aarburg* (Aarau, September 14, 1911), and *Bauvertrag, Brücke über die Aare bei Aarburg* (Aarau, October 7, 1911). The total cost of 112,000 francs for bridge and approaches was to be assessed as follows:

Canton Solothurn	15,000 francs
Olten-Aarburg Power Plant	12,000
Canton Aargau	30,000
Community of Aarburg	50,000
Nine small adjacent communities a total of	5,000
TOTAL	112,000 francs

33. Maillart speculated that his superintendent perhaps would be put in prison for some time because of negligence. Thus, unlike present-day practice in the United States, Maillart was not held liable. See letters from Robert in Zurich to Maria in Bern, October 13 and 15, 1910.

34. Letter from Robert in Zurich to Maria in Karlsbad, June 8, 1911, and letter from Robert in Zurich to Maria in Parpan, July 4, 1911. The bridge was opened ahead of schedule on January 15, 1912; see Mommsen, *Drei Generationen*, 367.

35. *Bauvertrag.* In early April 1912, one of the scaffold supports had settled after concreting the arch and much of the concrete had to be removed and the scaffold jacked up. The complication meant that he would lose money on the bridge. But more than that, it was yet another instance of the difficulty in carrying out Maillart's designs without Maillart to supervise.

36. Card from Robert in Fribourg to Maria in St. Margherita, March 2, 1908, and letter from Robert in Zurich to Maria in St. Margherita, March 3, 1908; see also *SBZ* 46(21) (November 18, 1905): 259–60.

37. "Pont de Pérolles in Fribourg," *SBZ* 51(9) (February 29, 1908): 116.

38. Maillart submitted his design on May 1; see letters from Robert in Zurich to Maria in St. Margherita, April 28, 30, May 2 and 4, 1908. In early June, he traveled to Fribourg to see the results. Card from Robert in Fribourg to Maria in Kilchberg, June 6, 1908.

39. "Pont de Pérolles," *SBZ* 51(23) (June 6, 1908): 301. For the jury report, see "Ideenwettbewerb, für den Pont de Pérolles in Freiburg," *SBZ* 52(6) (August 8, 1908): 77–81, and 52(7) (August 15, 1908): 89, 92–4. For Moser's ideas, see, as an example, R. Moser, "Betonbrücke der Gürbethalbahn," *SBZ* 38(24) (December 14, 1901): 257–8. For his love

of stone, see R. Moser, "Ueber Steinerne Brücken," *SBZ* 25(21) (May 25, 1895): 146–9.

40. One could interpret the criticism about architecture in two ways. First, there was no architect involved and hence no architectural renderings or decorative details typical of bridge design of the time. Second, the lack of architecture could also have meant that the aesthetic side was neglected whether an architect was involved or not. The jury did criticize one of the other designs for its aesthetics and did indicate how it could be easily improved. But with Maillart's, it said only that an architectural study was missing from the bridge.

41. F. von Emperger, "Ein Wettbewerb um Ideen für eine Brücke in Freiburg," *Beton und Eisen* 7(14 and 15) (1908): 343–5, 353–8. The only document surviving from Maillart's files is an undated and unsigned drawing entitled *Pont de Pérolles "Hohlbau" Blatt 1, Ansicht und Grundriss*. The drawing is a print on which color has been added; also pencil sketches of scaffolding are lightly visible (Fig. 53).

42. Another failed competition for Maillart was that for the Walche Bridge in Zurich; see "Zürich, Walchenbrücke," *Beton- und Eisen-Konstruktion* 15 (September 23, 1910): 144. The Maillart designs appear in the jury report, "Wettbewerb zur Erlangung von Plänen für die Walchebrücke in Zürich," *SBZ* 57(10) (March 11, 1911): 138–45. His own records for this bridge have not survived.

43. For the competition announcement, see "Lorrainebrücke in Bern," *SBZ* 56(23) (December 3, 1910): 314–15. The other members of the jury were the canton engineer of Basel, Ed. Joos of Bern as architect, and the city building commissioner of Bern.

44. On July 14, 1911, Maillart received a patent on the concrete block idea; see Hauptpatent 56981.

45. For the awards, see *SBZ* 57(15) (April 15, 1911): 213. The first place winner was the same Albert Buss against whom Maillart had competed for the Rheinfelden Bridge. Buss, combined with several other engineers and an architect, received 3,000 francs, and each of the three second place contestants won 1,750 francs. Professor Melan was a designer in one of those. For the Jury report, "Wettbewerb für eine Lorrainebrücke in Bern in Eisenbeton oder Stein," *SBZ* 57 (24 and 25) (June 17 and 24, 1911): 323–30, 344–51.

46. Letter from Robert in Zurich to Maria in Weggis, April 19, 1911.

47. Letter from Robert in Zurich to Maria in Karlsbad, May 26, 1911.

48. Letter from Robert in Zurich to Maria in Karlsbad, June 1, 1911.

49. "Der Wettbewerb eines Aareüberganges von der Stadt Bern nach dem Lorrainequartier," *SBZ* 30(7, 8, 9, and 10) (August 14, 21, 28, September 4, 1897): 50–2, 58–61, 67–9.

50. "Ueber die Lorrainebrücke Konkurrenz," *Die Schweizerische Baukunst* 12 (June 16, 1911): 161–9. The article itself, while praising the Maillart design and responding to the jury criticism of it, does present an architect's view of bridge design that was far from that of Maillart. Of his design, it said, for example: "Here clearly the architect has first given the bridge its proper form and then the engineer has worked out the structural problems." In the case of this design, of course, the architects did play an important role and it is just that role in the final built structure that hid the fact that it is a Maillart bridge. At the same time, an article on technical issues taken from Maillart's competition report appeared in "Zum Wettbewerb der Lorrainebrücke," *Beton- und Eisen-Konstruktionen*, 6 (June 16, 1911): 69–72.

51. "Wettbewerb für eine Lorrainebrücke in Bern in Eisenbeton oder Stein III," *SBZ* 58(3) (July 15, 1911): 33–9. The *Bauzeitung* also showed the similarity between the first place winner and the 1897 design and emphasized the loss to the profession of not seeing newer ideas rewarded. The *Bauzeitung* next described two designs that were not given prizes: one a reinforced-concrete rigid frame of surprisingly advanced appearance and a very low-cost proposal and the other the Maillart design. The first one showed quite new ideas both visually and technically; the second one, while also based on new ideas, was given primarily to show the weakness of the jury's report.

52. On November 21, 1912, the city council of Bern voted 43 to 32 to postpone the Lorraine Bridge question, *SBZ* 60(22) (November 30, 1912): 302, and on December 6 of that year, at a meeting of the Bern branch of the Swiss Society of Engineers and Architects, Hans Herzog, Maillart's former

employer and then on the city council, noted that the new bridge could not be built without new taxes. *SBZ* 61(1) (January 4, 1913): 14. Thus, the "Lorraine Bridge question," as it was called, found no prewar answer, but it did serve to make public the problems of competitions and it further brought closer together the views of the *Bauzeitung* with those of Maillart.

There were five points at issue and two are enough to give the sense of the debate. The jury had criticized the vertical walls of the Maillart design as being cladding, and thus too heavy in appearance by making the bridge appear as one of fake stone. On the contrary, the article stressed that the walls actually carry the loads and are part of the structure. Clearly, the jury did not understand Maillart's basic design, which the *Bauzeitung* was at pains to illustrate with drawings to make the point that a study of the report and calculations should have made clear to any engineer how the bridge was designed.

As a second example, the jury criticized Maillart's proposal as being very costly for an 11-meter-wide bridge; the other bridges were 15 meters wide. But here again the jury failed to study the plans, as the *Bauzeitung* emphasized. Maillart's arch is 11 meters wide but the deck is 15 and even over a major part of the length 2.5 meters wider than 15 meters. Hence, if compared properly with the others, his cost, as the *Bauzeitung* computed it, should be reduced by over 7 percent.

53. *SBZ* 71(5) (February 2, 1918): 58–9.
54. Of the six juried bridge competitions between 1908 and 1911, Moser's role was the engineer on four (Pérolles, Luzern, Laufenburg, and Lorraine).
55. Letter from Maria in Zurich to Bertha in Bern, November 10 and 12, 1903.
56. Letter from Robert in Zurich to Maria in Arosa, February 1, 1906, and a letter from Robert in Zurich to Maria in Italy, October 17, 1907.
57. Her travels are recorded in letters from Maillart; for example, letter from Robert in Zurich to Maria in Zuoz, July 16, 1907. He asked, "You did not tell me how you found the bridge. Have you seen the cracks?" Also letter from Robert in Zurich to Maria in Paraggi, Italy, October 17, 1907, and card from Robert in Bern to Maria in St. Margherita Ligure, Italy, November 11, 1907.

58. Letter from Robert in Kilchberg to Maria in Santa Margherita, December 30, 1907.
59. Aspects of separation are revealed in a letter from Bertha in Bern to Maria in Zurich, December 30, 1907, and in letters from Robert in Kilchberg and Zurich to Maria in St. Margherita, January 26 and February 24, 1908.
60. Letter from Robert in Zurich to Maria in St. Margherita Ligure, April 28, 1908.
61. Letters from Robert in Zurich to Maria in St. Margherita, April 28, 30, and May 2, 1908.
62. Letter from Robert in Zurich to Maria in Airolo, July 29, 30, 31, and August 2, 1908. See also letter from Bertha in Geneva to Maria in Airolo, August 1, 1908. Bertha was astonished that Robert had made the decision to take an apartment after Maria had left for Airolo.
63. For the sudden departure of Maria, see letter from Robert in Zurich to Maria in Görwihl, July 29, 1909, card from Robert in Albbruck (Germany) to Maria in Görwihl, August 1, 1909, and letter from Robert in Zurich to Maria in Görwihl, August 9, 1909.
64. Letters and cards from Robert in Copenhagen to Maria in Zurich, September 2 to 13, 1909.
65. Letter from Robert in Zurich to Maria in Karlsbad, June 5, 1911.
66. Letters from Bertha in Geneva to Maria in Zurich, October 5 and 19, 1912, and letter from Edmond in Geneva to Maria in Zurich, October 12, 1912. On December 30, 1912, Maria received a card from a good friend in Genoa who wrote, "I have heard from Rosa [Maillart's sister] that your move [to Voltastrasse 30] went well. Our congratulations to 'Madame la proprietaire.' It appears that the works in Russia are very lucrative!"
67. Charles Lehr, *Recollections of Robert Maillart* (Manasquan, New Jersey, April 10, 1978). I interviewed Lehr at his apartment in Manasquan in early March 1978 when he was already over 90 and in failing health. However, during the interview, he remembered many details of his time with Maillart between 1913–14 and 1916–17. I wrote these details in a document dated March 8, 1978, Princeton Maillart Archive.
68. But by far the most important record of that time comes from the recollections of Maillart's only daughter, whose memory of that time is vivid, in

spite of her having been not quite 8 years old when they left the house for Russia in 1914. See M.-C. Blumer-Maillart, *Recollections De Mon Père* (Princeton, 1978): 14.

69. Ibid., 15. For Maillart on the dance floor, little Marie-Claire would sometimes watch and one evening she recalls, "As I watched from the living room door, I saw my father suddenly stop, his eyes riveted to the floor worriedly. 'Let's see,' said the dancing master, 'one, two, three, four, . . . ' 'one, two, three, four,' my father repeated mechanically without raising his head. 'What should I do? Where is four?' "

70. Ibid., 15–16.

71. Ibid., 16–17.

72. F. Schüle, *Mitteilungen,* der Eidgen. Prüfungsanstalt am Schweiz. Polytechnikum in Zürich, Vol. 10, 1906. For a detailed review of this publication, see F. von Emperger, "Bücherschau," *Beton und Eisen* 5(7) (July, 1907): 183–4.

73. See, for example, Maillart's design for the 1905 Demutstrasse Bridge in St. Gallen, "Brücke an der Demutstrasse," drawing dated January 12, 1905, stamped by Maillart and Company, Zurich. See also E. Mörsch, *Concrete-Steel Construction* (English translation of the 1908 3rd ed., New York, 1910), 165. There Mörsch notes that the test of a beam with Hennebique-type shear reinforcement gave only three-quarters of that for a similar one with shear reinforcement placed perpendicular to the direction of the diagonal crack. Arthur N. Talbot at Illinois University used the same Hennebique-type shear reinforcement for his beam tests from 1904 to at least as late as 1908; see A. N. Talbot, *University of Illinois Engineering Experiment Station*, Bulletin Nos. 1 (September 1904), 4 (April 1906), 12 (February 1907), 28 (October 1908), and 29 (January 1909).

74. F. Schüle, *Protokoll der zweiten Sitzung: Schweizerische Commission für die Untersuchung des armierten Beton* (Zurich, October 31, 1906), from the Basel City Archive. The Commission consisted of Schüle as president; Mörsch as vice president; Fritz Locher (1842–1906), head of one of Switzerland's oldest construction companies; Maillart; the Basel City engineer; the Swiss Railway Director; and another structural engineer. Locher and Maillart represented the Swiss Society of Engineers and

Architects. When Locher died, Ed. Elskes replaced him.

75. Ibid., 4.

76. Ibid., 6.

77. Elskes was a student of Ritter in the sense of being influenced strongly by his ideas. He was a younger colleague of Ritter's on consulting projects such as the Coulouvrenière Bridge, but he graduated from the Polytechnikum in 1880, 2 years before Ritter returned from Riga.

78. F. Schüle, "Ueber vorschriften für armierten Beton," *SBZ* 49(1) (January 5, 1907): 29–34, and A. Considère, Schüle's counterpart in France responded in A. Considère, "Observations relative à deux articles de Monsieur le Professor Schüle," *SBZ,* 49(10) (March 9, 1907): 121–4.

79. Following the October 1906 meeting, Schüle called a series of subsequent ones in 1907 and 1908, and, finally, the commission agreed to a new code on April 30, 1909. *Vorschriften über Bauten in armiertem Beton,* aufgestellt von der Schweizerischen Kommission des armierten Beton nebst Erläuterungen von Professor F. Schüle, President und Berichterstatter der Kommission, June, 1909, Verlag der Eidg. Materialprüfungsanstalt in Zürich.

80. Letter from Bertha in Geneva to Maria in Zurich, February 6, 1911. "I did not know that it was an official affair – Paul simply told me that they had asked Robert to give fifteen lectures – in any case I congratulate him and you also because you get half of the honor that he receives. Paul has been astounded at the great works that Robert must direct. He truly needs force and health."

81. Letter from Robert in Zurich to Maria in St. Margherita, March 23, 1908, and to Maria in Adelboden, January 9, 1911. "The result [for the Walche Bridge competition] is not very satisfying," he wrote Maria, but he found it intriguing that one of the second place winners was Froté and another was Westermann, each acting as a builder and each combined with different engineers. Westermann's architect was Pfleghard and Haefeli and his enginer was Max Ritter, the brother-in-law of Mr. Haefeli. Ritter would become Maillart's most persistent detractor after the war.

82. Maillart's course expanded in 1912 to two class hours per week and three hours of computational exercises per week. Theophil Wyss, in his last year

at Zurich, took complete notes of both Maillart's and Ritter's courses during the winter semester of 1912/13 and left his manuscript in the archives of the Technical University. R. Maillart, *Armierter Beton* (Zurich, n.d., probably 1912/13, notes taken by T. Wyss). In addition, Hans Missbach kept a set of notes and sent them to Princeton at the suggestion of Professor Stüssi: R. Maillart, *Eisenbeton*, Vol. I, No. 1, 1913; Vol. II, February 1913; Vol. III, October 1913; Vol. IV, February/March 1914.

83. Max Ritter, *Statik der Eisenbetonbauten* (Zurich, n.d. and unsigned). The handwriting is the same as in the Maillart notes and were together with them; it is therefore reasonable to assume that Wyss took them both at about the same time.

84. For example, torsion is illustrated by Ritter with a pure, idealized cantilever beam somehow twisted, whereas Maillart shows two actual cases: the exterior rib of a ribbed floor and an electrification pole after wires on one side have broken.

85. Von Emperger had published surveys of types of courses taught and his data show the Zurich Institute to have been reasonably typical. See *Beton und Eisen* 7(3) (February 19, 1900): 68; 8(2) (January 28, 1909): 46; 9(2) (January 28, 1910): 42; 10(4) (February 13, 1911): 73; 11(3) (February 2, 1912): 69; and 13(7) (April 21, 1914): 153. The survey included the following Technical Universities: Aachen, Berlin, Braunschweig, Brünn, Danzig, Darmstadt, Dresden, Graz, Hanover, Karlsruhe, Lausanne, Lemberg, Munich, Prague, Stuttgart, Vienna, Zurich, and Hamburg.

86. *Canton Book* (Schwyz, 1910), 176–83, and "Competition Documents. Strassenbrücke über die Muota in Schwyz-Ibach," Akten No. 15, Staatsarchiv (Schwyz, March 30, 1912). Gubelmann's bridge consisted of a solid arch assumed to carry all the bridge load much as does the Stauffacher Bridge.

87. Oron J. Hale, *The Great Illusion: 1900–1914* (New York: Harper & Row, 1971), 76.

88. James H. Bater, *St. Petersburg: Industrialism and Change* (London: Edward Arnold, 1976), 1–16.

89. Ibid., 213–28.

90. For Maillart's first days in St. Petersburg, see letter from Robert in St. Petersburg to Maria in Zurich, April 17, 1912 (by the Russian calendar at that time, it was April 4, 1912).

91. Letter from Robert in St. Petersburg to Maria in Zurich, July 12, 1912.

92. Letter from Robert in St. Petersburg to Maria in Zurich, July 15, 1912, and letters from Robert in Bucharest and in Zurich to Maria in Rimini, July 24 and 30, 1912, resp.

93. The foundation materials on either bank were found to be uncertain, and after further study, Maillart decided to abandon the arch with its dependence upon horizontal as well as vertical foundation resistance. Instead, he proposed a cantilever bridge in which the large concrete masses under the approaches would serve as counterweights to the cantilevers, which in turn support a 5.84-meter-span beam at midspan. This system required only vertical resistance by the foundations, even though it led to more material and probably a higher cost than the arches. Maillart submitted his new proposal on August 23, 1912, for 75,142 francs. H. Gubelmann, "Die Muotabrücke in Vorder-Ibach," *SBZ* 62(26) (December 27, 1913): 355–8. Maillart's original hollow-box design appears to have been modeled after Tavanasa, although no drawings have been found. See letter from Maillart in Zurich to the canton engineer in Schwyz, March 30, 1912, with which he enclosed a report, a cost breakdown, and a drawing (now lost). The report refers to the description of Tavanasa in von Emperger's 1908 *Handbuch* on bridges. The new design was ready at least as early as July 29, 1912, and the complete drawing of the new design is dated August 15, 1912, in Zurich. On April 15, 1913, the bridge passed its load test. Letter from Gubelmann in Schwyz to Ochsner in Einsiedeln, April 10, 1913, and a report sent from E. Stettler in Bern to Gubelmann in Schwyz, April 29, 1913. Stettler was the father of Ernst Stettler, who would become Maillart's chief engineer in Bern from 1926 to 1940. The bridge was rehabilitated in 1979. Hans Weber and F. Pfister, "Die Sanierung der Muotabrücke in Ibach," *Strasse und Verkehr* 65(5) (May 1979): 183–8.

94. Letter from Robert in Zurich to Maria in Rimini, August 3, 1912.

95. Letter from Robert in Zurich to Maria in Rimini, August 8, 1912.

96. For some of these movements, see letters from Robert in Zurich to Maria in Rimini, July 30, August 3, 6, 8, 10, 22, 23, and 27, 1912.

97. See letters from Bertha in Geneva to Maria in Zurich, undated but found with those of January–February, 1913, and almost certainly from early February 1913. See also cards from Maria in Italy to Robert in Nice, March 17 and 19, 1913, and letter from Robert in Zurich to Maria in Genoa, March 29, 1913.

98. Letters from Robert in St. Petersburg to Maria in Zurich, April 11 and 15, 1913. He went to Peterhof on Saturday, April 19, to talk about the appearance of a bridge with some old generals of the "crown." There is no further mention of these bridges in his letter the following week nor in any further surviving letters from him.

99. Letter from Maria in Rimini to Robert in St. Petersburg, July 14, 1913. There has been speculation that Maillart did design several bridges in St. Petersburg, between 1912 and 1914; see André Corboz, "Un Pont de Robert Maillart Leningrad," *Archithese*, Lausanne, 2 (1971): 42–4. However, in the *Story of Leningrad's Bridges* (Leningrad, 1972; in Russian), there appears on p. 161 a photo of the same bridge shown by Corboz with the statement about the "Malo-Krestovskii Bridge, built in 1963, after the project by the engineer, J. L. Turkov and architect, L. A. Noskov. The span is 46 m." I am indebted to Gregory P. Tschebotarioff for finding this information and for M. S. Troitsky, who confirmed it independently. Also, Frederick Starr came to the conclusion after some investigation in Leningrad that the bridge shown by Corboz "is definitely not the work of Maillart"; see Frederick Starr, "Notes on Robert Maillart in Russia." Unfortunately, Maillart's July correspondence to Maria has not survived. However, in mid-July, Maria wrote him in St. Petersburg and commented that "There you are at the work, and I am pleased to know that the bridges have begun." It would seem, therefore, that some Maillart bridge design was built in 1913, but no surviving documents have yet been found to establish what types of bridges they were. Because the aesthetics had to be passed on by the court, it is unlikely that they would have been of the radical Tavanasa type. In a list of works Maillart made up in 1931, he included "Brücken Oranienbaum," St. Petersburg, 1913–15. This was item number 2 in a list of works beginning in 1913. Many items on this list are shown crossed off by Maillart and item 2 is one of these. It never appears on later lists. Maillart clearly did not think these bridges to be significant. See Chapter Four, n. 3.

100. Letters from Robert in St. Petersburg to Maria in Zurich, April 17 and 18, 1913.

101. On Monday, September 1, Maillart left Zurich once more for 2 weeks in Russia, followed by a week with Max in Barcelona, where he had established the branch office and was soon to begin a series of large works. See card and letters from Maria in Zurich to Robert in Riga, September 2, and September 6 and 8, 1913, resp. The card is addressed to Maillart in care of Mr. L. Schneider, Betonbau-Gesellschaft, Schulenstrasse, Riga.

102. Hale, *The Great Illusion*, 384, and Marc Ferro, *The Great War: 1914–1918* (London: Routledge and Kegan Paul, 1973; originally published in French, 1969), 22.

103. Cards from Robert at Eydtkuhnen to Maria in Adelboden, December 27, 1913, and from Maria in Adelboden to Robert in Riga, December 30, 1913.

104. Letters and one card from Robert in Zurich to Maria in Adelboden, January 6, 9, 10, 11, and 13, 1914. He had no time to visit his family in Adelboden, so he wrote to Maria that: "M. Nissen will bring you therefore my greetings and he will replace me for the children in order to play sport with them." Paul Nissen had by then become an intimate friend of the entire family, and along with Max von Müller and Otto Tschanz, Maillart's closest companion.

105. "Souvenir du 14 Février 1914," written by Maria Maillart.

Memories of the Month of February

1re	February, 1902 Firm of Maillart & Co.
2	Birth of the dear "Associate" [Maria]
6	Birth of the Signor Chief
14–1914	Harmonious evening! 1st Russian Million!
18–1909	Muscular contractions [pregnancy with René?]

| 25 | Successful birth of Anny [Niggli] |
| 27 | Birth of a "Rose" [Rosa Wicky] |

The other side began with the place and date and contained a word play on "Harmonie":

Odéone, the morning of 14 February 1914

Heureuse soirée!
Amis réunis
Revenue fidèles
Maintiendront
Oh! combien
Nouvelle et ancienne
Idéale "harmonie"
Et éternelle amitié!

Rosa Wicky
Maria Maillart
Maxli v Müller

Along with those signatures, there appeared, "Reviewed and corrected on the 16 of February 14 by Anny Niggli."

106. J. Rutkis (chief editor), *Latvia: Country and People* (Stockholm: Latvian National Foundation, 1967), 171–2.

107. Ibid., 180–1.

108. In 1930, the rubber factory was partly destroyed by fire, and an article in *Beton und Eisen* attributed this destruction partly to questionable design practice. Maillart responded in writing.

He made two important points: first, that the basic design, including the "architecture," was done by the Russo-Franco Rubber Company, "Prowodnik," and, second, that the collapsed part was built not by him but by a German firm. The main cause of failure was not because of design mistakes, but because of overloading and the lack of fire walls, conditions beyond his control. One of the main concerns in the article was the "feeling of an unsafe lightness." Maillart stoutly defended his light design in a reasonable enough way to convince the author, who ended his closure with, "I am entirely in agreement with Herr Maillart that the deflection of the floor [under heavy overload] is the primary cause of the catastrophe." See Ed. Weiss, "Die Brandschaden Prowodnik in Riga," *Beton und Eisen* 29(19) (October 5, 1930): 344–

8; R. Maillart, "Der Brand eines Fabrikgebäudes der Gummifabrik 'Prowodnik,' in Riga," *Beton und Eisen* 30(11) (June 5, 1931), 206–7; and Ed. Weiss, "Erwiderung," 207.

109. *Maillart Geneva Brochure* (Geneva, 1920), p. 33.

110. Card from Robert in Basel to Maria in Zurich, April 7, 1914.

111. Cards from Robert in Berlin and St. Petersburg to Maria in Zurich, April 8 and 9, 1914.

112. Letter from Robert in St. Petersburg to Maria in Gersau, April 10, 1914; letters from Maria in Gersau to Robert in Riga, April 12 and 14, 1914; and letters from Robert in Riga to Maria in Gersau, April 12 and 13, 1914.

113. Letter from Maria in Monsummano to Robert in Zurich, undated but probably May 18 or 19, 1914. She was in Italy for her health. René came down with scarlet fever on Sunday, May 17, 1914. See also letter from Maria in Zurich to Robert in Edinburg, May 30, 1914. He went by train to Berlin and thence north to Stettin, from where he took the 2-day boat trip to Riga. Telegram from Robert in Riga to Maria in Zurich, June 1, 1914: "bella traversata buona salute bacio te rené – robert." For this trip, see K. Baedecker, *Russia* (Leipzig, 1914), 46; see also cards from Maillart in Basel, Berlin, Stettin, and Riga to Maria in Zurich, May 29, 30, and June 1, 1914.

114. Rutkis, *Latvia*, 202–5.

115. Baedecker, *Russia*, 63–4. Edinburg is called Dzintari in Latvian and the Riga coast is called Rigas Jurmala.

116. Letter from Robert in Edinburg to Maria in Zurich, June 1, 1914. Maria's trip was planned for June 26; letter from Maria in Zurich to Robert in Edinburg, June 1, 1914, undated, but noted as "Monday of Pentacost."

117. Letter from Robert in Edinburg to Maria in Zurich, June 4, 1914, undated, but noted as "Thursday evening."

118. Sidney B. Fay, *The Origins of the World War*, 2nd ed. (New York: The Free Press, 1966), 49, 126.

119. Hale, *The Great Illusion*, 313–14.

120. Plans for the family's trip appear in a letter from Robert in St. Petersburg to Maria in Zurich, June 9, 1914; cards from Robert in Pskov and Riga to Maria in Zurich, June 9 and 11, 1914; card from Robert in Riga to Maria in Zurich, June 16, 1914.

121. Blumer-Maillart, *Recollections*, 19–21.
122. Hale, *The Great Illusion*, 293–312.

Chapter Four

1. Cards from Erica Baerwolff in Zurich to Edmond in Riga, July 29, 1914, and from Daisy Haag in Davos to René in Riga, July 31, 1914.
2. Alan Clark, *Suicide of the Empires: The Battles on the Eastern Front 1914–18* (New York: American Heritage Press, 1971), 5–47.
3. Eight of the Russian projects are given in a list Maillart made in 1913. When he revised that list in 1933, only two of the eight were left in but three others were added. In the 1940 list published by Mirko Ros, only a total of four projects remained. A composite list shows that Maillart completed the following eleven works in Russia between 1912 and 1916.

1.	1912	Refrigeration warehouse (Gerhard & Hey)	St. Petersburg
2.	1913	Refrigeration warehouse (S.I.A.E.C.)	St. Petersburg
3.	1913–15	Bridges	Oranienbaum–St. Petersburg
4.	1913–14	Cellar reinforcement (Gerhard & Hey)	St. Petersburg
5.	1913–14	Rubber factory (Prowodnik)	Riga
6.	1913–14	Stairs and caissons (Alexandrowski)	St. Petersburg
7.	1914–15	Warehouse (Prowodnik)	Riga
8.	1914–15	Warehouse (Stock Exchange)	Riga
9.	1914–15	Restaurant (Prowodnik)	Riga
10.	1915–16	Factory and buildings (Russian General Electric)	Karkov
11.	1916	Steel mills (Russian Government)	Kamenskaja

The 1931 list gives numbers 3, 4, 5, 6, 7, 8, 9, and 10; the 1933 list gives 1, 2, 5, 10, and 11; and the 1940 list gives 1, 2, 5, and 10.

4. Blumer-Maillart, *Recollections*, 20–1.
5. Cards from Bircher in St. Petersburg, Stockholm, and Malmo to the Maillart family at Industriestrasse 3a in Riga, October 3, 8, and 8, 1914. The St. Petersburg card is dated September 21, which was by the Russian calendar and was 13 days behind the western date. In the text, I have consistently used only the western dates.
6. Card from Robert in Moscow to Marie-Claire in Riga, November 16, 1914.
7. Card from Max Maillart in Geneva to Maria in Riga (forwarded to Moscow), December 30, 1914.
8. Blumer-Maillart, *Recollections*, 22.
9. Hindenburg later wrote, "are not these marches in the winter nights, that camp in the icy snowstorm . . . but the creation of an inspired human fancy?" He knew the horrible truth, not only that the fancy was fact, but also that "in spite of the great tactical [German] success . . . we failed . . . strategically." The quotations are taken from Winston S. Churchill, *The Unknown War: The Eastern Front* (New York: Scribner's, 1931), 298–9.
10. Blumer-Maillart, *Recollections*, 23. In these recollections, Madame Blumer-Maillart puts this scene with the Czar at the beginning of the 1915–16 winter. However, the family had moved from St. Petersburg to Kharkov in September, 1915 and there is no record of Marie-Claire having made the long trip to St. Petersburg after that time. Therefore, I assume the events occurred in the late winter of 1914–15.
11. Cards from Bircher in Switzerland to Marie-Claire in Riga, from Maria in Riga to Edmond in Edinburg, and from Bertha in Geneva (three cards) to Edmond, Marie-Claire, and René in Riga, April 12, 18, and 20, 1915 resp.; and also letter from Bertha in Bern to Edmond in Riga, May 19, 1915.
12. Blumer-Maillart, *Recollections*, 23. For the German advances, see Churchill, *Unknown War*, 309–15, 319–26.
13. Maillart described this region in his letter from St. Petersburg to Maria, July 15, 1912, and the trip to Oranienbaum is described in Baedeker, *Russia*, 178–183. The family arrived there sometime in early July; see card from Edmond in Petrograd to Walter Bisig in Zurich, July 19, 1915.
14. Baedeker, *Russia*, 183.
15. Churchill, *Unknown War*, 325–30.

16. Card from Robert in Kharkov to Viktor Tschiffely in Oerlikon, August 25, 1915. Maillart tells Tschiffely not to lose courage and holds out to him the hope that there may soon be work for him with Maillart in Russia.

17. Letter from Maria in Kharkov to Edmond in Geneva, September 21, 1915. Edmund arrived in Switzerland in mid-October.

18. Letter from Maria in Kharkov to Edmond in Glarisegg, October 14, 1915. The letter (which has not survived) from Maria in French was sent to Bertha in Geneva, who sent a German translation to Edmond in Glarisegg on November 1, 1915. See also Blumer-Maillart, *Recollections*, 28.

19. Clark, *Suicide of the Empires,* 91–3.

20. Letters from Bertha in Geneva to Edmond in Geneva, February 4, March 2 and 13, 1916.

21. Clark, *Suicide of the Empires.*

22. Blumer-Maillart, *Recollections,* 25, 26.

23. *The Russian Year-Book,* comp. and ed. N. Peacock (London; September, 1916), 460. This was the last edition of a work on business that had begun in 1911 and thus coincided with the intense commercial and industrial activity in Russia during the period of Maillart's building there.

24. Letter from Bertha in Creux de Genthod to Edmond in Zurich, June 13, 1916. Bertha tells of getting a letter from Maillart, which took only 18 days from Russia and in which he tells of Max's presence and of his hopes to get more work for his newly assembled construction force. Maillart's letter must have been sent from Kharkov on or about May 26, but it has not survived.

25. Clark, *Suicide of the Empires,* 103, and Churchill, *Unknown War,* 363–7.

26. Lehr, *Recollections.* For Tschiffely's position with Maillart, see Document, Kharkov, June 16, 1919 which states that he began work with Maillart as a manager of construction in 1911.

27. Maillart & Cie, Ingenieurs civils, *Enterprise de travaux en béton armé* (Geneva, c. 1920), 34.

28. Card from Robert in Moscow to the children in Edinburg, January 12, 1915; card from Maria in Moscow to Edmond in Edinburg, January 26, 1915; card from Edinburg friends to Marie-Claire in Riga, February 13, 1915; and card from Nissen in Zurich to Robert in Riga, February 19, 1915. Paul Nissen spent 3 months living in their Voltasstrasse house before going off to military service in late February.

29. Blumer-Maillart, *Recollections,* 22.

30. Ibid., 26–9.

31. Ibid., 35–6. See also letter from Maria in Kharkov to Edmond in Glarisegg, January 20, 1916.

32. Baedeker, *Russia,* 461–2, and cards from Maria in Kislovodsk to Edmond in Glarisegg, January 28, 19, and February 4, 1916, and also a telegram to Edmond on February 2, 1916.

33. Letter from Maria in Kharkov to Edmond in Glarisegg, February 22, 1916.

34. Blumer-Maillart, *Recollections,* 28–9.

35. Letters from Maria in Kharkov, March 14, 1916, and Bertha in Geneva, March 23, 1916, both to Edmond in Glarisegg.

36. Letter from Maria in Kharkov to Edmond in Glarisegg, April 12, 1916. This heartfelt letter is the last one from Maria to survive. Edmond also wrote a touching letter at this time in which he reassured his mother of his good condition; letter from Edmond in Zurich to the Maillart family in Kharkov, April 8, 1916. In her letter, Maria states that the Datcha was rented, whereas in her *Recollections,* Marie-Claire recalls that Maillart bought it (see Blumer-Maillart, *Recollections,* 29, 31.).

37. Blumer-Maillart, *Recollections,* 31–3.

38. The photo shows Olga Greter, Viktor Tschiffely, Ernst Eigenheer, even Adolph Zarn from Zurich, Vova Greter (Olga's brother), Mr. Oettli (director of the company for whom Maillart was building the factory), Carl Lehr, Eva, Maria at the center, Hans Bircher, Marie-Claire, Mrs. Wyss, Mr. Greter, Robert Maillart, Mrs. Novitska, Richard Wyss, Max von Müller, and René.

39. Blumer-Maillart, *Recollections,* 38–40.

40. René Maillart, *My Recollections of My Father Robert Maillart* (Princeton, 1974), 1.

41. August 9 (August 22 Swiss date), 1916, 1.

42. Letter from Robert in Kharkov to Edmond in Glarisegg, September 3, 1916.

43. Letter from Robert in Kharkov to Edmond in Geneva, October 3, 1916.

44. Churchill, *Unknown War,* 367–70.

45. René Maillart, *My Recollections,* 1.

46. Blumer-Maillart, *Recollections,* 40.

47. J. H. Billington, *The Icon and the Axe* (New York, 1966), 451–2.

48. Lehr, *Recollections*, 5. In these recollections, Lehr does not explicitly say that he exchanged his money into Swiss francs, but in my interview with him, he stated that he did.

49. Alfred Erich Lehn, *The Russian Revolution in Switzerland: 1914–1917* (Madison: University of Wisconsin Press, 1971), 228. For Lenin's ideas, see J. H. Billington, *The Icon and the Axe*, 524–32.

50. Cyril Falls, *The Great War* (New York, 1959), 282–6.

51. Blumer-Maillart, *Recollections*, 41.

52. Ibid., 42–5.

53. René Maillart, *My Recollections*, 2.

54. Falls, *The Great War*, 286–7.

55. Blumer-Maillart, *Recollections*, 45.

56. Arthur E. Adams, *Bolsheviks in the Ukraine: The Second Campaign 1918–1919* (New Haven: Yale University Press, 1963), 84. The Kharkov events following the armistice are given on pp. 82–93. See also George A. Brinkley, *The Volunteer Army and Allied Intervention in South Russia, 1917–1921* (South Bend, Indiana: University of Notre Dame Press, 1966), 79–87; and John Bradley, *Allied Intervention in Russia* (London: Weidenfeld and Nicolson, 1968), 137–42.

57. The story of the foreman and the Maillarts' subsequent escape is told in both Blumer-Maillart, *Recollections*, 45–54, and René Maillart, *My Recollections*, 2–4. The date of departure, December 30, 1918, appears in *Attestation* (Odessa, January 20, 1919), written by the Swiss Consulate to state that Maillart had left under duress and that "he must return to Kharkov as soon as possible."

58. V. Tschiffely, *Attestation* (Kharkov, June 26, 1919). See also the *Journal de Genève*, July 11, 1919, p. 2, where Tschiffely is listed as still being in Kharkov.

59. For mileages, rail and steamer lines, and descriptions of routes at that time, see Baedeker, *Russia*, 387–92.

60. Bradley, *Allied Intervention*, 141–2.

61. Blumer-Maillart, *Recollections*, 48.

62. Bradley, *Allied Intervention*, 141.

63. Blumer-Maillart, *Recollections*, 48–9.

64. Passport of Robert Maillart, issued April 21, 1917, in Petrograd. See also *Attestations* from the Swiss Consulate, January 18 and 20, 1919.

65. Fortunately, the passports and tickets were safe in his vest pocket and thus they finally were able to board their steamer. Blumer-Maillart, *Recollections*, 49–50.

66. René Maillart, *My Recollections*, 3–4.

67. Blumer-Maillart, *Recollections*, 51–3.

68. Ibid., 53–4. For the family's movements from Odessa to Geneva, see passport of Robert Maillart, April 20, 1917. For Maillart's account of his Russian experience, see "Robert Maillart, Ing. E.T.H.," *Schweizer Echo*, 14(2) (February 1934): 5.

Chapter Five

1. Letter from Robert in Zurich to Edmond in Geneva, April 11, 1919. See also letter to Marie-Claire on the same date.

2. R. Maillart, "Schuldschein" (note of debt) (Geneva, January 1, 1928). This document records his debt of 38,197.25 francs to his brother Paul out of a total debt of 237,464.20 francs. It states that the money will be paid back out of the funds he hopes to receive from the Russian government, which, he states, owes him 3,311,620 francs. The rate of exchange in 1919 was $1 = 5 francs and the ENR Construction Cost Index for 1919 was 198 compared to 5537 in March 1996, so the loan was about $(237,000/5) \times (5,537/198) = \$1,320,000$ in 1996 dollars.

3. Maillart tried unsuccessfully to secure two major works during this period. First, Pont Butin is the second project (no. 4001) listed in the Geneva Maillart Archive; Maillart made drawings and calculations for it intermittently between 1920 and 1923. The design chosen resembles a Roman aqueduct; see M. Brémond, "Le Pont Butin sur le Rhône près Genève," *Bulletin technique de la suisse romande* 46(14) (July 10, 1920): 157.

The second project Maillart sought was at the power plant Kloster-Küblis, approved for design in June, 1919; see *SBZ* 74(2) (July 12, 1919): 22. The plant construction appears in "Die Landquartwerke der Bündner Kraftwerke," *SBZ* 77(12) (March 19, 1921): 127–30. Maillart's first drawing for this project is a cover sheet dated September 18, 1920 and numbered 4006/1. The Geneva office projects begin with 4000, so that this one was the seventh; the number following the slash denotes the drawing number within any project. See James Chiu and David P. Billington, *Geneva Archive of Robert Maillart: Commentary and Catalogue* (Princeton: Princeton University, Department of Civil Engineering, April 16, 1974). The documents referred to in this catalogue are in the Maillart Archive at the Federal Technical Institute in Zurich with copies in the Princeton Maillart Archive.

4. Cards from Robert in Milan, one to René, one to Edmond, and one to Marie-Claire, all in Geneva, October 9, 1919. Cards from Robert in Bologna and in Barcelona to René in Geneva, October 12 and 16, 1919 resp. Cards from Robert in Barcelona, one to Marie-Claire and one to René, both in Geneva, October 20, 1919; letter from Robert in Barcelona to René in Geneva, October 27, 1919; and card from Robert in Barcelona to Marie-Claire in Geneva, October 31, 1919.

5. Cards from Robert in Oviedo, one to René and one to Marie-Claire both in Geneva, November 8 and 9, 1919, resp.; in Madrid, one to Marie-Claire, and in Cordoba and in Seville to Marie-Claire in Geneva, November 16 and 19, 1919, resp.

6. C. E. Black and E. C. Helmreich, *A History: Twentieth Century Europe,* 4th ed. (New York: Alfred A. Knopf, 1972), 261.

7. Before the war, he had not built a single structure in the French part of Switzerland. He had, however, built one factory in Lyon in 1914 and designed (but not built) a 1913 chocolate factory in Broc (near Gruyère) and five factories in France between 1913 and 1918. Probably Maillart had designed those last works before leaving Switzerland in 1914.

8. Cards from Robert in Paris to René in Chesières, January 21, 1920, and February 15, 1920; and letter from Robert in Geneva to René in Chesières, February 3, 1920. Cards from Robert in Lausanne to René in Chesières, February 27, 1910, and cards from Robert in Paris to René in Chesières and Marie-Claire in Geneva, March 7, 1920.

9. Letter from Robert in Paris to Marie-Claire in Geneva, March 27, 1920. Max had, like Maillart, lost his money in Russia; he was afflicted with poor health and was married to a Russian woman whom his family never accepted.

10. Letters from Robert in Geneva to Marie-Claire in Arosa, January 7, 1920, to René in Chesières, Jan-

uary 7 and 13, 1920, and to Edmond in Arosa, January 14, 1920.

11. Card from Maillart in Zurich to Marie-Claire in Geneva, March 31, 1922. Card from Robert in Stein am Rhein to René in Geneva, summer 1922.

12. Letter and card from Robert in Geneva to Marie-Claire in Lens, August 11 and 17, 1922, resp. At the Zarns, he felt well-treated. When they had a fish on Friday, he was given beefsteak.

13. Letter from Maillart in Zurich to Marie-Claire in Geneva, June 8, 1923. By this time, she was living with Hector Maillart; and Bertha and Rosa and Max now lived in a smaller place; see card from Robert in Geneva to René in La Rippe, June 26, 1922.

14. According to the Zurich Technical Institute Alumni lists of 1937, there were only Albert Huber and Alois Keller working as engineers for Maillart in 1923. Based on the drawings made in 1923, I estimate that only one draftsman was needed.

15. For Maillart's feelings during the summer of 1923, see letter from Maillart in Zurich to Marie-Claire in Geneva, June 8, 1923; and cards from Maillart: in Geneva to René in La Rippe, June 26, 1923; in Zurich to René in Levron, July 6, 10, and 18, 1923.

16. R. Sontag, *A Broken World: 1919–1939* (New York: Harper & Row, 1971), 113–17.

17. This trip is detailed by Maillart in the following: letter from Geneva to René in Levron, July 31, 1923; card from Munich to René in Levron, August 8, 1923; letter from Stettin to Marie-Claire in Bern, August 11, 1923; and card from Riga to René in Levron, August 12, 1923.

18. For his Riga visit, see the following: from Robert in Riga: letter to Marie-Claire in Bern, August 14, 1923; card to the Zarn family in Geneva, August 20, 1923; and card to René in Geneva, August 25, 1923. Also a card from Robert in Tallin (Estonia) to René in Geneva, August 28, 1923.

19. Sontag, *Broken World,* 152–3. All Swiss who had such losses had founded a society called *Secrusse,* which put their case before the Swiss government. *Secrusse* only accepted an amount that could be documented. Maillart could document that he was owed more than 3,300,000 francs. The actual sum was most probably far greater, perhaps 10 to 20

times as much, but the documents justifying those amounts were stolen from him at Odessa.

20. Letter from Robert in Geneva to Marie-Claire in Bern, January 3, 1924.

21. These concerns appear in an undated letter, probably in late 1923 from Maillart in Geneva to Edmond. Maillart writes in it, for example, that "they say that you do these things [sudden departures and not answering letters] expressly to exasperate your parents [he means himself and his mother, Bertha]."

22. The events of May to July 1924 are recorded in Maillart's letters from Geneva to Marie-Claire in Worb, May 25 (?), 30, June 3, 11, 18, July 17, 24, 27, and August 1, 1924. During this time, Maillart fell from a bicycle and broke his right arm, which some in the family intimated resulted from his worry over her.

23. These events are described in letters from Maillart in Geneva to Marie-Claire: in Zurich, August 12; in Worb, August 21, 24, 26, September 8, 9; and in Zurich, September 18, 1924; and in letters from Maillart in Geneva to Marie-Claire in Montmirail, October 3, 17, 27, November 25, December 13 and 20, 1924. All the fall and early winter there was the question of where Marie-Claire could go for the weekend and who would take her for the holidays. See also letters from Maillart in Geneva to Marie-Claire in Aigles, December 28, 1924; in Bern, December 31, 1924, and January 11, 1925.

24. Letter from Maillart in Geneva to Marie-Claire in Montmirail, January 21, 1925.

25. Letter from Maillart in Geneva to Marie-Claire in Montmirail, January 28, 1925.

26. Maillart & Cie, *Entreprise de travaux en béton armé.* For example, he noted that "this photograph [of the Barcelona Plant] of one area covered by our ribless floor shows a somewhat different form that gives a remarkable architectural effect."

27. There are two documents in the Geneva Maillart Archive from Bonneville, France, dated September 21, 1911. These show the original design that Maillart modified in the fall of 1920 and drew up in the winter. Six drawings of his survive in the Geneva Archive; they are dated December 2, 2, 14, 1920, and January 18, 28, and February 7, 1921. Maillart reduced the concrete dimensions slightly and made substantial reductions in some of the reinforcement.

28. He did considerable work on a concrete wall for a steel works at Homécourt near Verdun. Drawings for this retaining wall are in the Geneva Maillart Archive, Project No. 4000.

29. "Gesellschaft ehemaliger Studierender: Protokoll der Ausschuss-Sitzung, 4 Juli 1920," *SBZ* 76(7) (August 14, 1920): 81–2. Rohn and Maillart clashed at a meeting of the central committee on July 4, 1920, when the subcommittee report on curriculum was approved with only four (out of twenty) negative votes. Maillart spoke out strongly in its favor, and Rohn, who opposed it, was one of the four dissenters and the only member of the subcommittee to vote against it.

30. Rohn lectured on bridge aesthetics in Bern on January 24, 1919, see *SBZ* 73(10) (March 8, 1919): 113; in Zurich on November 5, 1919, see *SBZ* 76(17) (October 23, 1920): 202; and on November 26, 1919, he gave the same talk in Basel, see *SBZ* 75(24) (June 12, 1920): 272. He also presided at and took part in a discussion on engineering aesthetics in Zurich on January 29, 1919; see *SBZ* 73(8) (February 22, 1919): 87–8.

31. Between 1914 and 1926, Rohn was the principal advisor to ten successful Ph.D. students and coadvisor of one more. *Katalog der ETH-Dissertationen* (Zurich, January 30, 1981), 210–11.

32. C. Andreae, "Arthur Rohn obituary," *SBZ* 74(48) (December 1, 1956): 729–30. Following World War II, *SBZ* changed its volume numbering from two per year to one per year, so that old Vol. 74 was for the second half of 1919. The new numbering thus corresponds directly to the year of publication since 1883.

33. Le Corbusier, *Towards a New Architecture,* transl. F. Etchells (New York: Praeger, 1927), 16.

34. Arnold Keller, "Carl von Bach," *Neue Deutsche Biographie* (Berlin, 1952), 1:491–2. Bach was born on March 8, 1847, at Stolberg (Saxony) and died in Stuttgart on October 10, 1931. He graduated from Karlsruhe Technical Institute in 1867, and after 5 years in practice as a mechanical engineer, he was called to be a professor at Stuttgart. He was for a long time a member of the upper house in the Württemberg parliament. See also C. Bach, *Mein*

Lebensweg und meine Tätigkeit, 1926. Maillart's first critical writing was "Review of a Report: Versuche mit eingespannten Eisenbetonbalken," *SBZ* 76(23) (December 4, 1920): 267. The report was authored by C. Bach and O. Graf.

35. R. Maillart, "Review of *Elastizität und Festigkeit* by C. Bach, Berlin, 1920," *SBZ* 77(18) (April 30, 1921): 203–4.

36. In addition to the reviews given in Refs. 9 and 10, see reviews by R. Maillart: "*Der Eisenbeton* by R. Saliger," *SBZ* 79(14) (April 8, 1922): 186–7; "*Eisenbahnbau,* by R. Bastian et al.," *SBZ* 79(15) (April 15, 1922): 200; and "*Bogenbrücken* by Th. Gesteschi," *SBZ* 79(17) (April 29, 1922): 227.

37. Bach, *Elastizität,* 267–71.

38. For a review of Maillart's role in the center of shear development, see Eric Reissner, "A History of the Center-of-Shear Concept: Maillart's Work and Ramifications," *Maillart Papers* (Princeton: Princeton University, Department of Civil Engineering, 1973), 77–96.

39. A. Eggenschwyler, "Zur Festigkeitslehre," *SBZ* 76(23) (December 4, 1920): 266.

40. R. Maillart, "Zur Frage der Biegung," *SBZ* 77(18) (April 30, 1921): 195–7. Bach had tested his channels by placing loads first over the web center [see Fig. 79(a)] and second over the center of gravity of the cross-section [see Fig. 79(b)]. In both cases, the stresses he measured in the top flange varied from the tip of that flange to its junction with the web, whereas classical theory predicted a constant stress of 273 kg/cm². As Bach moved the load from over the web to 2.2 cm to the left (to the center of gravity point), the stress increased from a rise of 53 percent (or 144 kg/cm²) above the theory (to 417 kg/cm²) to a rise of 90 percent (or 245 kg/cm²) above to 518 kg/cm². Maillart looked at the results and observed that it is not so far fetched to assume that if a movement of the load to the left of the web center gives an extra increase of 90 − 53 = 37 percent above the classical stresses, then that increase might disappear if one moved the load to the right of the web center by 2.2 × 53/37 = 3.2 cm [see Fig. 79(c)].

That simple insight opened up the problem to a clear solution arrived at initially by physical insight. Maillart then proceeded to develop a proof of this idea, again working first from physical in-

sight before formulating the solution mathematically. He proved that the required distance to the right is actually 3.1 cm and he called this point the center of shear (*Schubmittelpunkt*); next he showed that by considering the loadings in Bach's tests as deviations from the center of shear, classical theory can be used to predict closely the experimental results. Maillart emphasized the desirability of designing tests in which channels are loaded at the center of shear (by some small but stiff device to connect the load point to the channel) to give more accurate results than his computations gave. Maillart did not call for tests to confirm his theory, about which he had no doubt, but rather to give further physical insight into the appropriate simplified calculations, because he had made a number of approximations to get results without complex mathematical methods.

Bach himself never returned to this problem and probably never saw Maillart's writing. When the ninth edition of his book appeared in 1924, the part on channel beams remained unchanged, with no reference at all to the vigorous discussion that followed Maillart's 1921 paper. Others, both in Germany and Switzerland, took up the issue, however, because it was, as Maillart had correctly noted, a surprising gap in the otherwise well-developed theory of bending in beams.

41. R. Maillart, "Bemerkungen zur Frage der Biegung," *SBZ* 78(2) (July 9, 1921): 18–19.

42. A. Rohn, "Diskussion," *SBZ* 82(6) (August 11, 1923): 79–80.

43. R. Maillart, "Der Schubmittelpunkt," *SBZ* 83(10) (March 8, 1924): 109–11.

44. A. Föppl and O. Föppl, *Grundzuge der Festigkeitslehre (Fundamentals of the Strength of Materials)* (Berlin, 1923), 81, 133. The first author was a renowned professor in Munich.

45. A. Rohn, "Zur Frage des Schubmittelpunktes," *SBZ* 83(12) (March 22, 1924): 131–2.

46. R. Maillart, "Zur Frage des Schubmittelpunktes," *SBZ* 83(15) (April 12, 1924): 174–7.

47. E. P. Popov, *Introduction to Mechanics of Solids* (Englewood Cliffs, N.J.: Prentice Hall, 1968), 245.

48. See "Eidgen. Technische Hochschule," *SBZ* 82(1) (July 7, 1923): 20, where Rohn's election is announced.

49. A. Eggenschwyler, "Zur Frage des Schubmittel-

punktes," *SBZ* 83(22) (May 31, 1924): 259–61, and Maillart's response in the same issue, pp. 261–2. Eggenschwyler objected to Maillart's arbitrary placement (in his July 9, 1921, article) of the center of shear for the angle section and noted its correct placement at the intersection of the center line of the two legs. Maillart at the end of the July 9, 1921, article noted that the center of shear for the angle section "must lie somewhere near the intersection of the inner surfaces of the two legs and thus in a different vertical line than the center of gravity." He was here discussing the erroneous conclusion of Sonntag in his work *Biegung, Schub, und Scherung* (Berlin, 1909), 96, that the angle section beam will have no twist if it is loaded in the vertical plane along its center of gravity. Clearly, Maillart did not work out the problem because his offhand remark, whereas correct in its principle, is not correct in its exact location of the center of shear, as Eggenschwyler pointed out.

50. S. Timoshenko, *The History of the Strength of Materials* (New York: Dover, 1983), 401 (first published by McGraw-Hill, New York, 1953).

51. Reissner, "Maillart's Work," 83.

52. For example, 64 of the 332 pages of the *Bauzeitung* 78 (July to December 1921) dealt with hydroelectric power plants.

53. Rothpletz, a classmate of Maillart's in Zurich, had worked on the Simplon Tunnel and the Lötschberg Tunnel before setting up his own office in Bern in 1918. From 1919 to 1921, he was a member of the National Assembly in Bern. Büchi, graduated from the Polytechnical Institute in Zurich in 1902 and by 1913 had opened his own office as consulting engineer for hydroelectric plants. In 1930, he was made an Honorary Doctor by the Institute.

54. M. Rothpletz, "Rapport des experts Rothpletz, Rohn et Büchi sur la formation des fissures dans la galerie sous pression de l'usine de Ritom des CFF," *Bulletin technique de la suisse romande* 47(5–11) (March 5 to May 28, 1921): 49–52, 61–5, 77–81, 88–91, 102–5, 109–19, 121–4.

55. C. Jegher, "Von Ritom-Kraftwek der S.B.B.," *SBZ* 76(2) (July 10, 1920): 19–20, a brief note in 76(1) (July 3, 1920): 10; and a further note in 76(4) (July 31, 1920): 56–7.

56. A. Jegher, "Von Ritomwerk der S.B.B.," *SBZ* 76(8) (August 21, 1920): 91–2.

57. R. Maillart, "Le béton armé à la galerie du Ritom," *Bulletin technique de la suisse romande* 47(17) (August 20, 1921): 198–201.

58. "Rapport," 104–5.

59. Maillart described this contrast himself later in his article, "De la construction de Galeries sous pression intérieure," *Bulletin technique de la suisse romande* 48(22, 23, and 25) (October 28, November 11 and December 9, 1922): 256–60, 271–4, and 290–3, resp., and 49(4 and 5) (February 17 and March 3, 1923): 41–5 and 53–8, resp. Maillart had the *Bulletin* bind his five-part article into one reprint, which he sent to a wide circle of leading engineers, both in Switzerland and abroad. The list of people, June 1923, is in the Zurich Maillart Archive.

60. Johann Schmid, *Statische Grenzprobleme in kreisförmig durchörtertem Gebirge* (Berlin: Springer, 1926).

61. This comparison, which identified Maillart's design as the most practical one then devised, appears in the *Bulletin technique de la suisse romande*, 49(26) (December 22, 1923): 318–19. Construction of the Maillart design at Amsteg is shown in "Brochure" (Holderbank-Wildegg, n.d.), 51. The comparison also appears in Hans Studer, "Das Kraftwerk Amsteg der Schweizerischen Bundesbahnen," *SBZ* 86(23) (December 5, 1924): 285–9.

62. A. Schrafl, "Kurzer Bericht über die Druckstollen-Versuche der S.B.B.," *SBZ* 83(1 and 3) (January 5 and 19, 1924): 7–10 and 27–30. The excellent experience at the Klosters-Küblis plant is noted on p. 30.

63. "Die Landquartwerke der Bündner Kraftwerke," *SBZ* 77(12) (March 19, 1921): 127–30.

64. R. Maillart, "Ueber Gebirgsdruck," *SBZ* 81(14) (April 7, 1923): 168–71. Heim had been a professor at the Zurich Polytechnical Institute from 1873 to 1911 and at Zurich University from 1887 to 1911.

65. Card from Albert Heim in Zurich to Maillart in Geneva, April 29, 1923.

66. C. Andreae, "Der Einfluss der Ueberlagerungshöhe auf die Bemessung des Mauerwerks tiefliegender Tunnel," *SBZ* 85(6) (February 7, 1925): 71–3. See also the letter from C. Andreae in Zurich to Maillart in Geneva, January 27, 1923. Andreae was professor of railways and tunnel construction in

Zurich from 1921–8, rector of the Institute from 1926–8, and president of the Swiss Society of Engineers and Architects from 1924–8. In 1928, he left Zurich to become director of the Technical Institute in Cairo, from which he returned in 1937.

67. Maillart based the paper (see n. 68) on his patent first taken out in Germany in 1921 (DPR No. 348219), then in Switzerland in 1922, and finally published in Bern on May 1, 1923. Swiss Patent No. 98968, *Stoss-verbindung*.

68. R. Maillart, "Zum Vernietungs-Problem," *SBZ* 82(4) (July 28, 1923): 43–5. In his paper, Maillart showed how the standard method of multirivet connections led to high overstressing in some rivets and hardly any stress in others. The problem as Maillart sketched it was that when two elements or bars of steel were connected in the standard way by a line of rivets [Maillart showed a connection with five rivets, as in Fig. 81(a)], the design tension force P in one bar is transferred to the other bar only through the two outermost rivets [Fig. 81(a)], with the three interior ones carrying none of the load. He proceeded to show by simple mathematical argument that the force could be transferred by all rivets in a line if the two bars were reduced in thickness continuously throughout the connection [see Fig. 81(b)]. In this way, Maillart argued, each rivet would carry the same load, in this case, $P/5$, rather than having two rivets carry $P/2$ and the other three zero, as in the case of the standard type of connection. Thus, he showed "the usual method of calculation (which assumed each rivet to carry the same, $P/5$, in this case) is fundamentally incorrect; the shear stress would be in our example not 800 kg/cm² ($P/5A$, where A is the cross sectional area of one rivet) but rather 2,000 kg/cm² ($P/2A$)." In other words, because the end rivets really carry $0.5P$ rather than the $0.2P$ assumed by the usual calculation rule found in practice, the stress on the end rivets is 2.5 times as great as is regularly calculated.

Maillart admitted readily that this does not lead directly to failure (otherwise, why would it have been the usual way for so long without ceaseless failures?) because once steel reaches its maximum stress (called yield stress), it simply gives (or yields) but does not break. In real structures, the overstressed rivets yield and the others begin to pick up the extra load. Then at failure, all rivets are at their yield stress and the safety against total failure is, as assumed in 1923 practice, measured by all rivets equally stressed.

69. Rohn's reaction appears in *SBZ* 82(6) (August 11, 1923): 79–80, along with Maillart's response.

70. The debate continued with another and shorter retort by Rohn that appeared in the September 15 *Bauzeitung* along with a much longer discussion by a Dr. A. Frieder of Bern, which had been written on August 6, 1923. Maillart responded this time only to Frieder and the *Bauzeitung* editor commented that he thought that the debate should now be closed; "Korrespondenz: Nochmals zum Vernietungs-Problem," *SBZ* 82(11) (September 15, 1923): 144–5.

In spite of the editor's attempt to close the debate on September 16, Frieder insisted on reopening it, and the *Bauzeitung* published a final exchange on December 8, 1923, between Frieder and Maillart; "Nochmals zum Vernietungsproblem," *SBZ* 82(23) (December 8, 1923): 304–5. Frieder's discussion was answered by one from Maillart written on September 28, 1923. Frieder focused on recent tests in Zurich by Dr. Wyss, a student of Rohn's, as a proof of his earlier position that Maillart had brought nothing new and that present practice was confirmed by tests. Th. Wyss, "Beitrag zur Spannungsuntersuchung an Knotenblechen eiserner Fachwerke," *SBZ* 82(11) (September 15, 1923): 133–6. In his reply, Maillart reiterated the significance of changing loads (fatigue).

71. See *Riveted Joints: A Critical Review of the Literature Covering their Development* (New York: American Society of Mechanical Engineers, 1945), 159.

72. "Konkurrenzen," *SBZ* 76(2) (November 27, 1920): 254, and A. Rohn, "Die Zähringerbrücke in Freiburg," *SBZ* 81(16) (April 21, 1923): 189–94. See also at this time A. Rohn, "Les débuts des travaux du Pont de Pérolles à Fribourg," *SBZ* 76(16) (October 16, 1920): 182–4.

73. The expert commission consisted of three engineers (A. Bühler, F. Hübner, and Rohn), three architects, a senator, and a professor of art. Rohn was proud of this design, as illustrated in his article and in a lead article of 1926 published in the volume of the

first International Congress on Bridges and Buildings, *Schweizerische Ingenieurbauten in Theorie und Praxis.* (Zurich, 1926). Moreover, the appearance of the Zähringer Bridge is very similar to that of the Eglisau Bridge for which Rohn was the designer.

74. "Nachschrift der Redaktion," *SBZ* 81(16) (April 21, 1923): 194.

75. For these other two competition announcements, see "Die Neue Sitterbrücke bei Bruggen," *SBZ* 81(10) (March 10, 1923): 125–6, and "Hochbrücke Baden-Wettingen," 81(11) (March 17, 1923): 132–4 and 139–140.

76. "Wettbewerb für eine Hochbrücke Baden-Wettingen," *SBZ* 82(24 and 26) (December 15 and 29, 1923): 307–11 and 333–6. For this bridge, the expert commission consisted of Rohn, Bonatz, and the Zurich architect Karl Moser.

77. For example, in late 1921, Maillart objected strongly when, at an alumni committee meeting in Zurich, Rohn proposed protecting Swiss engineers from the international competition of foreign professionals working in Switzerland. Such outbursts, unusual for Maillart, illustrate his obvious distaste for things connected with his young adversary. *SBZ* 79(9) (March 4, 1922): 119–20.

78. See Geneva Maillart Archive, Project No. 4016, 1922.

79. Gustav Kruck, "Das Kraftwerk Wäggital," *Neujahrsblatt* (Zurich: Naturforschenden Gesellschaft, 1925). Rohn was appointed consultant for the dam designs and reviewed two proposals in late 1920; see p. 56. Alfred Stucky had completed his thesis, "Arch Dams," in November 1919; it was published as *Etude sur les barrages arqués* (Lausanne, 1922).

80. "Diga di Rochemolles," Geneva Maillart Archive No. 4010/1, September 21, 1921, and Swiss Hauptpatent No. 101777, "Gewölbestaumauer," August 17, 1922 (priority: Germany, September 22, 1921).

81. W. Sattler, "Zwei Wasserkraft-Projekte in den französischen Alpen," *SBZ* 91(5) (February 4, 1928): 59–61. Robert Maillart, "Gewölbe-Staumauern mit abgestuften Druckhohen," *SBZ* 91(15) (April 14, 1928): 183–5, and Robert Maillart, "Die Wahl der Gewölbestärke bei Bogenstaumauern," *SBZ* 92(5) (August 4, 1928): 55–6.

82. The Maillart dam design was criticized by his friend, the distinguished Swiss dam designer, Fred Nötzli, "Laminated Arch Dams with Forked Abutments," *Proceedings of the ASCE*, paper no. 1774 (New York: February 1930): 558–9. For Maillart's dam projects, see Chiu and Billington, *Geneva Archive,* Nos. 29, 54, 220, and 221, 1974. In May 1929, Maillart went to Spain as a consultant on a cracked dam. See letters from Maillart in Bern and Zurich to the Blumers in Serang-Djaja, June 6 and 13, 1929.

83. Maillart even made contributions to concrete pavement design when late in August 1928, representatives of the Swiss cement industry asked him to examine cracks in the first 6-kilometer section of a concrete road built between Brugg and Schinznach-Bad. Maillart inspected the road and wrote his observations with simplified but extensive calculations in a report submitted on August 30. In the 8 days between his visit and the completion of his report, Maillart surveyed the literature on concrete roads, studied the latest technical work in the mathematical analysis of slabs on elastic foundations, and gave a clear and convincing interpretation of the cracks found on this early roadway. R. Maillart, *Bericht über Betonstrassenbau* (Geneva, August 30, 1928), 14 pages plus tables and diagrams. Maillart based his calculations on the book by F. Schleicher, *Kreisplatten auf elastischer Unterlage* (Berlin: Verlag von Julius Springer, 1926). Drawing on the results of his investigation, Maillart strongly urged the cement industry to conduct carefully measured, full-scale tests on that road.

At this time, the Swiss cement industry founded a separate consulting company, Betonstrassen A.G., for which, "no less than the highly gifted engineer Robert Maillart worked out design fundamentals which helped the company to brilliant achievements." This quote comes from the foreword to *Betonstrassen*'s 50-year-anniversary publication in which a full article on Maillart's pioneering work describes how it still provides a reasonable basis for modern design. W. Wilk, "Foreword," in *50 Jahre Betonstrassen A.G.* (Wildegg, 1980), 6, and A. Voellmy, "Die Maillartschen Ueberlegungen zur Bemessung der Betonstrasse Brugg-Schinznach-Bad im Lichte der Streifenmethode," in *Betonstrassen,* 37–42.

In the United States after World War II, the Bureau of Public Roads (taking the same approach Maillart had recommended in 1928) returned "to observation-oriented full-scale field tests to guide" roadway design, after abandoning them in the late 1920s in favor of "scientific laboratory tests." Bruce E. Seely, "The Scientific Mystique in Engineering: Highway Research at the Bureau of Public Roads, 1918–1940," *Technology and Culture* 25(4) (October, 1984): 827.

84. "Kraftwerk in Wäggital," *SBZ* 78(17) (October 22, 1921): 211, and Kruck, "Das Kraftwerk."

85. Maillart's 1921 drawings for the Marignier Bridge show a substantial reduction in reinforcement from the 1911 design. Maillart could reduce that steel because the deck-arch frame had much greater bending stiffness than the arch alone. That frame existed in the earlier design, but the previous designer did not recognize its potential.

86. See Chap. 1, n. 20.

87. See Billington, *Robert Maillart's Bridges,* Chap. 6.

88. See E. Stettler, "Reflections on Maillart," *Maillart Papers* (Princeton, 1973), 129–36.

89. C. J. Whitney, "Design of Symmetrical Concrete Arches," *Transactions of the ASCE* 88 (1925): 931–1029.

90. C. T. Morris et al., "Final Report of the Special ASCE Committee on Concrete and Reinforced Concrete Arches," *Transactions of the ASCE* 100 (1935): 1431.

91. N. Newmark, "Interaction Between Rib and Superstructure in Concrete Arch Bridges," *Transactions of the ASCE*, Paper No. 1981, 103 (October 1938): 62–80.

92. H. Cross, "The Relation of Analysis to Structural Design," *Proceedings of the ASCE* 61 (October 1935): 1119–30 (see *Arches, Continuous Frames, Columns and Conduits,* selected papers of Hardy Cross, ed. N. Newmark [Urbana: University of Illinois, 1963], 91–2, 113–14).

93. D. P. Billington, "Deck-Stiffened Arch Bridges of Robert Maillart," *Journal of the Structural Division of the ASCE* 99(ST7) (July 1973): 1534–6.

94. "L'Aqueduc sur le Trebsenbach des Forces motrices du Wäggital," *Hoch-und Tiefbau* 24(28) (July 11, 1925): 270–2. The author merely signed M.; it seems possible that it was written by Maillart or perhaps by his newly employed engineer Lucien

Meisser. For a description of the uniqueness of the Trebsenbach Aqueduct, see E. Marquardt, "Rohrleitungen und geschlossene Kanäle," *Handbuch für Eisenbetonbau,* 4th ed. (Berlin, 1934), 9:651–5.

95. Felix Favetto, "Ein Stück, Seestrasse des Wäggitalwerkes," *Hoch - und Tiefbau* 24(24) (June 13, 1925): 235–8.

96. For a summary of these ideas, see R. Mark, J. K. Chiu, and J. F. Abel, "Stress Analysis of Historic Structures: Maillart's Warehouse at Chiasso," *Technology and Culture* 5(1) (January 1974): 49–53.

97. Ibid., 53–64.

98. For a detailed discussion of how function follows form, see D. P. Billington, *The Tower and the Bridge* (New York: Basic Books, 1983), Chap. 6.

99. This stimulus to the Chiasso form is discussed in Billington, *Robert Maillart's Bridges,* 64–79. Also in June 1924, Maillart sent a card to René showing a photo of one of these strengthened Gotthard Bridges; card from Maillart [in Geneva] to René [La Rippe], June 3 (?), 1924.

100. Ibid., 68.

101. For a discussion of this deck-stiffened, archlike behavior, see D. P. Billington, *Robert Maillart and the Art of Reinforced Concrete* (Cambridge, Massachusetts, 1990).

102. It is unlikely that Gaudi's tiled brick columns influenced Maillart. There is no record of his reaction to those buildings except in one postcard to René showing the Sagrada Familia Church; Maillart commented, "Doesn't this make you want to leave the hotel business and become an architect." This mildly satirical remark hardly implies a significant influence from Gaudi. See card from Maillart in Barcelona to René in Geneva, February 19, 1925.

103. See Maillart's calculations of 1929 in which he bases the form for the Sihlhölzli gym roof on the hanging string and refers to Chiasso as a model. R. Maillart, "Submissions Berechnung: Turnhallen Sihlhölzli" (Zurich: Zurich Maillart Archive, October 15, to November 1, 1929). The sketch appears on p. 1, which is undated. Page 5 has the date of October 15, 1929.

104. *Brücke über den Valtschielbach: Statische Berechnung,* No. 4094/4 (Geneva: Maillart and Company, April 20, 1925), 4. The complete

calculations appear in a mere 3½ pages, of which only the last half-page is for deck stiffening.

105. Mirko Ros, "Neuere schweizerische Eisenbeton-Brückentypen," *SBZ* 90(14) (October 1, 1927): 172–7. The load test took place on September 29, 1926, in the presence of Maillart and Ros (Fig. 102). For a negative criticism, see H. Ruckwied, *Brückenästhetik* (Berlin, 1933), 75.

106. Ros, "Neuere schweizerische Eisenbeton," 175–7.

107. Robert Maillart, "Note explicative concernant une variante pour L'AQUEDUC SUR L'EAU NOIRE" (submitted to the Federal Railroad Office in Bern) (Zurich and Geneva, October 3, 1924).

108. Maillart's arrangement of supports leads to bending moments of 186 ton-meters at the supports and 94 ton-meters at the midspans. This means that the maximum moments are about the same at *all* the supports and hence the reinforcement can be standardized; it also means that the midspan moments, that is, the ones leading to tension under water and hence possibly cracking, are only half the support moments and therefore the danger of leaks is reduced. These values are all for Maillart's preliminary design and were slightly changed in the final dimensioning.

109. Max Bill, *Robert Maillart,* 3rd ed. (Zurich: Verlag für Architektur, 1969), 52.

110. "Aqueduc sur l'eau noire: Calcul statique," No. 4079/9 (Geneva, Maillart and Company, January 22, 1925), 4–8. The main gravity forces are calculated on one page, but he does go into more detail in these final design sheets to include temperature and to calculate detailed stresses in the critical cross-sections.

111. Adolphe Bühler, "Der Umbau des Grandfey – Viaduktes der Schweizerischen Bundesbahnen," *SBZ* 88(16, 17, and 18) (October 16, 23, and November 6, 1926): 217–21, 231–7, 267. See also Adophe Bühler, "La reconstruction du Viaduc de Grandfey," *L'Entreprise,* Zurich, 15–20 (1928).

112. The actual design was made by engineers of the Swiss Railroads under the direction of Adolphe Bühler. Maillart was hired first to study the scaffolding and to advise on the concrete design. He met frequently with the railway engineers from late December 1924 until the spring of 1926. Maillart made only seven engineering drawings (4086/1–4086/7) and no detailed arch calculations. However, he was at this time beginning to see most arch-bridge designs as deck-stiffened arches. He was designing Valtschielbach in early 1925, and through 1925 he was working on the Lorraine Bridge for which he made a deck-stiffened design as well. Bühler in his articles wrote about the interactions of the deck and the arches at Grand Fey; he does not credit Maillart with the idea, but he does list him prominently as the consulting engineer on the title page of both major articles. It seems obvious, therefore, that the idea for deck stiffening is Maillart's.

113. The visual quality of the design was clearly recognized by one of the most perceptive engineers in the United States, George Beggs, professor at Princeton University, who wrote in 1932 that in

> the design of the Grand Fey Viaduct, a double-track railway bridge built in Switzerland about ten years ago [*sic*], full advantage of deck participation was taken in order to reduce the amount of material, thus securing economy and at the same time reducing the foundation load, which otherwise would have been excessive. . . . Although the specified load is heavy and the unit working stresses low according to American standards, the resulting design is impressively light in its proportions. A recent examination made by me revealed that after ten years of service no damage to the superstructure is discoverable. All the advantages of deck participation have been realized, and by carefully studied design the disadvantages of continuity have been avoided.

From George E. Beggs, "Letter to the Editor," *Civil Engineering* 3(1) (January 1933): 35–6.

114. Maillart left for Barcelona on February 10, 1925, and returned in early March. See letters and cards from Maillart to René and to Marie-Claire, February 6, 17, 19, 24, 27, and March 1, 1925.

Chapter Six

1. C. Jegher, "Eidgen. Materialprüfungsanstalt," *SBZ* 86(2) (July 11, 1925): 28.

2. Ros, "Neuere schweizerische Eisenbeton."

3. Robert Maillart, "Die Brücke in Villeneuve-sur-lot, nebst Betrachtungen zum Gewölbebau," *SBZ* 85(12 and 13) (March 21, 28, 1925): 151–4, 169–70.

4. Robert Maillart, "Zur Entwicklung der unterzuglosen Decken in der Scheweiz and in Amerika," *SBZ* 87(21) (May 22, 1926): 263–5.

5. Johan Huizinga, *Homo Ludens* (Boston, 1955; written in 1938), 1–13.
6. *Bulletin technique de la suisse romand* 52(11) (May 22, 1926): pages 135–136; also see letter from Maillart in Geneva to René in Thonon, October 6, 1925.
7. *SBZ* 86(21) (November 21, 1925): 265, and letter from Maillart in Geneva to Marie-Claire in Montmirail, October 25, 1925.
8. Letter from Maillart in Geneva to Marie-Claire in Tring, June 11, 1927.
9. Letter from Maillart in Geneva to Marie-Claire in Tring, June 22, 1927.
10. Letter from Maillart in Geneva to Marie-Claire in Tisbury, January 11, 1927.
11. R. Maillart, promissory note, January 1, 1928. The note was in German and is as follows (see also n. 2, Chap. 5):

The undersigned attests to Mr. Paul Maillart in Geneva to be in his debt for 38,197.25 francs.
This debt will be taken from the total amount which the Hilfe und Kreditoren Genossenschaft für Russland im Genf (Secrusse) will deliver to the account of the undersigned the outstanding debt in the total admitted sum of 3,311,620.95 francs.
Since the total outstanding debt of the undersigned amounts to 237,464.20 francs, thus Mr. Paul Maillart will receive from each payment by Secrusse a fraction of the total equal to: 38,197.25/237,464.20 until the total amount of his loan is paid.
R. Maillart
Geneva the 1 January, 1928. Eingesehen

This note was given to me by Ella Maillart in June 1980. It is not in Maillart's hand, but was copied from the original by someone in the family after Maillart's death. He had to pay over 6 percent interest on his debt. This amounted to 15,000 francs per year or over half his total income. See letters from Maillart in Geneva to Marie-Claire in Montmirail, May 4, and to René in Thonon, May 12, 1926.
12. Sontag, *Broken World,* 134 and 202.
13. Card from Maillart in Geneva to Marie-Claire in Tisbury, February 6, 1927.
14. Cards from Maillart in Bern to Marie-Claire first in Geneva, February 1, and second in Zurich, February 7, 1927.
15. Blumer-Maillart, *Recollections,* 65.
16. Ernst Stettler, "Reflections on Maillart," in *Maillart Papers,* eds. D. Billington, R. Mark, and J. Abel (Princeton: Princeton University, Department of Civil Engineering, 1973), 129–36.
17. The Zurich office began work in May 1925 on the "Freilager" (Geneva catalogue 4075), which had been sketched out in November 1924 in the Geneva office. The main bulk of this project was drawn up in 1926 (94 drawings of the 118 total).
18. Zarn was still coming in to the Geneva office in 1928, but he had little influence on Maillart's work. During this period, Zarn had been working on a material for covering concrete to give it an attractive appearance and some protection. Called contex, it was not a commercial success. Written up in "Das Contex-Verfahren zur Behandlung von Beton-Sichtflächen," *SBZ* 89 (11) (March 12, 1927): 145.
19. Stettler, "Reflections," 131. Much of the detail for Maillart's weekly tours in 1928 comes from Blumer-Maillart, *Recollections,* 63–86.
20. Stettler, "Reflections," 129.
21. Blumer-Maillart, *Recollections,* 65.
22. "Nekrologe, Eduard H. Blumer-Maillart," *GEP Bulletin* 122 (March 1981): 7.
23. Letter from Maillart in Geneva to Marie-Claire in Zurich, January 9, 1929.
24. Letter from Maillart in Geneva to Marie-Claire in Serang-Djaja, November 27, 1929. Also letter from Maillart in Geneva to Marie-Claire in Serang-Djaja, May 1, 1929.
25. Letter from Maillart in Bern to Marie-Claire in Serang-Djaja, July 4, 1929.
26. Letters from Maillart in Geneva and Zurich to Marie-Claire in Serang-Djaja, July, 1929, and May 7, 1930, resp.
27. Letter from Maillart in Geneva to the Blumers in Serang-Djaja, December 10, 1929, and letter from Maillart in Geneva to Marie-Claire in Serang-Djaja, January 7, 1930.
28. See letter from Maillart in Geneva to the Blumers in Serang-Djaja, April 21, 1930.
29. Ros, "Neuere schweizerische Eisenbeton," 177.
30. Carl Jegher, "Zur Zerstörung der Rheinbrücke bei Tavanasa," *SBZ* 90(15) (October 8, 1927): 194–5. For a more general description of the storm, see "Zur Rhein-Hochwasserkatastrophe von 25. September 1927," *SBZ* 90(16) (October 15, 1927): 206–9. For the details of the catastrophe at

Tavanasa, see Cavegn Eusebi, Gemeindepräsident Brigels, "über die Naturkatastrophe," abstract from *Community Chronicle,* September 25, 1927.

31. Mirko Ros, "Zur Zerstörung der Rheinbrücke bei Tavanasa," *SBZ* 90(18) (October 29, 1927): 232–6.

32. This evolution is described in detail in Billington, *Robert Maillart's Bridges,* Chap. 8.

33. "Schuders," *Dictionnaire Géographique De La Suisse* (Neuchâtel, 1906), 4:543.

34. Ingenieurbureau Maillart, "Salginatobel," Drawing Nos. 168/1, 169/2, 168/3 (Geneva, August 21, 1928). Drawing No. 168/1 from this date does not survive, but a version dated March 5, 1929, does.

35. R. G. Collingwood, "A Philosophy of Progress," in *Essays in the Philosophy of History* (New York, 1965; essay first published in 1929 in *The Realist*), 116.

36. The September 15, 1928, submission by Prader included an explanatory report in addition to the three engineering drawings, a binding offer for price, and a perspective drawing. This last does not survive, but from his numerous other projects, we can see that Maillart did not use perspective drawings in his designs; rather, his sketches were always in profile and section.

37. As we shall see later with Maillart's Schwandbach Bridge, where the bridge deck itself makes a radical horizontal curve, the integration of deck and arch can lead to a dramatic spatial view from one bank and the design vision must move out of the pure planer mode. This more spatial vision is very rare in structural art and the Schwandbach Bridge is a rare case. Only an artist of Maillart's stature could make such a work. Nevertheless, the form comes out of the discipline of planar thinking and resulted from Maillart's 8-year, intensive study of deck-stiffened, arch-bridge designs. See, for example, D. P. Billington, "Bridge Aesthetics: 1925–1933," in *Proceedings of the 11th Congress of the International Association for Bridge and Structural Engineering,* Vienna (Zurich, 1980), 47–52.

38. Peter Lorenz, *Report* (Filisur, September 24, 1928). See also letter from P. Lorenz in Filisur to J. Solca in Chur, October 19, 1928.

39. Stettler, "Reflections," 134.

40. See Billington, *Robert Maillart's Bridges,* 77–81.

41. Stettler, "Reflections," 134.

42. *Salginatobel Brücke, Statische Berechnung,* Drawing No. 168/11 (Geneva, June 1, 1929), 31–44.

43. The form does serve to keep dead-load stresses low; see the detailed study by Laurent Maillart, *The Salginatobel Bridge* (Princeton: Princeton University, Department of Civil Engineering and Operations Research, April 1992).

44. David P. Billington, "Explanation of the Arch Analysis," in *Background Papers for the Second National Conference on Civil Engineering: History, Heritage, and the Humanities,* commemorating the hundredth anniversary of the birth of Robert Maillart, ed. John F. Abel (Princeton, October 4–6, 1972), 91–102.

45. Ibid., List of Documents, 90.

46. Letter from Maillart in Bern to Marie-Claire in Serang-Djaja, July 4, 1929.

47. Letter from Maillart in Geneva to the Blumers in Serang-Djaja, July 23, 1929. See also Jürg Conzett, "Richard Coray," *Fünf Schweizer Brückenbauer* (Zürich, 1985), 32–57.

48. Letters from Maillart in Geneva to Marie-Claire and to the Blumers in Serang-Djaja, September 25 and October 23, 1929, resp.

49. Letter from Maillart in Geneva to Marie-Claire in Serang-Djaja, August 13, 1929.

50. Letter from Maillart in Geneva to the Blumers in Serang-Djaja, August 20, 1930.

51. "Einweihung der Salginabrücke," *Der Freie Rätier,* August 20, 1930.

52. Mirko Ros, "Bericht über die Ergebnisse der Belastungsversuche mit der Strassenbrücke über das Salgina-Tobel von 18 August 1930" (Zurich, December 1930). Slightly revised, this report appeared in M. Ros, *Bericht 99* (1937): 110–22, with added figures and test diagrams.

53. Blumer-Maillart, *Recollections,* 92–3.

54. R. Maillart, "Construction and Aesthetics of Bridges," *The Concrete Way* (May–June, 1935): 305–7.

55. "Die Eröffnung der Lorrainebrücke," *Berner Tagblatt,* May 19, 1930. Maillart's address appeared in several other newspapers as well. Letter from Maillart in Geneva to the Blumers in Serang-Djaja, May 21, 1930. Maillart referred to his talk as a sermon; see letter from Maillart in Geneva to the Blumers in Serang-Djaja, May 27, 1930. See also *Program for the Inauguration of the New Lorraine*

Bridge in Bern, Saturday, May 17, 1930. The bridge opening was widely reported in newspapers all over Switzerland and even Paris editions of the *Chicago Tribune,* May 19, 1930, and the *Daily Mail,* May 18, 1930, carried the opening under large headlines and with a photo.

56. "Einweihung der Salginabrücke." See also "Graubünden," *Neue Zürcher Zeitung,* week of August 18, 1930.

57. "Un beau travail technique dans le "Prättigau," *Journal de Genève,* Thursday, August 21, 1930.

58. Ibid.

59. "Einweihung der Salginabrücke."

60. "Un beau travail technique." For the cost of the Lorraine Bridge, R. Maillart, "Die Lorraine-Brücke über die Aare in Bern," *SBZ* 97(5) (January 31, 1931): 49. The full article appears in *SBZ* 97(1, 2, 3, and 5) (January 3, 10, 17, and 31, 1931): 1–3, 17–20, 23–7, and 47–9.

61. For a note about the Maillart works at the Paris Fair, S. Giedion on the Paris Fair, see *Neue Zürcher Zeitung: Literatur und Kunst* (Zurich, August 5, 1937), and *L'Europe Nouvelle* (Paris, September 4, 1937), Part V, Suisse. For the 1991 event, see D. P. Billington, "Maillart and the Salginatobel Bridge, Switzerland," *Structural Engineering International* 1(4) (November 1991): 46–7.

62. A. Reber, "Berner Brückenfragen," *SBZ* 83(19) (March 8, 1924): 120.

63. A. Reber, "Lorrainebrücke," *SBZ* 86(23) (December 5, 1925): 295, and "Die Lorraine-Brücke in Bern," 86(25) (December 19, 1925): 323. For Maillart's design work see "Hand Calculations of Robert Maillart," Lorraine Bridge, March 10–17, 10 pages; and March 15–17, 11 pages (Geneva, 1926). Letters from Maillart in Geneva to Marie-Claire in Bursins, July 27, 1926, and November 7, 1926; to Marie-Claire in Tisbury, December 19, 1926; to Marie-Claire in Tring, May 25, 1927; and to René in Vittel, June 15, 1927.

64. The referendum was for 3.8 million Swiss francs. See C. Jegher, "Das Projekt der neuen Lorrainebrücke in Bern," *SBZ* 90(11) (September 10, 1927): 141–3.

65. Maillart, "Die Lorraine-Brücke," 1–3, 17–20, 23–4, and 47–9. See also letters from Maillart in Geneva to the Blumers in Serang-Djaja, October 9 and December 25, 1929.

66. Letters from Maillart in Geneva to the Blumers in Serang-Djaja, March 3 and 11, 1930.

67. The official design called for an arch 4.8 meters wide and an average of about 75 centimeters thick, whereas Maillart projected a 30-centimeter-thick arch 3.7 meters wide at midspan, increasing to 5.56 meters at either support. In turn, the deck proposed by Maillart had two rather than four beams, and its total weight was about 5 percent less than that of the official design.

68. R. Maillart, "Leichte Eisenbeton-Brücken in der Schweiz," *Der Bauingenieur* 12(10) (October 1930): 167.

69. Ibid., 170.

70. R. Maillart, *Brücke über die Landquart bei Klosters: Erläuterungen,* (Geneva, February 11, 1930), 1.

71. Ibid., 2.

72. Ibid., 3. See also P. J. B., "Die Linienverlegung der Rhätischen Bahn in Klosters," *SBZ* 96(25) (December 20, 1930): 337–41. In 1930, however, no one suspected that the foundations would move in unforeseen ways and would have to be repaired; but that would occur in 1944 after Maillart's death. For the rehabilitation of this bridge, see L. Meisser, "Umbau der Landquartbrücke der Rhätischen Bahn in Klosters: III. Technische Einzelheiten der Rekonstruktion," *SBZ* 65(3) (January 18, 1947): 32–4. The bridge was taken down in 1992 in a thoughtless reconstruction program by the railways.

73. Letter from Maillart in Geneva to the Blumers in Serang-Djaja, December 9, 1930. For the description of the competition, see "Dreirosenbrücke in Basel," *SBZ* 96(23) (December 4, 1930): 322, and 97(11, 12 and 14) (March 14, 21, and April 4, 1931): 129–33, 144–46, and 173–5.

74. For Maillart's design, see Project No. 209 from the Maillart Geneva Archive, Drawings 3 and 4. Maillart's comment on the concrete designs appear in his letter from Geneva to the Blumers in Serang-Djaja, December 17, 1930.

75. On November 1, he spoke in Bern on a proposal to pave the Lorraine Bridge with concrete rather than asphalt, a new idea in Switzerland; on November 9, he represented Geneva at the meeting of Swiss engineers and architects at Sion; and on Saturday, November 24, he went to Basel to see the

site of a major bridge competition for which he had been named to the jury (as an alternate). See letters from Maillart in Geneva to the Blumers in Serang-Djaja, November 13, 20, and 27, 1929. See also brief notices in *SBZ* 94(24 and 25) (December 14 and 21, 1929): 309, 323, and 95(1) (January 4, 1930): 14. For the Basel Bridge, see C. Jegher, "Die Dreirosenbrücke in Basel," *SBZ* 95(4 and 5) (January 25 and February 1, 1930): 58 and 72. In December, he gave a lecture at Basel where he showed a movie about the construction of the Lorraine Bridge. See letter from Maillart in Geneva to the Blumers in Serang-Djaja, December 8, 1929, and brief notices in *SBZ* 94(24) (December 14, 1929): 312, and *SBZ* 94(26) (December 28, 1929): 335. See also letter from Maillart in Geneva to the Blumers in Serang-Djaja, December 15, 1929.

76. He concluded his letter of December 27, 1930, with a half page written on the last day of the year. It was thus on December 31, 1930, that he spoke of his good health.

77. Letter from Maillart in Geneva to the Blumers in Serang-Djaja, May 5, 1931.

78. For the post office, see "Die Eisenbeton-Pilzdecken des Sihlpost-Gebäudes, *SBZ* 97(13) (March 28, 1930): 165–8. Ad for contex and the Lorrainebrücke, *SBZ* 96(1) (July 5, 1930): facing the title page.

79. Stettler, "Reflections," 133.

80. Letter from Maillart in Geneva to the Blumers in Serang-Djaja, November 8–11, 1931. The article was published first by Ros in November 1930 and then in R. Maillart, "Masse oder Qualität im Betonbau?" *SBZ* 98(12) (September 19, 1931): 149–50. A translation by P. M. Shand appears in Max Bill, *Robert Maillart*, 1st ed. (Zurich, 1949), p. 26.

81. Letter from Maillart in Geneva to Marie-Claire in Serang-Djaja, November 12, 1930. Once again, on October 29, 1937, Maillart would dine with Ammann who spoke warmly of Maillart's work. Ammann was in Zurich to give a lecture; see brief notice in *SBZ* 111(20) (May 14, 1938): 260. Maillart described his meeting with Ammann in a letter from Geneva-Zurich to Edmond in America, January 15–16, 1938.

82. Hans Kruck, "Recollections on Robert Maillart – Part One," Zurich, Autumn 1973, six pages.

Kruck's recollections were written to D. P. Billington and are in the Princeton Maillart Archive.

83. Kruck's interview with Maillart comes from Hans Kruck, "Recollections on Robert Maillart – Part Two," Zurich, December 1974.

84. Gustav Kruck, "Das Kraftwerk Wäggital," *Neujahrsblatt* Naturforschenden Gesellschaft in Zürich, Report No. 127 (1925).

85. He even turned down some of these requests because they took time, were not very interesting, and did not allow him to use his engineers. Letter from Maillart in Geneva to the Blumers in Serang-Djaja, June 15, 1931.

86. Letter from Maillart in Geneva to the Blumers in Serang-Djaja, April 14, 1931.

87. Letter from Maillart in Geneva to the Blumers in Serang-Djaja, July 28, 1931.

88. Letter from Maillart in Geneva to Marie-Claire in Serang-Djaja, November 25, 1930.

89. Letter from Maillart in Geneva to the Blumers in Serang-Djaja, January 6, 1931.

90. Marcel Fornerod, "Reflections on Maillart," in *Maillart Papers*, 137–42.

91. The simplified analysis method, called the trial-load method, combined the cantilever and arch behavior of the dam in the same way Maillart had done for the cantilever and ring behavior of his fixed-base water container of 1902.

92. Letter from Maillart in Zurich to the Blumers in Serang-Djaja, October 10, 1931.

93. Letter from Maillart in Geneva to the Blumers in Serang-Djaja, September 1, 1931. See also letter from Maillart in Geneva to Eduard Blumer in Serang-Djaja, February 25, 1931. Maillart was responding to a January 29, 1931, letter from Blumer, who had included an engineering drawing, with statistics, of the new Sidney Harbor steel-arch bridge with a central span of 1650 feet. Ammann's Bayonne Bridge, just completed, had an arch span of 1652 feet. Maillart observed that such works made him feel giddy.

94. Robert Maillart, "Einige Neuere Eisenbetonbrücken," *SBZ* 107(15) (April 11, 1936): 157–63.

95. *Bulletin of the International Association for Bridge and Structural Engineering*, Zurich, 1 (October 1, 1933): 27. This bulletin gave illustrations of twenty-one recent bridges worldwide, of which the Bohlbach is by far the smallest one shown. See also

"Gekrümmte Eisenbeton-Bogenbrücken," *SBZ* 102(13) (October 28, 1933): 218–19, an unsigned article on the Bohlbach and Schwandbach bridges but certainly written by Werner Jegher.

96. Letter from Maillart in Geneva to the Blumers in Serang-Djaja, September 11, 1930.

97. Letter from Maillart in Geneva to the Blumers in Serang-Djaja, July 14, 1930.

98. Op. cit., letter September 11, 1930.

99. S. Giedion, "Lumière et Construction: Réflexions à propos des ateliers de chemins de fer de Freyssinet," *Cahiers d'Art* 4 (1929): 275ff.

100. S. Giedion, "Maillart, Constructeur des dalles aux Champignons et leur utilization dans l'architecture," *Cahiers d'Art* 5 (1930): 146–52.

101. Op. cit., letter July 14, 1930.

102. C. Zervos, "Contemporary Lyricism," *Cahiers d'Art* 1 (1926): 56–7. My discussion of the modern movement and Maillart is largely taken from George Collins, "The Discovery of Maillart as Artist," in *Maillart Papers,* 35–60. Hans Kruck spoke of Giedion's visit to Maillart's office (in conversations with D. P. Billington in November 1974), but Maillart never referred to the visit in his surviving letters.

103. P. Morton Shand, "Two Remarkable Concrete Arch Bridges," *The Concrete Way* 4(2) (September 1931): 61–73. This article dealt with the Lorraine Bridge and with the Echelsbach Bridge in Germany near Oberammergau by the Melan system. Shand mixed up the history of the Lorraine Bridge competition, but he did understand Maillart's concrete-block idea. Later in the same issue, there appeared on pp. 84–8 a short article, "More Bridges," unsigned but certainly written by Shand in which the other Maillart works were illustrated.

104. P. Morton Shand, "On Getting Back to England," *The Concrete Way* 5(3) (November 1932): 151–5. Maillart is also noted in the *Architects' Journal,* Wednesday, April 6, 1932, 455.

105. P. Morton Shand, "Robert Maillart: The Architecture of a Great Swiss Engineer," *Journal of the Royal Institute of British Architects* 45, Series 3, 19 (September 12, 1938): 957–69. There is also a frontispiece photograph of the Salginatobel with the scaffold in place. The quotations in this section are from Shand's article.

106. For example, Shand says Maillart set up his independent business in 1901 (it was 1902).

107. He did make mistakes such as in the Zuoz and Aarburg Bridge designs, and in that sense Shand was correct even though neither structure was in danger of failure.

108. M. S. Ketchum, Jr., "Thin-Section Concrete Arches as Built in Switzerland," *Engineering News Record,* January 11, 1934, pp. 44–5.

109. "Architektur-Ausstellung in Paris," *Neue Zürcher Zeitung,* March 22, 1934. Nelson and Maillart were getting ready to enter a hospital competition in Zurich. See letter of Maillart in Geneva to the Blumers in Pladjoe, November 29, 1934.

110. These events are described by George Collins, "The Discovery of Maillart," 43–58. See S. Giedion, "New Bridges by Maillart," *Cahiers d'Art* 9 (January 1934): 66–9, and Herbert Read, *Art and Industry* (London, 1934), Fig. 11.

111. "Rigid-Girder Arch Bridges in Switzerland," *Australian Engineer,* Sydney, April 7, 1934.

112. Letter from Maillart in Geneva to the Blumers in Serang-Djaja, December 23, 1930. Edmond's largess probably consisted of his taking Maillart to dinner in Zurich.

113. Letter from Maillart in Geneva to the Blumers in Serang-Djaja, December 27, 1930.

114. Letter from Maillart in Geneva to Marie-Claire in Serang-Djaja, November 25, 1930.

115. The story of this trip comes from Stettler, "Reflections" in *Maillart Papers,* 135–6, and from a letter of Maillart in Bern to the Blumers in Serang-Djaja, October 16, 1931.

116. Letter from Maillart in Geneva to the Blumers in Serang-Djaja, October 23, 1931.

117. Letters from Maillart in Geneva to the Blumers in Serang-Djaja, October 20, November 20, 26, and December 3, 1931. Letter from Maillart in Geneva to the Blumers in Serang-Djaja, December 11, 1931. Maillart dated this November 11, 1931, but the text makes clear that it is 1 month later.

118. Letter from Maillart in Geneva to the Blumers in Serang-Djaja, November 8–11, 1931.

119. Letters from Maillart in Geneva to the Blumers in Serang-Djaja, December 17, 1931, and January 13, 1932.

120. Letter from Maillart in Geneva to the Blumers in Serang-Djaja, February 3, 1932.

121. Letter from Maillart in Bern to Marie-Claire in Ser-ang-Djaja, May 26, 1932.

122. Letter from Maillart in Geneva and Zurich to the Blumers in Serang-Djaja, June 1, 1932.

123. Letter from Maillart in Bern to the Blumers in Ser-ang-Djaja, June 29, 1932.

124. Letter from Maillart in Bern to the Blumers in Ser-ang-Djaja, July 5, 1932.

Chapter Seven

1. For the meeting in Zurich on September 29–30, 1922; see "Technische Kommission des Verbandes Schweiz. Brückenbau- und Eisenhochbau-Fabri-kanten," *SBZ* 81(7) (February 17, 1923): 82–4.

2. R. Maillart, "Eine schweizerische Ausführungs-form der unterzuglosen Decke – Pilzdecke" ("A Swiss-developed Form of Beamless Slab – The Mushroom Slab"), in *Schweizerische Ingenieur-bauten in Theroie und Praxis, Internationaler Kon-gress für Brückenbau und Hochbau,* ed. M. Ros (Zurich, 1926).

3. R. Maillart, "Druckbeanspruchung bei Biegung," in *First International Congress for Testing Mate-rials* (Amsterdam, 1927), 2:13–17.

4. Blumer-Maillart, *Recollections,* 88–9.

5. *Report of the Second International Congress for Bridge and Structural Engineering,* Vienna, Sep-tember 24–8, 1928 (Vienna: Verlag von Julius Springer, 1929), 6.

6. Blumer-Maillart, *Recollections,* 88.

7. R. Maillart, "Weitgespannte Wölbbrücken, Dis-kussion," *Report of the Second International Con-gress,* 417–419.

8. Letter from A. Kleinlogel in Darmstadt to Maillart in Geneva, December 3, 1928. Kleinlogel was the editor of *Beton und Eisen,* which had been founded by Fritz Emperger. The speech appeared rather in *Report of the Second International Congress.* (n. 7).

9. Eugène Freyssinet, "Discussions," in *Report,* 419–21, and "Les arcs du Pont de Plougastel. Les ex-pèriences et l'éxecution de l'ouvrage," in *Report,* 669–78.

10. Robert Maillart, "Report on Sessions about Re-inforced Concrete," *SBZ* 92(21) (November 24, 1928): 262–3.

11. Over 50 years later, Anton Tedesko, who had seen the Plougastel Bridge in 1930, recalled that it made him "proud to be a civil engineer just to see such a work." Meanwhile in 1928 and unknown to all but a few, Freyssinet had just conceived the idea of prestressing stimulated by the Plougastel design; right after the Congress, he would return to Paris and take out his historic patent. Also at the Con-gress, a young Spanish engineer, Eduardo Torroja, presented a paper describing a concrete aqueduct in which, perhaps for the first time, prestressing was used as a major part of the design; see Eduardo Torroja, "L'emploi des cables d'acier dans les con-structions en béton armé," *Report,* 683–8.

12. The keynote was given by Professor Hartmann, whose book appeared that same year, *Aesthetik im Brückenbau* (Leipzig and Vienna, 1928).

13. "Concrete and Reinforced Concrete Architecture," in *Proceedings of the First International Congress for Concrete and Reinforced Concrete* (Liège, 1930): 2:1–73.

14. R. Maillart, "Note sur les ponts voûtés en Suisse," in *Proceedings of the First International Congress for Concrete and Reinforced Concrete* (Liège, 1930): 3:4.

15. Letter from Maillart in Geneva to the Blumers in Serang-Djaja, September 11, 1930.

16. Letters from the International Association of Bridge and Structural Engineering in Zurich to Maillart in Geneva, June 19, 1931, and April 18, 1932. Maillart attended the special meeting of the Central Committee on May 18, the day before the Congress opened.

17. "Programme of the Working Meetings," *First Con-gress, International Association of Bridge and Structural Engineering,* Paris, May 19–25, 1932, 1–5.

18. R. Maillart, "Note sur les dalles sans nervures, sys-tème Maillart," *Premier Congres de l'A.I.P.C.* [In-ternational Association of Bridge and Structural Engineering], typescript original, May 1932.

19. Maillart's paper was published as "Discussion," *Final Report,* First Congress of the International Association for Bridge and Structural Engineering, Paris, 1932 (Zurich, December 1933): 197–208.

20. Rohn expressed the goal of Congress and of the Association in his opening address as being one of "applying always better the laws of mathematics and physics on which are based their research and their studies and to achieve thereby an adaptation

more and more precise by their structures to these laws . . ."; *Final Report,* 29.

21. "Brücke über den Tessin," No. 247, nine separate documents, November 17, 1931, to February 2, 1932. See also Bill, *Robert Maillart,* 86. Bill shows the final design taken from Maillart's February 2, 1932, Drawing No. 247/4, except that he has incorrectly labeled the sections A and C and has a different span at the left end of the longitudinal section.

22. The columns are still not so high that they need to be thickened because of the resulting increase in height.

23. Rather than Maillart's design for the Giubiasco Bridge, a steel arch was constructed.

24. Letters from Maillart in Geneva to the Blumers in Serang-Djaja, March 9 and 16, 1932.

25. Drawing No. 503/3, Ingenieurbureau W. Pfeiffer, Winterthur, April 13, 1932. I am indebted to P. Pfeiffer, who sent me this and other documents concerning his father's work on this bridge. Letter from P. Pfeiffer in Winterthur, October 28, 1992, with documents included. Mistakenly, in my 1990 book, *Robert Maillart and the Art of Reinforced Concrete,* I left out Mr. W. Pfeiffer's name from the Töss Bridge; Maillart did not do that.

26. Drawing with no title, Ingenieurbureau Maillart, Geneva, April 21, 1932. This drawing shows an elevation and one cross-section for the Töss River Bridge.

27. Letter from Maillart in Geneva to the Blumers in Serang-Djaja, April 27, 1932. He put Töss aside to work on his Paris papers.

28. "Neue Tössbrücke," Drawing No. 254/1, Ingenieurbureau Maillart, Geneva, June 24, 1932.

29. The final changes are sketched on the June 24, 1932, drawing and are shown in Bill, *Robert Maillart,* 106.

30. "Neue Tössbrücke: Statische Berechnung," Drawing No. 254/2, Ingenieurbureau Maillart, Geneva, June 24, 1932, corrected August 1, 1932.

31. Ibid. These calculations are for the deck slab and he referred directly in a footnote to R. Maillart, "Zum Entwurf der neuen schweizerischen Vorschriften für Eisenbetonbauten," *SBZ* 99(5) (January 30, 1932): 58, which is his article on the proposed new Swiss code. See also R. Maillart, "Diskussion," *Beton und Eisen* 31(1) (January 1932): 10.

32. P. E. Souter, "Fussgängersteg über die Töss bei Winterthur," *SBZ* 105(3) (January 19, 1935): 32–4. This bridge cost 32 percent more than Maillart's design for the same crossing.

33. *Bericht für die Rossgrabenbrücke,* August 3, 1931. This four-page report was Maillart's first proposal made in great haste and without any drawings. Six weeks later he submitted a second report, *Rossgrabenbrücke über das Schwarzwasser: Bericht zu der Projekten A und B,* September 16, 1931, three pages and four drawings: Nos. 522/1, 522/2, 522/3, and 522/3a (522/1 is missing in the Bern Maillart Archive, but I assume it was included only in the September submission).

34. Maillart, Drawing No. 522/4, January 12, 1932, Bern. The contract is referred to in the final accounting: *Abrechnung über den Bau der Rossgrabenbrücke bei Schwarzenburg* (Schwarzenburg and Bern, December 22, 1932).

35. On August 6, just prior to construction, Maillart made a quick business trip to Paris. See letter from Maillart in Bern to the Blumers in Serang-Djaja, August 10, 1932.

36. *Berner Tagblatt,* November 21, 1932, 3; *Neue Berner Zeitung,* Bern, November 22, 1932, and Berner *Volkszeitung,* Herzogenbuchsee, November 22, 1932. All these accounts are closely similar and report Maillart's talk by stressing his claim for a world's record.

37. Letter from Maillart in Bern to the Blumers in Serang-Djaja, November 22, 1932. For a description of the load test, see M. Ros, "Ergebnisse der Belastungsversuche an der Rossgrabenbrücke über das Schwarzwasser, Kt. Bern," *Bericht 99* (Zurich, 1937), 182–91. For the celebration, see also Stettler "Reflections," 135. A detailed description of the bridge appeared in E. Stettler, "Die Rossgrabenbrücke über das Schwarzwasser (Kt. Bern)," *Schweizerische Technische Zeitschrift* 11 (March 16, 1933): 153–7.

38. For the canton design, see "Thurbrücke, Detailplan," Kanton St. Gallen, Plan No. 1001, No. 6, September 1932, and for Maillart's first proposal, "Thurbrücke," Drawing Nos. 738/1 and 738/3, Ingenieurbureau Maillart, Zurich, October 20, 1932. Maillart writes of the opposition from the canton engineer in a letter to the Blumers from Geneva, January 4, 1933.

39. Letter from Maillart in Geneva to the Blumers in Serang-Djaja, December 21, 1932.

40. Letter from Maillart in Geneva to the Blumers in Serang-Djaja, January 4, 1933.

41. Ibid.

42. Fornerod, "Recollections," 140–1.

43. "Thurbrücke, Uebersichtsplan," Maillart, Drawing No. 738/10/4, January 31, 1933. On this copy, there is a stamp of the canton of St. Gallen, approving it on March 31, 1933.

44. In Bern on June 10, Maillart approved the first drawing for the new Schwandbach Bridge and began to write a report.

45. R. Maillart, *Bergweg Hinterfultigen-Schönentannen: Schwandbachbrücke. Bericht* (Bern, June 27, 1933). Along with this report, Maillart sent a five-page construction outline in the nature of a contract, which was signed by the two contractors, Binggeli and Losinger, but was on Maillart's own stationery. This contract stated that the bridge is to be done by the middle of November 1933. It was actually opened on November 29. See M. Ros, "Belastungsversuche an der gebogenen Eisenbetonbrücke über den Schwandbach, Kt. Bern," *SBZ* 113(5) (February 4, 1939): 53.

46. Max Bill reports a total cost of 47,300 francs, which may not include the design fee; Bill, *Robert Maillart*, 90.

47. Maillart offered Kruck 600 francs per month in 1930, Fornerod the same in 1931, and he must have paid Keller about 1,200 francs a month to run the office. I assume the two draftsmen received about 300 francs per month each, so that the total is about 3,000 francs. Probably Stettler and Daetwyler in Bern cost him about 1,500 francs per month and the largest office, Geneva, about 4,000 francs. If the office rent and expenses were about 20 percent of salaries, then his total monthly expenses, exclusive of anything for himself, would have been about 10,000 francs per month.

48. "Gekrümmte Eisenbeton-Bogenbrücken," *SBZ* 102 (18) (October 28, 1933): 218–19. Jegher's review of H. Rukwied's book appeared in W. Jegher, "Review of *Brückenästhetik*," *SBZ* 101(10) (March 11, 1933): 120, and Mirko Ros gives the dates of construction for the Schwandbach Bridge in "Belastungsversuche," 53–8.

49. "Interview with Karl Hofacker," Gockhausen, July 11, 1978. I am indebted to Prof. Hans Hauri, who organized this interview at his home.

50. Ros, "Belastungsversuche," 53–8. Letters from Maillart in Geneva to the Blumers in Badjoebang, June 19 and 26, 1935. See also the test announcement in "Wissenschaftliche Belastungsversuche an der Schwandbachbrücke, Kt. Bern," *SBZ* 105(23) (June 8, 1935): 270. Truck loading gave stresses so small that they could be hardly distinguished from the influence of the sun.

51. Along with the Salginatobel Bridge, the Schwandbach Bridge is frequently shown and discussed by art historians. See especially the article by Collins, "Discovery of Maillart."

52. "Gekrümmte Eisenbeton-Bogenbrücken," *SBZ* 103 (11) (March 17, 1934): 132–3, discussion by F. Bohny and response by R. Maillart. Maillart's response is in reality a major paper.

53. Billington, *Robert Maillart's Bridges*, 141, has these calculations.

54. Letters from Maillart in Geneva to the Blumers in Serang-Djaja, February 1 and 28, 1933.

55. Letter from Maillart in Geneva to the Blumers in Serang-Djaja, May 3, 1933. The actual production work did not begin until early 1934 even though some work must have gone on in 1933. No business records exist from Maillart's Zurich office, but the official drawings numbered 745/11 to 745/21 in the Zurich city archive are dated from June 14, 1934, to October 4, 1935.

56. Fornerod, "Reflections," 141.

57. Ibid., 142; the article was "Berechnung mehrstöckiger kontinuierlicher Rahmen durch die Methode der algebraischen Momentenverteilung," *SBZ* 102(19) (November 4, 1933): 223–7.

58. M. Turrettini and R. Maillart, "Neues Bankgebäude der Schweizer. Kreditanstalt an der Place Bel Air in Genf," *SBZ* 101(4) (January 28, 1933): 47.

59. "L'inauguration du nouveau Quai de Vevey," *Feuille d'avis*, Lausanne, October 22, 1934, and "L'inauguration du nouveau Quai," *La Suisse*, October 21, 1934. See also E. Meyer-Peter and Henry Favre, "Experimentelle Bestimmung der Beanspruchung von Bauwerken . . . ," *SBZ* 101(4) (January 28, 1933): 48–51.

60. R. Maillart, "Der Ausbau des Quai Perdonnet in Vevey," *SBZ* 108(15) (October 10, 1936): 159–61.

61. "Reconstruction du quai Perdonnet, Projet General," Maillart, Geneva, Drawing No. 262/4 September 14, 1932. The design work began in July 1932 and was mostly completed in early 1933. He also did a minor quai project in Vevey at the same time. "Port de Plaisance," Maillart, Bern, Drawing No. 539/6, December 12, 1932.

62. Letter from Maillart in Geneva to the Blumers in Serang-Djaja, March 28, 1933. The story of the collapse appeared in *SBZ* 101(20) (May 20, 1933): 231–3. The final report, *Ville de Vevey: Reconstruction du quai Perdonnet* by R. Maillart, 11 pages, went to the city on November 28, 1933.

63. "Les essais de résistance du nouveau quai," *Feuille d'avis*, Vevey, October 1, 1934. Losinger was proud enough of the project to place an advertisement: "Losinger and Co.," *SBZ* 104(15) (October 13, 1934): facing p. 163.

64. "L'Inauguration du nouveau Quai," *Le Courrier de Vevey*, October 22, 1934.

65. Letter from Maillart in Geneva to the Blumers in Serang-Djaja, June 27, 1933.

66. "Aarebrücke in Wangen," Maillart, Bern, Drawing Nos. 284/1 and 284/2, August 26, 1933. Although the bridge is in the canton of Bern, Maillart did the project in his almost dormant Geneva office.

67. "Aarebrücke in Innertkirchen," Maillart, Bern, Drawing Nos. 545/7, 545/8, and 545/15, September 20, 1933.

68. Letter from Maillart in Geneva to Milo Ketchum in Illinois, September 18, 1933.

69. "Aarebrücke bei Wangen," Maillart, Bern, Drawing No. 284/5, October 10, 1933. The story about Bösiger comes from Stettler, "Reflections," 130–1.

70. See "Sirakovo Viaduct," Maillart, Geneva, Drawings No. 305, May 7, 17, and July 7, 18, 1934. They are reproduced in Bill, *Robert Maillart,* 99–101.

71. Letter from Maillart in Bern to the Blumers in Pladjoe, November 22, 1934. See "Sitter Bridge," Maillart, Geneva, Drawings No. 310, September 13, and October 1, 1934. The Sitter design appears in Bill, *Robert Maillart,* 102–3.

72. See "Russeinertobel Bridge," Maillart, Geneva, Drawings No. 311, September 18–21, 1934. On February 22, 1935, Maillart traveled to Chur with Florian Prader most probably to discuss this project with the new canton engineer; see the letter from Maillart in Geneva to the Blumers in Pladjoe, February 26, 1935. Maillart also began his design of the Twannbach Bridge in September 1934 although it would not be built until 1936; see "Twannbach Bridge," Maillart, Bern, Drawings No. 561, September 15, 1934.

73. See "Lancy Bridge," Maillart, Geneva, Drawings No. 308, November 3, 1934; "Vessy Bridge," Maillart, Geneva, Drawings No. 319, initial report December 11, 1934. The Lancy project and its subsequent redesign by M. Tremblet were described in Bill, *Robert Maillart,* 120–5.

74. This contrast between Plougastel and Sitter helps us to understand Maillart's ideas. It is not, however, a fair comparison because Maillart's Sitter design was never built and therefore a true visual comparison is not possible.

75. Letters from Maillart in Geneva to the Blumers in Pladjoe, November 29 and December 4, 1934.

76. Sontag, *Broken World,* 163.

77. Letter from Maillart in Geneva to the Blumers in Serang-Djaja, April 4, 1933.

78. Sontag, *Broken World,* 155–7.

79. Letter from Maillart in Geneva to the Blumers in Serang-Djaja, February 14, 1933. See also Sontag, *Broken World,* 164–5.

80. Letters from Maillart in Geneva to the Blumers in Serang-Djaja, April 26 and June 20, 1933.

81. Contract bridge, invented in 1926 by Harold Vanderbilt of New York, rapidly replaced auction bridge itself, which was introduced around 1903. By 1932, bridge became an immensely popular game called "a social phenomenon unparalleled in the history of games," *Encyclopedia Britannica* (Chicago, 1963), 4:180. Newspapers in 1932, such as the *Journal de Genève,* began to include bridge problems and discussions. Letter from Maillart in Geneva to the Blumers in Serang-Djaja, January 11–14, 1933. See also letter of March 7, 1933.

82. Letter from Maillart in Geneva to the Blumers in Serang-Djaja, January 17, 1933.

83. Stettler, "Reflections," 131–2.

84. Letter from Maillart, June 20, 1933. See also letter from Maillart in Geneva to the Blumers in Serang-Djaja, June 17, 1933.

85. Letters from Maillart in Geneva to the Blumers in Serang-Djaja, July 18 and 26, 1933.

86. Sontag, *Broken World,* 171, 237, 249.

87. Letter from Maillart in Geneva to the Blumers in Serang-Djaja, August 2, 1933.

88. Letter from Marie-Claire in Sumatra to Maillart in Geneva, September 12, 1933.

89. Letter from Maillart in Geneva to the Blumers in Serang-Djaja, January 20, 1932.

90. Letter from Maillart in Geneva to the Blumers in Serang-Djaja, January 24–7, 1932. Maillart would sleep in the smaller room and use the larger one as a sitting room or as a guest room for when his children visited. It was very inexpensive, only 1,260 francs per year (about $250 per year!), and he would have the mail and laundry service of the hotel.

91. Blumer-Maillart, *Recollections,* 90–1.

92. Ibid., 66–7, 71. See Chap. Seven, n. 70.

93. Letters to Marie-Claire in Pladjoe from Maillart in Geneva, October 30 and November 8 and 14, 1934.

Chapter Eight

1. Letter from Maillart in Geneva to the Blumers in Badjoebang, January 2, 1935. The official claim against Maillart by the city of Zurich was dated December 27, 1934. Neither in this letter nor any others does he mention the gym case.

2. R. Maillart, "Die Zürcher Sport-und Grünanlagen im neuen Sihlhölzli: Konstruktives," *SBZ* 101(9) (March 4, 1933): 103–5. The actual drawing appears as Fig. 19 on p. 104. The reinforcement in the central vertical member was shown there to be four bars each 16 mm in diameter, an area of steel only 38 percent of that required by design. See letter from Vorstand des Bauwesens II, Zurich, May 20, 1933, and letter from Maillart in Zurich to Hochbauamt Zurich, May 30, 1933.

3. R. Maillart, *Turnhallen Sihlhölzli: Statische Untersuchung der Dachbinder* (Zurich, April 26, 1933).

4. Letter from Hochbauamt Zurich to Bauvorstand I, June 10, 1933. Ritter's name first appears in reference to a meeting on September 15, 1933, in a letter from Maillart in Zurich to Herter in Zurich, October 4, 1933.

5. On June 10 in Neuchâtel, against Maillart's strong objections, the Swiss Society of Engineers and Architects approved the new concrete code prepared under Ritter's leadership. See P. E. Soutter, "Mitteilungen der Vereine," *SBZ* 102(9 and 11) (August 26 and September 9, 1933): 112–14, 138–40, for a detailed description of the Neuchâtel meeting.

6. Letter from Maillart in Zurich to Herter in Zurich, October 4, 1933. See also letter from Herter in Zurich to the city building officials, October 17, 1933, and the minutes of an official meeting in Zurich "Vorstand des Bauwesens I," October 20, 1933, following Maillart's suggestion that Ritter accepted in a letter of October 24, 1933.

7. This professional situation parallels the private problem he faced after Marie-Claire's romantic "error" of nearly 10 years earlier. The question for many Genevans was how to punish her for her error, whereas for Maillart, it was how to make sure that the past would not impair her future. Once a mistake was made, for Maillart, the issue was to make certain that all would be well thereafter, not that the error itself be reason for chastisement. He did not immediately say that the broken rule was irrelevant, rather he first studied the error on concrete physical behavior to determine how his creation would work under future stress.

8. This controversy appears in letters of: Maillart to Ritter, November 13, 1933; Ritter to the city, November 15, 1933; Herter to the city, November 20, 1933; Maillart to Ritter, November 24, 1933; and the city to Ritter, December 13, 1933. See also letter from the city to Ritter, December 26, 1933.

9. Letter from Ritter in Zurich to the city, February 12, 1934. With this letter, Ritter included a nine-page report and a detailed set of calculations.

10. Memorandum from Maillart in Zurich to the city, March 2, 1934.

11. "Protokoll, Vorstand des Bauamtes II," Zurich, Sitzung (meeting on), June 19, 1934. The chairman of the meeting was a city councilman, Dr. J. Hefti.

12. Letter from the city to Maillart in Zurich, June 27, 1934.

13. Maillart did make one drawing on July 3, but he did not send it right away. There were other things to do such as help plan and attend the June 29 load test for the Felsegg Bridge. M. Ros, *Bericht über die Ergebnisse der Belastungsversuche, Thurbruecke Gossau-Will* (Zurich, February 1936).

14. Letter from the city to Maillart in Zurich, August 24, 1934, and Maillart's reply on August 29, 1934.

15. *Protokoll, Meeting at the Bureau des Vorstandes des Bauamtes II* (Zurich, September 5, 1934).

16. R. Maillart, "Verstärkung einer Eisenbetonkonstruktion," *SBZ* 105(11) (March 6, 1935): 130–2.

17. Abstract from the Zurich city protocol, February 23, 1935.

18. Letter from Carl Jegher in Zurich to Councilman J. Hefti in Zurich, March 16, 1935.

19. Maillart, "Verstärkung."

20. Letter to the Editor of the *Schweizerische Bauzeitung* from Dr. E. Ammann in Zurich, May 15, 1935. Jegher responded by observing that Maillart's rebuttal would be published with the city's letter; see letter from Carl Jegher in Zurich to councilman J. Hefti in Zurich, May 31, 1935.

21. R. Maillart, "Reply to the Letter from the Building Authority II of 15 May 1935," Zurich, May 31, 1935. Maillart was here criticizing not only the building department of Zurich, but the new Swiss code, which, over his strong public objection, had just been announced on May 14, 1935. See "Mitteilungen der Vereine," *SBZ* 105(20) (May 18, 1935): 234.

    On the first point in Ammann's letter, Maillart explained that the city had failed to understand the distinction between assumed values and measured results. On the second point, Maillart chastised the city for concealing the fact that the peak stress in the eaves only occurred when the hanger stress was considerably less than the maximum. Moreover, the high concrete stress in the eaves that the city found unacceptable still allowed a factor of safety of 3 in the concrete (the concrete stresses were only one-third of those required to crush the material).

    Against the third objection that he had placed all the load above the hangers and columns, Maillart replied sarcastically: "The fact that the forces [are] shown above [is] for one familiar with the elements of statics, without significance when he examines the force diagram even superficially."

22. Maillart, "Reply," 2.

23. See Billington, *Robert Maillart's Bridges,* Chaps. 9 and 10.

24. Letter from the city of Zurich to Maillart, February 6, 1935.

25. Letters from Maillart in Geneva to the Blumers in Badjoebang, February 4 and 13, 1935. Maillart's first drawing for the Huttwil Bridge is dated February 6, 1935, and his "Hausarrest" was on February 4 in Geneva, so that he most likely sketched out the design then.

26. In 1907, Maillart had designed and built two continuous-beam bridges over the railroad in St. Gallen near Aach, but their forms were probably dictated by the railway engineers and do not have the interest of Liesberg. See article by Adophe Bühler in the 1926 conference volume at Zurich.

27. See "Generelles Projekt, Birsbrücke in Liesberg," Maillart, Geneva, Drawing No. 315/1, October 29, 1934. Subsequently there appeared drawings dated November 6, 8, 26, December 12, 19, 1934, and January 3, 15, 24, 29, 1935.

28. Letter from Maillart in Geneva to the Blumers in Badjoebang, May 22, 1935.

29. Robert Maillart, "Einige neuere Eisenbetonbrücken," *SBZ* 107(15) (April 11, 1936): 157–63.

30. Letter from Maillart in Geneva and Zurich to the Blumers in Badjoebang, June 11 and 12, 1935. See also letters from Maillart in Geneva to the Blumers in Badjoebang, May 22, and June 19, 1935.

31. Maillart, "Einige neuere," 162.

32. Card from Maillart included in a letter from Geneva to the Blumers in Badjoebang, June 19, 1935. For Shand's lecture, see "Tradition und neues Bauen in England," *SBZ* 105(25) (June 22, 1935): 295. The Liesberg Bridge appeared with Maillart's photos in P. Shand, "Flat-Arch Bridge over the River Birs at Liesberg," *The Concrete Way* (July–August, 1935): 26–7, with a text written by Shand.

33. Letter from Maillart in Geneva to the Blumers in Badjoebang, August 6, 1935.

34. "Passage supérieur de la route cantonale à son croisement au Km. 1.603 avec la ligne Huttwil-Wolhusen," *L'Entreprise* 35(19) (May 9, 1936): 160–2.

35. For the bridges of this period that were praised for their appearance, see Wilbur Watson, *A Decade of Bridges* (Cleveland, 1937). Maillart began a number of other small bridge designs in 1935. See catalogue of Bern Maillart Archive for dates of bridge drawings.

36. Letter from Maillart in Geneva to the Blumers in Badjoebang, February 26, 1935. See also letters

from Maillart in Geneva to the Blumers in Badjoebang, February 19 and March 4, 1935.

37. Marcal Cayla, "Les ponts en béton armé de Laifour et d'Anchamps, sur la Meuse," *Le Génie Civil* 105(26) (December 29, 1934): 602–4.

38. R. Maillart, "La Construction des Ponts en béton armé envisagée au point de vue esthetique," *Le Génie Civil* 106(11) (March 16, 1935): 262–3.

39. R. L. Bidwell, *Conversion Tables: A Hundred Years of Change* (London: Rex Collings, 1970), 20 and 49. The French claimed their bridges were economical and they were correct. Their two bridges cost about 1,000 French francs per square meter of roadway surface, which in terms of Swiss francs (ratio 5 to 1 between 1930 and 1935) was about 200 Swiss francs per square meter. Maillart's costs ranged from 298 francs for the much more difficult Saligina to 103 for the Vessy. For the hollow-box designs:

Name	Date	Span	Cost per Square Meter
Laifour	1934	100 m	200 Swiss francs
Salgina	1930	90 m	298 Swiss francs
Rossgraben	1932	82 m	196 Swiss francs
Felsegg	1933	72 m	138 Swiss francs
Vessy	1936	56 m	103 Swiss francs

Because the cost increases with span and also with the difficulty of scaffolding, we can see that these bridges are comparable. Further distinctions are probably as much related to site differences as to design differences.

40. R. Maillart, "Ponts-voûtes en béton armé: de leur développement et de quelques constructions spéciales exécutées en suisse," *Travaux* 19(26) (February 1935): 64–71. An earlier German version appeared as "Leichte Eisenbeton-Brücken in der Schweiz," *Der Bauingenieur* 12(10) (1931): 165–71. For the *Travaux* article, Maillart simply translated his 1931 *Bauingenieur* article, deleting his little Ladholz Bridge and adding the Rossgraben, Schwandbach, Töss, and Felsegg designs. The French version does have one major difference: a table of all the bridges giving their principal dimensions, their dates, articles written about them, and their costs. In the 1931 German version, he had given a much smaller table comparing only the amount of concrete in each of five bridges. Maillart insisted on including costs, so that aesthetics could be judged within the context of economics. He did not compare his costs with those of the French designs.

41. Letters from Maillart in Geneva to the Blumers in Badjoebang, May 8, 14, and 29, 1935. His views on Germany at this time were succinctly expressed in a letter to the Blumers on July 21–4, 1935.

42. The translated article in a slightly expanded version appeared in R. Maillart, "The Construction and Aesthetic of Bridges," *The Concrete Way* 7(6) (May–June 1935): 303–9.

43. In late September, he attended the general biannual meeting of his Alumni Association (G.E.P., Gesellschaft Ehemaliger Polytekniker), where Basel's leading consulting engineer, M. Gruner, told him of Japanese interest in his bridges. Letter from Maillart in Geneva to the Blumers in Badjoebang, October 1, 1935.

44. Letters from Maillart in Geneva to the Blumers in Badjoebang, August 26 and September 4, 1935.

45. "Wettbewerbe," *SBZ* 106(11) (September 14, 1935): 130. The jury report appeared in, "Ideenwettbewerbe," *SBZ* 106(21) (November 23, 1935): 244–50.

46. Letter from Maillart in Geneva to the Blumers in Badjoebang, September 23, 1935.

47. Letters from Maillart in Geneva to the Blumers in Badjoebang, October 9 and 15, 1935.

48. Letter from Maillart in Zurich to the Blumers in Badjoebang, October 31, 1935. The lecture was announced in *SBZ* 106(17) (October 26, 1935): 203–4, to be on 25 years of mushroom slab construction (1910–35). Later it was referred to as "My Experience with Reinforced Concrete over 40 Years," *SBZ* 108(16) 16 (October 17, 1936): 178, and A. Roth, "Ehrenabend für Ing. Maillart," *Weiterbauen* 6 (December 1936): 47.

49. The recollections of Alfred Roth came in conversation with D. P. Billington and letters from Roth in Zurich to Billington in Princeton, October 21, 1974, and March 5, 1975.

50. Letter from Maillart in Bern to the Blumers in Badjoebang, November 20, 1935, and Stettler, "Reflections," 130–1. Years later, Stettler met Bösiger, then retired, who told him, "I advise you to specialize in granite bridges."

51. Letter from Maillart in Geneva to the Blumers in Badjoebang, November 11, 1935.

52. Maillart, "Vessy Bridge" (Chapter Seven, n. 73).

53. Christian Menn, "Aesthetics in Bridge Design," *Bulletin of the IASS,* Madrid (August 1985): 53–62.

54. Letter from Maillart in Geneva to the Blumers in Badjoebang, July 14, 1935.

55. Maillart's cost estimate of December 11, 1934, was 140,000 francs, whereas on August 26, 1935, after substantial design effort, his estimate dropped to 91,444.40 francs. On September 30, the canton sent him the eight bids, the lowest of which was 74,819.10 francs and the highest 87,637.10 francs. He recommended that the canton accept the next lowest bid of 78,222.20 francs. Letter from Maillart in Geneva to the Canton Department of Public Works, October 2, 1935.

56. Letters from Maillart in Geneva to the Blumers in Badjoebang, May 12 and June 2, 1936.

57. Letter from Maillart in Geneva to the Department of Public Works of the Canton of Geneva, February 15, 1937. Here he observes that his fee should have been 11.5 percent of 85,000 francs or 9,800 francs according to the rules of the Swiss Society of Engineers and Architects. Rather they had bargained him down to 8,300 francs, but with his condition that the construction be completed by October 1, 1936. Because the bridge still dragged on into 1937, Maillart's expenses rose because he was responsible for field supervision. By December 1936, he wrote he had total expenses well above 8,300 francs and requested some compensation. He sent the canton a complete listing of his direct costs to the end of May 1937 (11,000 francs) and the payments of the syndicate (8,300 francs). See letters from Maillart in Geneva to the Canton Department of Public Works in Geneva, May 25 and June 21, 1937.

58. Letter from Maillart in Geneva to the Blumers in Tarakan, December 8, 1937. Letter from Maillart in Geneva to the Blumers in Tarakan, July 21–2, 1937. The *Tribune de Genève* reported on the successful load test, including a photo showing Maillart.

59. Letter from Maillart to the Canton Department of Public Works in Geneva, December 7, 1937.

60. N. Newmark, "Interaction between Rib and Superstructure in Concrete Arch Bridges," *Transactions of the ASCE,* Paper No. 1981, 103 (October 1938): 62–80.

61. Letter from Maillart in Geneva to the Blumers in Badjoebang, January 20–2, 1936.

62. Letter from Maillart in Geneva to the Blumers in Badjoebang, January 27, 1936.

63. Letters from Maillart in Geneva to the Blumers in Badjoebang, January 7, 14, and 15, 1936.

64. Letter from Maillart in Geneva to the Blumers in Badjoebang, February 9, 1936.

65. Letter from Maillart in Geneva to the Blumers in Badjoebang, February 18, 1936.

66. M. Ros, *Bericht über die Ergebnisse der Belastungsversuche, Thurbrücke Gossau-Wil* (Zurich, February 12, 1936). This report later appeared in Ros, *Bericht 99.*

67. See letter from Maillart, October 1, 1935, and Raymond J. Sontag, *A Broken World: 1919–1939* (New York, Harper & Row, 1971), 288. See also letter from Maillart, January 7, 1936.

68. Sontag, *A Broken World,* 292–4.

69. Johann Huizinga, "Im Schatten von Morgen," review by C. Jegher, *SBZ* 107(1) (January 4, 1936): 10.

70. Letter from Maillart in Geneva to the Blumers in Badjoebang, March 30, 1936.

71. Letter from Maillart in Geneva to the Blumers in Badjoebang, March 17, 1936.

72. Letters from Maillart in Zurich to the Blumers in Badjoebang, July 8 and 13, 1936.

73. Letters from Maillart in Zurich to the Blumers in Badjoebang, July 20, 29, and August 5, 1936.

74. M. A. Huber, "Le Nouveau pont sur l'Arve à Vessy près de Genève," *Strasse und Verkehr* 24(9) (April 29, 1936): 143–8. Maillart warned the owner in a letter to A. Claret of May 29, 1936. See also letters from Maillart in Zurich to the Blumers in Baedjoebang, August 11, 12, and 19, 1936. For the confusion at the Vessy Bridge site, see letters from Huber in Geneva to Société Romande des Ciments Portland S.A., September 14, 1936, and to M. Mouchet, September 17, 1936.

75. Letter from Maillart in Zurich to the Blumers in Badjoebang, August 26, 1936.

76. R. Maillart, "Der Ausbau des Quai Perdonnet in Vevey: I Allgemeines und Projektierung," *SBZ* 108(15) (October 10, 1936): 159–61. The construction is described by Kurt Egli on pp. 161–3.

77. Letter from Maillart in Geneva to the Blumers in Badjoebang, November 24, 1936.

78. "Neue Strassenbrücke über das Sittertobel ('Krä-zernbrücke') bei St. Gallen-Bruggen," *SBZ* 108(13) (September 26, 1936): 148.

79. "Wettbewerbe," *SBZ* 107(7) (February 15, 1936): 75.

80. Letter from Maillart in Geneva to the Blumers in Badjoebang, March 24, 1936. For Jehger's comments, see "Zur Bauausschreibung der SBB," *SBZ* 107(10) (March 7, 1936): 108–9, and 107(13) (March 28, 1936): 142.

81. Letter from Maillart in Geneva to the Blumers in Badjoebang, October 15, 1936. The official design by Bühler had already appeared in *SBZ* (June 9, 1936): V. 103, 270.

82. Letter from Maillart in Baden to the Blumers in Badjoebang, October 7, 1936.

83. Letter from Maillart in Geneva to the Blumers in Badjoebang, October 28, 1936. The prize appeared in "Wettbewerbe," *SBZ* 108(18) (October 31, 1936): 199–200, and the critiques in 108(20) (November 14, 1936): 211–16.

84. Letters from Maillart in Zurich to the Blumers in Badjoebang, November 6, and 11, 1936, and letter from Maillart in Geneva of November 18, 1936.

85. For the lost Kräzernbrücke, see "Projekt-Wettbe-werbe für eine Strassenbrücke, die 'Kräzernbrücke' über die Sitter bei St. Gallen," *SBZ* 108(24 and 25) (December 12 and 19, 1936): 266–9, 272–6. For the Schaffhausen discussion, see "Sitzung," *SBZ* 108(24) (December 12, 1936): 270. See also Chap. Eight, n. 118, for Maillart's reflections on competitions.

86. Sontag, *A Broken World*: 294.

87. R. Maillart, "Zum Entwurf der neuen schweizer-ischen Vorschriften für Eisenbetonbauten," *SBZ* 99(5) (January 30, 1932): 55–9.

88. R. Maillart, "Diskussion," *Beton und Eisen* 31(1) (January 1932): 10.

89. Letters from Maillart in Geneva to the Blumers in Serang-Djaja, February 24 and March 2, 1932.

90. A. Paris, "Projet de revision des normes suisses de béton," *SBZ* 99(10) (March 5, 1932): 119–20. Maillart's response appeared in the same number on pp. 125–6.

91. Letter from Maillart in Geneva-Bern to the Blumers in Serang-Djaja, April 27, 1932.

92. Letter from Maillart in Geneva to the Blumers in Serang-Djaja, May 5, 1932. The Zurich code committee meeting must have been held on Saturday, April 30, according to Maillart's letters, although the *Bauzeitung* reported it to have been on April 23; see "6. Fachgruppen," *SBZ* 101(25) (June 24, 1933): 304.

93. Letters from Maillart in Geneva to the Blumers in Serang-Djaja, November 9, 30, and December 6, 1932. The lecture was noted in *SBZ* 101(25) (June 24, 1933): 304.

94. Maillart was essentially arguing for a simple means of determining the minimum depth of beams or slabs and the reinforcement required for them. He proceeded to define the bending moment M in terms of a total section depth h by the formula $h = \sqrt{M}$, and the reinforcement, A_s, required by the equally simple formula $A_s = M/h$. See R. Maillart, "Zum Entwurf."

By the 1930s, the analysis of reinforced-concrete structures had become complicated for two reasons: one having to do with its monolithic nature and the other with its peculiar material properties. Maillart had analyzed monolithic concrete flat slabs and columns on the basis of substantial large-scale tests in his construction yard; for monolitic deck-stiffened arch bridges, he had developed a method of analysis that reduced the problem to very simple formulas whose accuracy was proven by full-scale tests.

The material properties of reinforced concrete presented a different problem to designers. Unlike steel or wood, concrete is far weaker in tension than in compression. Under downward bending, a beam will crack near the midspan on its lower half. Reinforcing steel bars embedded in the concrete will carry the tension and restrict cracking to hairline or nearly invisible cracks. Given the dimensions of the concrete, the designer's task is to calculate the amount of steel needed for the reinforcing bars. The composite material (steel and concrete) works well in practice but not in theory. It is difficult to determine accurately the compression stresses in the concrete and the tension stresses in the steel. By the 1930s, the standard way to determine these stresses was through a set of formulas in which one key parameter was $n = E_s/E_c$, where E_s is the material stiffness of steel and E_c is the material stiffness of concrete (E is the modulus of elasticity and defines the ratio of stress to strain).

The ratio n varies depending on the quality of the concrete and can be as low as 8 or as high as 20. The use of n implies that the values of E_s and E_c are constant for a given beam. E_c in fact is not a constant, as Maillart and others well knew, but code committees insisted in retaining n and assuming E_c constant.

Maillart believed that the academics who controlled the code committees kept analysis complex because they had devoted much research to develop it and because they believed that it led to greater accuracy. The standard method approved by codes and academic authorities in the 1930s and Maillart's method as applied to the Töss River footbridge deck design give essentially the same answer, but the contrast between the two approaches is striking. The standard method requires an assumption of n and then a series of calculations that seem to be scientific. These lead to a number j, which then appears in the formula for the area of steel (A_s). Maillart never used n, avoided all the formulas needed to get j, and simply used one elementary equation, the bending moment M divided by the total thickness of the slab (or height of the beam h). He introduced a single adjustment by adding 10 cm-tons to the calculated bending moment.

It is quite obvious that for this example, Maillart's method is fully equivalent to the standard method (they yield almost the same results), but it is not obvious why this is so. The highly simplified formula $A_s = M/h$ only works under two assumed conditions: first, that the allowable steel stress $f_s = 1{,}200$ kg/cm^2 = 1.2 tons/cm^2, and, second, that the quantity $jd = 5h/6$. Making these assumptions, Maillart converted the standard formula $A_s = M/(f_s \times jd) = M/1.2 \times 5h/6 = M/h$. In the 1920, the steel stress was usually taken to be 1.2 tons/cm^2, but his assumption for jd was only an estimate. For shallow slabs, this was too high (for the Töss Bridge deck for $n = 10$, $jd = 0.883 \times 2.5 = 2.2$in. $= 0.63h$ rather than $5h/6 = 0.833h$), so Maillart increased the moment by 10 cm-tons as compensation. Where the height of the beam is much greater, jd does approach $5h/6$ and the correspondingly larger bending moment is less influenced by the addition of 10 cm-tons.

One further observation that Maillart recognized was the insignificance of n. If, for example,

we recalculate the Töss deck for $n = 20$, then $A_s = 24.4$ cm^2. This result for A_s is larger than Maillart's 23.4 cm^2. Thus, doubling n increases A_s by only 4 percent. However, Maillart provided 28.2 cm^2 anyway because the sizes of reinforcing bars rarely give calculated areas, that is, he used four bars of 30 mm in diameter.

95. See P. Soutter, "S.I.A. Versammlung," *SBZ* 102(9 and 11) (August 26 and September 9, 1933): 112–14, 138–40, for a detailed description of the Neuchâtel meeting.

96. R. Maillart, "Ponts-Voûtes en béton armé," *Bulletin du Ciment* 2(8) (August 1934): 2–6. The bridge designed by Ritter was the Hundwiler-Tobel Bridge. See also Fritz Stüssi, "Nekrologe, Max Ritter (1884–1946)," *SBZ* 127(14) (April 6, 1946): 167–8.

97. Letters from Maillart in Geneva to the Blumers in Badjoebang, November 26 and December 5 and 11, 1935. The article appeared as R. Maillart, "Die Schweizerischen Normen für Eisenbeton von 1935," *Der Bauingenieur* 47/48 (November 22, 1935): 481–5.

98. Letter from Maillart in Geneva to the Blumers in Badjoebang, December 17, 1935 (completed on December 18).

99. The Concrete Building Code in the United States dropped the allowable stress method for design in its code revision of 1971.

100. Letter from Maillart in Geneva to the Blumers in Badjoebang, January 7, 1936.

101. Letter from Maillart in Geneva to the Blumers in Badjoebang, January 14–15, 1936.

102. Letter from Maillart in Geneva to the Blumers in Badjoebang, February 9, 1936.

103. R. Maillart, "The Modular Ratio," *Concrete and Constructional Engineering* (September 1937): 517–21. Partly Maillart was stimulated by earlier articles; see K. Hajnal-Konyi, "The Modular Ratio II," *Concrete and Constructional Engineering* (February 1937): 129. Maillart's discussion of Emperger's article appears in Ing. Maillart, Genf, "Diskussion," *Beton und Eisen* 31(1) (January 5, 1932): 10. For comments by Hajnal-Konyi on Maillart's paper, see Hajnal-Kónyi, "Reply," *Concrete*, 522, 528–30. See also letter from Maillart in Zurich to the Blumers in Tarakan, April 8, 1937.

104. R. Maillart, "Die Sicherheit der Eisenbetonbau-ten," *SBZ* 53(9) (February 27, 1909): 119–20, translated in Billington, *Robert Maillart's Bridges,* 46–7, 133–4.

105. "S.I.A. Diskussionstag," *SBZ* 110(19) (November 6, 1937): 240, gives the announcement of the two lectures.

106. R. Maillart, "Aktuelle Fragen des Eisenbeton-baues," *SBZ* 111(1) (January 1, 1938): 1–5. Max Bill included about half the text in his book *Robert Maillart,* but omitted the two diagrams and all of the last part, which included the formulas.

107. For this legacy, see D. P. Billington, *Robert Maillart and the Art of Reinforced Concrete* (New York: Architectural History Foundation, 1990), Chap. 9.

108. Letter from Maillart in Geneva to the Blumers in Tarakan, February 8, 1937 (continued on February 9 and 11 and completed February 12). For Maillart's written text, see "Lichtbilder Vortrag in Basel 1938" [the date is incorrect and should be 1937]. The nine-page typed copy is in the Zurich Maillart Archive. It was never published, but the ideas appeared in his two major *Bauzeitung* articles in 1938.

109. Letter from Maillart in Geneva to the Blumers in Takakan, February 4, 1937. The announcement for Maillart's talk appeared in "Sitzung," *SBZ* 109(6) (February 6, 1937): 72.

110. R. Maillart, "Einige neuere Eisenbetonbrücken," *SBZ* 107(15) (April 11, 1936): 157–63. For some of the bridges shown by Bonatz, see Walter Wirth, "Die Deutschen Reichsautobahnen," *SBZ* 108(21) (November 21, 1936): 223–9.

111. *Paul Bonatz: Work from the Years 1907–1937,* ed. Friedrich Tamms (Stuttgart: Verlag Julius Hoffmann, 1937), 94 pages with 102 illustra-tions.

112. Erna Lendvai-Dirchsen, *Reichsautobahn: Mensch und Werk* (Bayreuth: Gaulverlag, 1937). A new edition appeared in 1942 in memory of the recently killed Fritz Todt, general director of the Reichsau-tobahn. Todt was memorialized in 1943 by a book devoted to his career in which once again the mon-umental bridges reappear; Eduard Schönleben, *Fritz Todt: der Mensch, der Ingenieur, der Nation-alsozialist* (Oldenburg: Verlag Germard Stalling, 1943). Todt (1891–1942) appears throughout the book in pictures together with Hitler frequently visiting construction sites. He was Hitler's top en-gineer and was minister of armaments and muni-tions at his death; see Albert Speer, *Inside the Third Reich* (New York, 1970), 193–4.

113. For the Basel Art Museum, see R. Christ and P. Bonatz, "Das neue Basler Kunstmuseum," *SBZ* 109(4 and 5) (January 23 and 30, 1937): 42–3, 51–8.

114. For Bonatz's lecture announcement, see "Sitzung," *SBZ* 107(13) (March 28, 1936): 144. For a brief description of Bonatz as Autobahn architect, see Speer, *Inside the Third Reich,* 80.

115. Maillart, "Einige Neuere," 158.

116. *Paul Bonatz,* 41–7.

117. Wirth, "Die Deutschen Reichsautobahnen," 225. Letter from Maillart in Geneva to Eduard Blumer in Badjoebang, June 2, 1936, partly in response to a card from Eduard Blumer in Bandoeng (above Jakarta) to Maillart in Geneva, May 22, 1936.

118. Maillart, "Lichtbilder Vortrag," 7–9. The section on beauty includes Maillart's evaluation of com-petitions:

When it concerned a competition to win a prize, I could never participate successfully with the collabo-ration of an architect. Only when it was a matter of a pure utilitarian structure in an out-of-the-way place could I succeed with my two structural systems and indeed always only on the basis of economy which, in the face of the negative aesthetic feeling, settled the matter. With competitions, where I dared to propose these structural systems, they regularly dropped out after the first elimination.

There was strict refusal in certain circles, whereas country folk, remarkably free from the [dictates] of custom, agree quite enthusiastically on these structural types.

119. Letter from Maillart in Geneva to the Blumers in Tarakan, February 22, 1937.

120. Letter from Maillart in Geneva to the Blumers in Tarakan, July 7, 1937. For Vulpera, see Karl Bae-deker, *Switzerland* (Leipzig, 1938), 502–3.

121. Letter from Maillart in Vulpera to the Blumers in Tarakan, July 28, 1937. For Maillart's election as honorary corresponding member, see *Journal of RIBA* 154(N18) (August 14, 1937): 971.

122. Letter from Maillart in Vulpera to the Blumers in Tarakan (in German), August 11, 1937, and letter

from Maillart in Geneva to Marie-Claire in Tarakan, August 19, 1937.

123. Letters from Maillart in Geneva to Marie-Claire in
Tarakan, August 31 and September 15, 1937.

124. *SBZ* 110 (22 and 24) (November 27 and December
11, 1937): 277, 291-302. The previous year
(1936), the *Tribune de Genève* published an article
on the Vessy Bridge without mentioning Maillart.
His friend Braillard wrote the *Tribune* a strong reprimand. Although Maillart found it a bit "eulogis-
tic," the article helped appease his annoyance at
not being credited. Letters from Maillart in Baden
to the Blumers in Badjoebang, September 30 and
October 7, 1936. Similar things had happened in
the *Bauzeitung* in which a long article on deck-
stiffened arches by Fritz Stüssi, a former assistant
of Rohn's, failed to mention Maillart's designs, and
an article on pressurized tunnels referring to the
Ritom failures ignored Maillart's pioneering work.
Fritz Stüssi, "Der Formänderungseinfluss beim ver-
steiften Stabbogen," *SBZ* 108(6) (August 8, 1936):
57-9, and H. F. Kocher-Preiswerk, "Erfahrungen
aus dem Druckstollenbau," *SBZ* 108(9) (August
29, 1936): 98-101.

125. Maillart's design, made in 1935, is No. 569 in the
Bern Maillart Archive; see Ernst Stettler, *Projekte
Bureau Maillart Bern* (Bern, October 7, 1974). For
the Wooden Bridge, see *SBZ* 112(19) (November
5, 1938): 227-31.

126. P. Morton Shand, "The Changing Bridge" (Parts I
and II), *The Listener* (October 6 and 13, 1937):
722-4, 796-8.

127. Ibid. On October 31, 1937, Maillart welcomed
Shand and his wife in Geneva. Shand urged Maillart to visit London soon to lecture to the Institution of Civil Engineers. He could speak in either
French or German and Shand would translate.
"Can't you just imagine that!" exclaimed Maillart
to the Blumers. Letter from Maillart in Geneva to
the Blumers in Tarakan, November 1, 1937.

128. Letter from Maillart in Geneva to the Blumers in
Tarakan, October 5, 1937. He completed the last
page of this letter at midnight in Zurich.

129. Letter from Maillart in Geneva and Zurich to the
Blumers in Tarakan, November 16 and 17, 1937.
For Maillart's other visits at this time, see letter
from Maillart in Geneva to the Blumers in Tarakan, November 29, 1937.

130. Letters from Maillart in Geneva to the Blumers in
Tarakan, December 22, 1937, and January 3,
1938. In this last letter, he wrote of a new design
for the Jugolsav Bridge over the Tara River. See
Bill, *Robert Maillart,* 123. See also Geneva Archive
No. 380.

131. James Chiu and David P. Billington, *Geneva Archive of Robert Maillart: Commentary and Catalogue* (Princeton: Princeton University, Department
of Civil Engineering, April 16, 1974), 38-41.

132. Letter from Maillart in Zurich and Geneva to Edmond, Germantown, Pennsylvania, July 1 and 2,
1938. Later Maillart warned his son about
"Sharks" and "Hyenas" who would steal Edmond's patents just as Froté and Westermann had
stolen his Zuoz one; see the letter to Edmond of
August 16, 1938.

133. S. Giedion, "Construction and Aesthetics," *Transition,* translated from German by Eugène Jolas
(Fall 1936): 181-201. See letter from Maillart in
Geneva to the Blumers in Tarakan, March 1-2,
1937.

134. For the vote on the Swimming Hall, see "Drei
grosse Bauwerke," *SBZ* 111(9) (February 26,
1938): 111, and for Maillart's calculations, see Zurich Maillart Archive: Hs RM 1938-1, Projekt 777.
Maillart's drawing appears as No. 777/6/2, Hal-
lenschwimmbad, Maillart, Zurich, March 4, 1938.

135. Letter from Maillart in Geneva to Edmond in Or-
eland, Pennsylvania, February 24, 1938. See also
letter from Maillart in Geneva to Edmond in Or-
eland, January 15-16, 1938.

136. C. Jegher, "Schweizerische Landesausstellung Zürich: Bauen," *SBZ* 112(10) (September 3, 1938):
126-8. The Hall of Ceramics, shown next to the
Cement Hall, was designed by the architect Hans
Leuzinger of Glaris-Zurich with engineer F. Pfeiffer
of Zurich. Leuzinger had been responsible for laying out the entire building section for the fair, but
as the figures show in September 1938, Maillart
did the form of the Cement Hall alone.

137. See, for example, the extensive French coverage in
"Expositions," *Le Génie Civil* 65(7) (August 12,
1939): 133-8. By contrast, the Cement Hall was
featured in "Aus der Abteilung Bauen," *SBZ*
113(10) (March 11, 1939): 122-3. Again, the
Bauzeitung clearly identifies the design with
Maillart and not with Leuzinger. Below a photo of

the completed work in the *Bauzeitung* appears the caption "In the section 'building' [37]. Arch. H. Leuzinger, Ing. R. Maillart." The 37 refers to the exposition number for the building section, which includes both the Ceramic Hall designed by Leuzinger and the Cement Hall by Maillart. This confusing caption has contributed to the mistaken belief that Leuzinger collaborated with Maillart for the form of the Cement Hall.

138. Ulrich Fischer, "Die Mitwirkung des Aufbaues massiver Bogenbrücken," *Beton und Eisen* 37(19) (October 5, 1938): 310–15.

139. Arnold Moser, "Les Planchers-Champignons," *La Technique des Travaux,* Paris, 14 (March 1938): 137–52.

140. M. Ritter, "Aus dem Institut für Baustatik: Eine wirtschaftliche Bemessung von Eisenbetonquerschnitten," *SBZ* 111(14) (April 2, 1938): 165–6. One practicing engineer did discuss Maillart's ideas. See L. Bendel, "Bemerkungen zum Aufsatz von R. Maillart über 'Aktuelle Fragen des Eisenbetonbaues'," *SBZ* 111(21) (May 21, 1938): 261–2. He carried further Maillart's seminal ideas in "Aktuelle Fragen," but Bendel was not part of the establishment and Ritter never acknowledged his discussion.

141. M. A. Huber, "Le nouveau pont sur l'Arve à Vessy près de Genève," *Strasse & Verkehr* 24(9) (April 29, 1938): 143–8.

142. "Der Quai Turrettini," *Strasse & Verkehr* 24(9) (April 29, 1938): 149–52. See also the earlier article, "Les Travaux du quai Turrettini," *La Suisse* (October 13, 1937): 8.

143. "Die Arve-Brücke bei Vessy in Kanton Genf," *Hoch-und Tiefbau* 37(20) (May 21, 1938): 187–91.

144. Klett and Hummel, "Die Donaubrücke bei Leipheim im Zuge der Reichsautobahn Stuttgart-München," *Die Bautechnik* 16(40/41) (September 23, 1938): 521–35. The quote is on p. 524. The bridge had already been featured in Karl Schaechterle and Fritz Leonhardt, *Die Gestaltung der Brücken* (Berlin: Volk und Reich Verlag, 1937), 95–8.

145. From M.-C. Blumer-Maillart and E. Stettler.

146. R. Maillart, "Ueber Eisenbeton-Brücken mit Rippenbögen unter Mitwirkung des Aufbaues," *SBZ* 112(24) (December 10, 1938): 287–93.

147. "Oesterreichischer Ingenieur- und Architekten-Verein," *SBZ* 112(14) (October 1, 1938): 179.

148. Sontag, *A Broken World,* 354. On October 24, 1938, Hitler's foreign minister, Joachim von Ribbentrop, began discussion with Poland on the question of Danzig.

149. "Call for the Moral Rearmament of Switzerland," *SBZ* 112(18) (October 29, 1938): 215 (signed by 28 leading Swiss citizens). About the threat to Switzerland, see A. J. P. Taylor, *The Origins of the Second World War* (New York: Atheneum, 1962).

Chapter Nine

1. Blumer-Maillart, *Recollections,* 74–5. Maillart described some of the events of Marie-Claire's visit in a letter from Geneva to Edmond in Germantown, Pennsylvania, on August 16, 1938.

2. Blumer-Maillart, *Recollections,* addendum 1987.

3. Card from Maillart and Marie-Claire in Vulpera to Rosa in Geneva, July 15, 1938, and card from Maillart and Marie-Claire in Vulpera to Edmond in Germantown, Pennsylvania, July 15, 1938.

4. Blumer-Maillart, *Recollections,* 69–70. The 1930s was the bridge decade, and starting with Culbertson, many systems of bidding and playing conventions (i.e., signaling one's partner) were developed.

5. Letter from Maillart, August 16, 1938. Of course, Maillart had been reading the bridge columns in the *Journal de Genève* regularly and, in spite of not playing in strict competition, he was aware of new conventions.

6. Card from Maillart, Rosa, and Amélie in Geneva to Edmond in Germantown, Pennsylvania, August 1, 1938.

7. Letter from Maillart, August 16, 1938.

8. C. Jegher, "Karl Emil Hilgard," *SBZ* 112(6) (August 6, 1938): 70–1.

9. Card from Maillart, the Blumers, Nissen, and the Nigglis in Mollis, Canton Glarus, to Edmond in Philadelphia, October 7, 1938. The first photo in the *Cement Hall Photo Album* is dated October 13, 1938, and it shows the shell scaffold in place.

10. Blumer-Maillart, *Recollections,* 92–3. Photographs taken at the Salginatobel show a scene without snow and without summer foliage; therefore I assume the trip to be sometime in late fall, probably between September 25 and November 29 when

there are no surviving letters to or from Maillart. Maillart's November 29, 1938, card to the Blumers in Engi does not mention the trip, so I estimate that it had taken place well before the end of November.

11. Letter from Maillart in Geneva to Edmond in Philadelphia, December 10 and 11, 1938.

12. The last photo in the *Cement Hall Photo Album* shows the structure complete and free-standing; it is dated December 1, 1938.

13. Blumer-Maillart, *Recollections,* 79. The Ceylon project was an apartment house for Baur and Co., Geneva Maillart Archive, No. 390.

14. Card from Marie-Claire, Maillart, Eduard, Nissen, and Alfred and Maria Maillart from Genoa to Edmond in Philadelphia, February 16, 1939.

15. Blumer-Maillart, *Recollections,* 75 and 92–3.

16. F. Bolens, "Das Rhône-Kraftwerk Verbois," *SBZ* 114(27) (December 30, 1939): 318–21.

17. Geneva Maillart Archive, No. 386, Drawing Nos. 19–21, January 17, 1939.

18. Bern Maillart Archive, No. 597, Drawing Nos. 2, 6–7, 9–15, January 12–13, 1939.

19. Fr. Hübner, "Concours pour la construction du nouveau pont de Peney sur le Rhône," *Bulletin technique de la suisse romande* 65(20) (October 7, 1939): 257–61.

20. F. Bolens, "Das Rhône-Kraftwerk Verbois," 318–21. The article is in two parts: one on the power plant taken from Bolens' longer article in *Bulletin technique* 65(13) (July 1, 1939): 169–80, and devoted solely to the power plant; the other part on the bridge uses information from Hübner's article (n. 19 above), but is written by the *Bauzeitung* editors.

21. This predilection is illustrated in Elizabeth B. Mock, *The Architecture of Bridges* (New York: Museum of Modern Art, 1949), 65.

22. Hübner, "Concours," p. 259.

23. For photos of the Weissensteinstrasse Bridge compared to Peney Bridge, see Bill, *Robert Maillart,* 126–29.

24. Blumer-Maillart, *Recollections,* 67. See also letter from Marie-Claire in Pankalan-Brandan to Edmond in Philadelphia, May 30, 1939.

25. Letter from Maillart in Geneva to Edmond in Philadelphia, April 3, 1939. For the dangers to Switzerland, see Sontag, *A Broken World,* 351–9.

26. "Vom Stand der moralischen Aufrustung," *SBZ* 113(14) (April 8, 1939): 165–6 (unsigned but almost certainly written by C. Jegher).

27. Ros, "Belastungsversuche," *SBZ* 113(5): 53–8, and see also p. 65 (written in October 1938 in Lausanne) for a review of *Bericht 99* by Ed. Elskes.

28. "Aus der Abteilung Bauen," *SBZ* 113(10) (March 11, 1939): 122–23.

29. Letter from Maillart in Geneva to Edmond in Philadelphia, April 3, 1939.

30. R. Maillart, "Mass or Quality in Concrete Structures," *SBZ* 98(12) (September 19, 1931): 149–50.

31. Billington, *Robert Maillart and the Art of Reinforced Concrete,* 104.

32. "Das Zürcher Tonhalle- und Kongressgebäude," *SBZ* 113(17) (April 29, 1939): 209.

33. Letter from Maillart in Geneva to Edmond in Philadelphia, May 22, 1939. Maillart led a guided tour, announced in *SBZ* 113(N19) (May 13, 1939): 238.

34. Speer, *Inside the Third Reich,* 74, 152–4; the plates following p. 286 show the 1939 model. The irony of the contrast between Hitler's dome and Maillart's shell was that Germany pioneered thin-shell concrete structures and that German engineers like Franz Dischinger and Ulrich Finsterwalder were world figures in engineering by 1939.

35. Franz Dischinger had made a thin-shell concrete design for the Berlin Dome.

36. "Internat. Vereinigung für Brücken-und Hochbau," *SBZ* 114(5) (July 29, 1939): 58–9. See also card from Maillart en route to St. Gallen to Edmond in Philadelphia, June 4, 1939.

37. For the May activity, see Geneva Maillart Archive, Nos. 337, 400, 401, and 402, and the Bern Maillart Archive, Nos. 599, 565, and 566.

38. *Wettbewerb über die Erstellung von Zwei Ueberfahrtsbrücken Altendorf-Lachen,* Schweizerische Bundesbahnen Kreisdirektion III (Zurich, July 25, 1939).

39. Letters from Maillart in Gurnigelbad to Stettler in Bern, July 21 and Aug. 3, 1939, and card from Maillart in Gurnigelbad to Edmond in Philadelphia, July 28, 1939. For the Chateau d'Oex Bridge, he reduced the thickness of the columns while widening them so that they would appear more slender in profile. He eliminated some elements and changed the plan layout to achieve a greater sym-

metry and thus "get a better appearance." He indicated to Stettler how he would be able to lower the cost from the official design.

40. While Maillart was at Gurnigelbad in July 1939, the *Bauzeitung* published a brief note on the Cement Hall, apologizing for not having credited the architect Leuzinger with a role in its design. "The Cement Hall in the Building Section," *SBZ* 114(4) (July 22, 1939): 48.

41. *Report,* Section Chief, Region III, Swiss Railways, August 15, 1939. It consisted of five pages plus cost breakdowns for twenty-three bids. The official cost estimate of 99,000 francs appears in "Voranschlag" [cost estimate], December 28, 1938. Technically, the fine designs with minimum materials and visual elegance appealed to the railways, but the bridges' low cost was decisive. On August 23, a contract was signed in Zurich between Prader and the Swiss Railways. *Report,* Section Chief Region III, Swiss Railways, August 21, 1941.

42. Sontag, *A Broken World,* 380–1.

43. Letter from Marie-Claire in Branden to Maillart in Geneva, September 1–2, 1939, and letter from Marie-Claire in Brandan to Maillart in Geneva, September 27, 1939. See also letter from Maillart in Geneva to Stettler in Bern, September 8, 1939.

44. The description of Maillart's illness appears in cards from Rosa in Geneva to Edmond in Philadelphia, October 30 and November 7 and 13, 1939. Rosa's densely written cards were like letters written quickly and with evident passion.

45. Letter from Maillart in Geneva to Marie-Claire in Brandan, November 22, 1939.

46. Letter from Maillart in Geneva to the Blumers in Brandan, December 6, 1939.

47. Letter from Rosa in Geneva to the Blumers in Brandan, December 21, 1939.

48. Letter from Maillart in Zurich to Edmond in Philadelphia, December 24, 1939.

49. Letter from Maillart in Geneva to the Blumers in Brandan, January 2, 1940.

50. R. Maillart, "L'ingenieur et les autorités," *Vie, art et cité* (January–February, 1940).

51. Little did Maillart know that his pachiderm would one day become the inspiration for the world famous McDonald's logo. See Alan Gowans, "Reevaluation: Space Time, and Architecture," *Progressive Architecture* 71(5) (May 1990): 123.

52. "Die Zementhalle der LA," *SBZ* 115(4 and 5) (January 27 and February 3, 1940): 50, 62. For the detailed report, see M. Ros., "II. Ergebnisse der Belastungsversuche an der Zementhalle der Schweizerischen Landesaustellung Zürich 1939," *Bericht 99 EMPA,* Zurich, Zweite Ergänzung [second addendum] (August 1, 1940): 21–60.

53. Letter from Maillart in Geneva to the Blumers in Brandan, February 7, 1940.

54. Letter from Maillart in Geneva to the Blumers in Brandan, February 27, 1940. It was in this letter that he referred to the Cement Hall as the back of the elephant. In a postscript to this letter dated February 28, 1940, Maillart described Rosa's seventy-fifth birthday party.

55. Letter from Amélie in Geneva to Marie-Claire in Brandan, March 8, 1940.

56. "Persönliches," *SBZ* 115(12) (March 23, 1940): 142.

57. "S.I.A. Fachgruppe," *SBZ* 115(9) (March 2, 1940): 108.

58. Letter from Rosa in Geneva to the Blumers in Brandan, March 17–18, 1940.

59. Letter from Rosa in Geneva to Marie-Claire in Brandan, March 22, 1940, completed on March 26.

60. Letter from Rosa in Geneva to Marie-Claire in Brandan, April 1, 1940, continued on April 2, 3, and finished on April 4.

61. Letter from Rosa in Geneva to Marie-Claire in Brandan, April 9, 1940. This is a long, moving letter describing Maillart's last days, the arrangement for the business and the funeral, and the funeral itself. Rosa also wrote long letters to Edmond in Philadelphia on March 30, April 4, and May 5, 1940. In this last letter, she described the doctor's final diagnosis. Maillart did have a tumor near the bladder; it grew rapidly at the end.

Maillart's Major Bridges

Number	Date	Place (Name)	Crossing	Number	Date	Place (Name)	Crossing
1	1896	Pampigny	Le Veyron Brook	25	1931	Adelboden (Spital)	Engstligen River
2	1899	Zurich (Stauffacher)	Sihl River	26	1931	Frutigen (Ladholz)	Engstligen River
3	1901	Zurich[a]	Hadlaub Street	27	1931	Schangnau	Hombach
4	1901	Zuoz	Inn River	28	1931	Schangnau	Luterstalden Brook
5	1903	St. Gallen	Steinach Brook	29	1932	Nessental	Triftwasser Brook
6	1904	Billwil	Thur River	30	1932	Schwarzenburg	Rossgraben
7	1905	Tavanasa[b]	Rhine River	31	1932	Habkern	Traubach
8	1907	Aach	Railroad	32	1932	Habkern	Bohlbach
9	1909	Wattwil	Thur River	33	1933	Hinterfultigen	Schwandbach
10	1910	Wyhlen	Unterwasser Canal	34	1933	Felsegg	Thur River
11	1911	Laufenburg	Rhine River	35	1934	Innertkirchen	Aare River
12	1912	Aarburg	Aare River	36	1934	Wülflingen-Winterthur	Töss River
13	1912	Augst-Wyhlen	Rhine River	37	1935	Liesberg	Birs River
14	1912	Rheinfelden	Rhine River	38	1935	Huttwil	Railroad
15	1913	Ibach	Muota River	39	1936	Twann	Twannbach
16	1920	Marignier	Arve River	40	1936	Vessy	Arve River
17	1923	Wäggital[a]	Flienglibach	41	1937	Gündlischwand[a]	Lütschine River
18	1924	Wäggital[c]	Ziggenbach	42	1938	Bern	Weissenstein Street
19	1924	Wäggital	Schrähbach	43	1938	Wiler	Gadmerwasser
20	1924	Châtelard	Eau-Noire	44	1940	Laubegg	Simme River
21	1925	Donath	Valtschielbach	45	1940	Altendorf	Railroad
22	1930	Bern (Lorraine)	Aare River	46	1940	Lachen	Railroad
23	1930	Schiers	Saginatobel	47	1940	Garstatt	Simme River
24	1930	Klosters[a]	Landquart River				

[a] Bridge later replaced.
[b] Bridge destroyed by avalanche.
[c] Spelling later changed to Wägital.

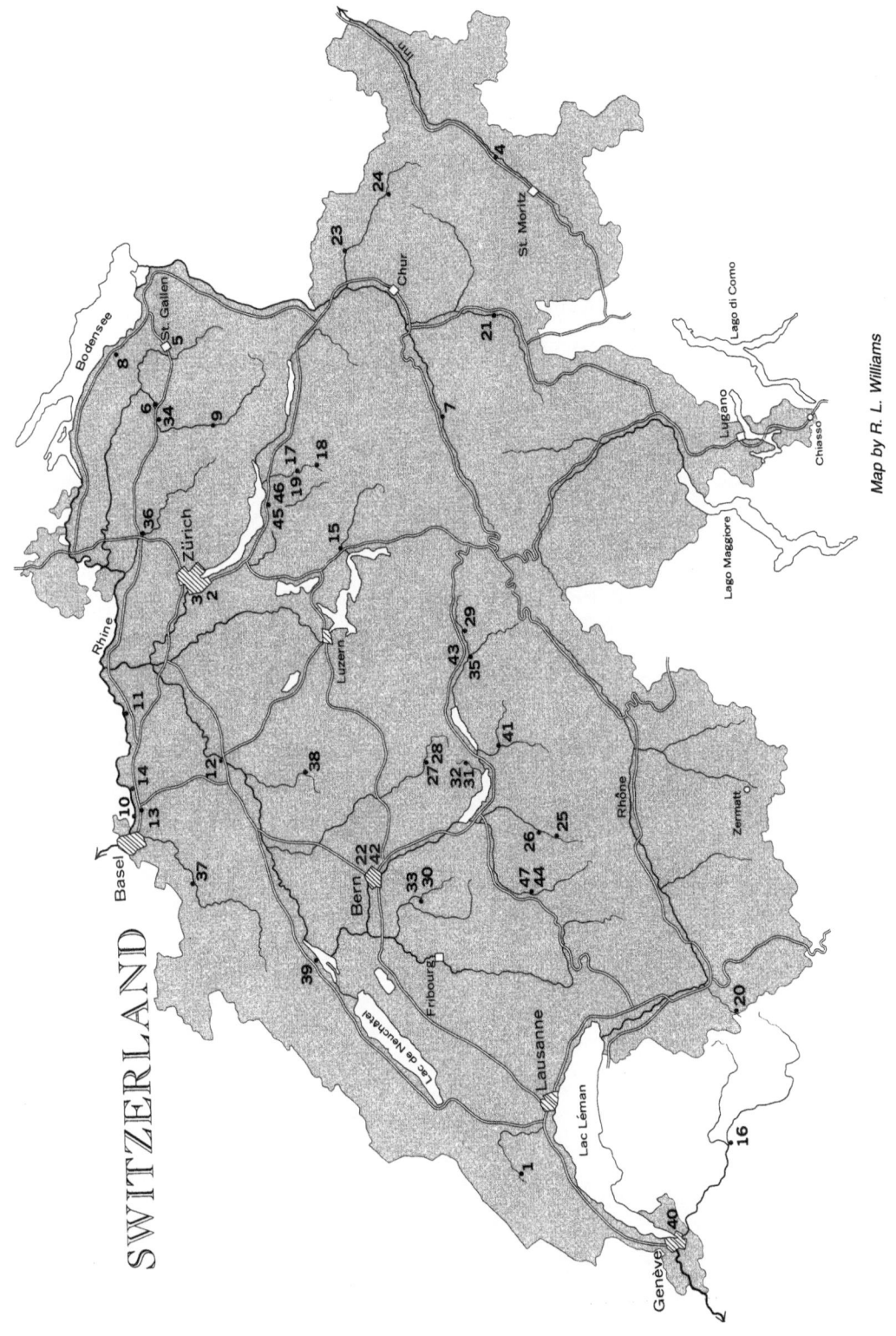

SWITZERLAND

Bodensee

Rhine

St. Gallen

Zürich

Basel

Luzern

Bern

Fribourg

Lac de Neuchâtel

Lausanne

Lac Léman

Genève

Chur

Inn

St. Moritz

Rhône

Zermatt

Lago Maggiore

Lago di Como

Lugano

Chiasso

8
6
34
9
5
36
3
2
11
14
12
13
10
37
38
22
42
33
30
39
45 46
19 17
18
15
27 28
32
31
41
35
43
29
7
21
23
24
44
25
26
47
44
20
16
40
1

Map by R. L. Williams

Maillart's Works

Date	Structure Title and/or Description	Location	Client
	DESIGNS MADE AS AN EMPLOYEE		
1896	Bridge crossing, Le Veyron Brook, hingeless arch	Pampigny	Chemin de Fer B. A. M. (Bière-Apples-Morges)
1899	Bridge over the Sihl River, nonreinforced concrete, three-hinged arch[a]	Stauffacher (Zürich)	Stadt Zürich
1901	Bridge over Hadlaub Street, continuous beam	Zürich	Stadt Zürich
	Bridge over the Inn River, hollow-box, three-hinged arch[a]	Zuoz	Kanton Graubünden
	STRUCTURES DESIGNED AND BUILT BY MAILLART AND COMPANY		
1902	Two gas tanks of about 10,000 m³ in volume apiece[b]	St. Gallen	Stadt St. Gallen
	Two water tanks with combined volume of 300 m³	Hundwil	Kanton Appenzell
	Two water tanks with combined volume of 300 m³	Döttingen	Kanton Aargau
1903	Bridge over the Steinach, concrete-block arch[b]	St. Gallen	Stadt St. Gallen
	Bridge over the Thur, hollow-box, three-hinged arch[b]	Billwil	Kanton St. Gallen
	Water storage reservoir of 250 m³ in volume	Rebstein	Gemeinde Rebstein
	Two water tanks with combined volume of 500 m³	Männedorf	Kanton Zürich
	Floors, columns, and roof construction for an embroidery building	St. Gallen	Herren Jäger & Brennwald
	Floors and columns over a storage space	St. Gallen	Stadt St. Gallen
	Two water tanks with combined volume of 300 m³	Balgach	Gemeinde Balgach
	Two water tanks with combined volume of 400 m³	Rüthi	Gemeinde Rüthi
	Roof for a factory building	Schönengrund (Appenzell)	Herren Locher & Cie.
	Church of Bruggen: foundations and floors as well as earth and wall work	St. Gallen	Gemeinde Straubenzell-Bruggen
1904	Protective cover against falling stones	Chur	Rhätische Bahnen
	Two water tanks with combined volume of 200 m³	Oberstetten (St. Gallen)	Gemeinde Oberstetten

Date	Structure Title and/or Description	Location	Client
	Forty-two transmission poles[b]	Zürich	Stadt Zürich
	Canal crossing	Schaffhausen	Schweizerische Bundesbahnen
	Factory building: columns, floors, stairs, and roof	Zürich	A. Waltisbühl
	Two water tanks with combined volume of 200 m³	Heiden	Kanton Appenzell
	Steam boiler and machine foundation	St. Gallen	Stadt St. Gallen
	Health resort – new building	Davos	Deutsche Heilstätte
1905	Bath house	Davos	Deutsche Heilstätte
	Cloth factory[b]	Wädenswil (Zürich)	Pfenninger & Cie.
	Thirty-two transmission poles	Zurich	Stadt Zürich
	Steinach Bridge for the Demutstrasse	St. Gallen	Stadt St. Gallen
	Factory enlargement	Zürich	A. Waltisbühl
	Hotel Valsana	Arosa	Jösler & Morgenthaler
	Hadwig School House, floors, etc.	St. Gallen	Stadt St. Gallen
	Two water tanks with combined volume of 200 m³	Kronbühl (St. Gallen)	Gemeinde Kronbühl
1906	Water tower	St. Gallen	Schweizerische Bundesbahnen
	Casino terrace	Bern	Bürgergemeinde Bern
	Floors, galleries, and cellars	Kubel (St. Gallen)	Elektrizitätswerk Kubel
	Wine building: foundations, floors, and columns	Wädenswil	Obst und Weinbaugenossenschaft vom Zürichsee
	Two water tanks with combined volume of 200 m³	Zuoz (Engadin)	Gemeinde Zuoz
	Two water tanks with combined volume of 200 m³	Sils (Engadin)	Gemeinde Sils
	Transmission poles	Baden	Elektrizitätsgesellschaft "Motor" A.G.
	Floors and columns together with foundations for the new concert hall	St. Gallen	Tonhallebaugesellschaft
	Gas tanks of 2,000 m³ volume	Baden (Switzerland)	Stadt Baden
	Bridge over the Rhine, hollow-box, three-hinged arch[b]	Tavanasa	Kanton Graubünden
1907	Sanitorium	Davos-Platz	The Queen Alexandra-Sanitorium
	Bank building	St. Gallen	Eidgenössische Bank, A.G.
	Siphon for the Riehenteich	Basel	Stadt Basel
	Two water tanks with combined volume of 1,200 m³	Kilchberg & Rüschlikon	Gemeinden Kilchberg und Rüschlikon

Date	Structure Title and/or Description	Location	Client
	Roof structure	Herisau	A.-G. Cilander
	Two overpass bridges	Linie Winterthur-Romanshorn	Schweizerische Bundesbahnen
	Two water tanks with combined volume of 400 m^3	Thalwil (Zürich)	Gemeinde Thalwil
	Floors, columns, and stairs	Herisau	Buff & Cie.
	Transformer station floors and roof construction	Letten	Stadt Zürich
	The steps in the school houses by Aemtlerstrasse	Zürich	Stadt Zürich
	Two watertanks with combined volume of 2,000 m^3	Sittertobel (St. Gallen)	Frischknecht & Cie.
	Two watertanks with combined volume of 300 m^3	Hemberg	Kanton St. Gallen
	Two watertanks with combined volume of 300 m^3	Wolfhalden-Aussertobel	Kanton Appenzell
	A watertank of 80 m^3	Interlaken	Harderbahngesellschaft
1908	Music pavillion, Bürkli Gardens	Zürich	Stadt Zürich
	Administration and machine building	Zürich	Neue Zürcher Zeitung
	Foundation slab at electricity plant	St. Gallen	Stadt St. Gallen
	200 transmission poles[b]	Rheinfelden	Kraftübertragungswerke
	Office building: floors of pumice concrete	St. Gallen	O. Münch, Goldschmied
	Office building on the Neugasse: floor construction	St. Gallen	Buzzi & Lutz
	Railway station	Teufen	Appenzeller Strassenbahn
1909	Bridge over the Thur, hollow-box, three-hinged arch[b]	Wattwil	Kanton St. Gallen
	Office building	Rorschach	Gebr. Witta
	Office building: floors of pumice concrete	St. Gallen	Konsumverein St. Gallen
	Office building (and retaining wall)	St. Gallen	Otto Alder & Cie.
	Coal silo	Uzwil	Gebr. Bühler
	Wine building	Zürich	Obstverwertungs-Genossenschaft Zürich III
	School house on the Münchhaldenstasse	Zürich	Stadt Zürich
	Pump foundations for a villa	Zürich	Dr. W. v. Muralt
	Reinforced concrete piles[b]	Venice	Prof. Conrad Zschokke, Engineer, Aarau
	Caissons[b]	Venice	Prof. Conrad Zschokke, Engineer, Aarau
	Moveable Stageset Warehouse	Zürich	Stadt Zürich

Date	Structure Title and/or Description	Location	Client
1910	Apartment house and office building	St. Gallen	Dr. Schönenberger
	Apartment house and office building	St. Gallen	Zünd-Bischoff
	Hat factory[b]	Wädenswil	Felber & Cie.
	Reinforced-concrete foundations, slabs and pumice concrete floors for two villas	Zürich	F. Egli-Schneider
	Bridge for cables crossing the Unterwasser Canal, continuous hollow-box beam[b]	Wyhlen	Kraftübertragungswerke Rheinfelden
	Car repair shop floors out of pumice concrete	Zürich	Schweizerische Bundesbahnen
	Warehouse at Gieshübelstrasse[b]	Zürich	Lagerhausgesellschaft A.-G.
	Fish pond	Bern	A. v. Stürler
	Factory building	Zürich	H. Weilenmann-Girsberger, Zürich III
1911	Bridge over the Rhine, concrete-block arch[b]	Laufenburg	Kraftwerk Laufenburg
	Church hospital: new building, floors out of pumice concrete	Chur	Kreuzspitalkommission
	Bridge over the Aare, hingeless arch[b]	Aarburg	Kanton Aargau
	Office building	Uzwil	Gebr. Bühler
	University building, student building with tower[b]	Zürich	Kanton Zürich
	Factory building	Zürich	Schweizerische Kohlensäurewerke
	Warehouse	Chur	Stiffler & Conrad
1912	Grain depot[b]	Altdorf	Schweizerische Eidgenossenschaft
	Warehouse for refrigeration[b]	St. Petersburg	A.-G. Gerhard & Hey
	Dam bridge over the Rhine by the power plant of Augst-Wyhlen, continuous beam	Aarau	A.-G. Conrad Zschokke
	Commercial building on Hohlstrasse[b]	Zürich	Konsumverein Zürich
	Bank and administration building[b]	Herisau	Kantonalbank
	Reinforced-concrete roof to replace iron roof destroyed by fire	Wangen a.d. Aare	Elektrizitätswerk Wangen a.d. Aare
	Apartment and business building	Herisau	E. Frehner
	Bridge over the Rhine, hingeless arches[b]	Rheinfelden	Stadt Rheinfelden
	Bridge over the Muota, hollow-box cantilever beam[b]	Ibach	Kanton Schwyz
	Armored acid tank, ventilation and connection duct systems	Kempttal	Fabrik von Maggi's Nahrungsmitteln
	Filtration plant[b]	Rorschach	Stadt St. Gallen
	Roofs for the railway station	Oerlikon	Schweizerische Bundesbahnen

Date	Structure Title and/or Description	Location	Client
1913	Sanitorium	Agri di Lugano (Davos)	Deutsche Heilstätte
	Reinforcement work in the Castle of Jegenstorf	Bern	A. v. Stürler
1913–14	Factory[b]	Riga	Gummifabrik "Prowodnik"
1914	Factory[b]	Villanueva y Geltrù (Spain)	Kabelfabrik Pirelli & Cie.
	Shoe factory[b]	Lyon	Bally-Camsat
1916	Factory, warehouse, and office building[b]	Charkov	Russische Allgemeine Elektrizitätsgesellschaft
	Steelworks[b]	Kamanskaja	Russian government
1919–20	Power plant building	Barcelona	Central Catalana de Electricidad
1920	Weaving Mill[b]	Sallent (Spain)	Mata y Pons
	Warehouse for screws	Barcelona	Union Metallurgica
	Spinning mill[b]	Badalona	Cotonoficio

LIST OF SECONDARY WORKS IN RUSSIA, 1913–18

Date	Structure Title and/or Description	Location	Client
1913–14	Cellar enclosure	St.Petersburg	—
	Steps and caissons	Alexandrowski, St. Petersburg	—
1913–15	Bridges Oranienbaum	St. Petersburg	—
1914–15	Warehouse	Prowodnik (Riga)	—
	Stock market warehouse	Riga	—
	Food storage house	Prowodnik (Riga)	—

WORKS DESIGNED BY MAILLART, BUILT BY OTHERS

Date	Structure Title and/or Description	Location	Client
1913	Warehouse[b]	Broc (Fribourg)	Schokoladefabrik Cailler
	Refrigerated storage building	St. Petersburg	Société S.I.A.E.C.
	Factory building	Lancey (Isère)	Papeteries Bergès & cartonneries du Sud-Est
1915	Factory building	Nanterre	Papeterie de la Seine
	Coal bunker	La-Chaux-de-Fonds	Städtisches Gaswerk
1916–17	Steam plant for generating electricity[b]	Barcelona	Catalana de Gas y Electricidad
1917	Factory building	Lyon	Carburateurs "Zenith"

Date	Structure Title and/or Description	Location	Client
1918	Factory building	Suresnes (France)	Flugzeugfabrik Bleriot
	Factory building	Lyon	Co. Electro-Mécanique
1920	Bridge over the Arve, hingeless arch	Marignier (France)	Department Hoch-Savoyen
1921	Cellar rooms	Zürich	Rentsch & Cie.
1922	Pier	Eaux-Vives	Kanton Genf
	Roof structure	Geneva	Banque de Genève
	Business building	Geneva	Société Citadine
	Clinic	Champel (Geneva)	Société La Colline
	Bank building, office block[b]	Geneva	Banque Nationale
	Water tower of 120 m³ in volume[b]	Chancy	Kraftwerk Chancy-Pougny
	Swimming pool	Barcelona	Schwimmklub
1922–4	Tunnel lining by the Maillart method, penstock, aquaduct over the Trebsenbach and bridges for the lakeside road, including the bridge over the Flienglibach, deck-stiffened arch, bridge over the Ziggenbach, bridge over the Schrähbach, deck-stiffened arch[b]	Wäggital	Kraftwerk Wäggital
1923	Factory building	Paris	Société "Pile Leclanché"
	Commercial building	Geneva	S.I. Centre C
1923–4	Office building[b]	Geneva	Bureau Internationale du Travail
	Warehouse[b]	Chiasso	Freilagergesellschaft
1923–30	Lorraine Bridge[b]	Bern	Stadt Bern
1924	Factory building	Lyon	Ateliers électriques de Delle
	Aquaduct over the Eau-Noire, continuous hollow-box beam and frame[b]	Châtelard	Schweizerische Bundesbahnen
1924–5	Warehouse	Geneva	Naef & Cie.
	Factory building	Geneva	Naef & Cie.
1924–6	Warehouse	Zürich-Albisrieden	Freilagergesellschaft
1924–7	Viaduct de Grandfey[c]	Fribourg	Schweizerische Bundesbahnen
1925	Bridge over the Valtschielbach, deck-stiffened arch[b]	Donath	Kanton Graubünden
1925–6	Drainage canal	Zürich	Stadt Zürich
	Exposition building[b]	Geneva	Societé du Palais des Expositions

Date	Structure Title and/or Description	Location	Client
1926	Administration building	Geneva	Naef & Cie.
	Institute de Jeunes Gens	Geneva	Institut Widemann
	Garage[b]	Vernier	L. Givaudan & Cie. S.A.
	Garage and workshop for the incineration plant	Zürich	Stadt Zürich
	Factory building	Geneva	S.I. Falaise-Pêcheries
1926–30	Office building by the Sihl River[b]	Zürich	Schweizerische Bundesbahnen und Postverwaltung
1927	School house	Hermance	Commune d'Hermance
1927–8	Apartment houses	Geneva	S. I. Quai Wilson 41
	Apartment control canals	Zürich	Stadt Zürich
	Dwelling houses	Geneva	S. I. Square des Tranchées
	Three dwelling houses and garage	Geneva	S. I. Auto-Avion
	Refrigeration warehouse[b]	Geneva	Schweizerische Bundesbahnen
	Commerical building	Eaux-Vives, Geneva	S. I. L'Anneau
	Home for the aged	Near Geneva	Kanton Genf
	Apartment house	Geneva	S. I. St.-Jean-Le Vuache
	Silos and elevated crane tracks[b]	Cairo (Egypt)	Fabrique de Ciment Tourah
1927–9	Three apartment blocks[b]	Geneva	S. I. Deux Parcs
	Cellar structure and rebuilding	Geneva	Brasserie de l'Avenir
	Milk distribution center, workshop, garage, pumping station[b]	Geneva	Laiteries Réunies
1928	Castle renovation	Dardagny (Geneva)	Commune de Dardagny
1928–9	County seat hall	Chêne-Bourgeries	Commune de Chêne-Bourgeries
	Sanatorium	Montana	Sanatorium populaire genevois
1928–30	Church	Chêne-Bourgeries	Société St. François de Sales
1929–30	Bridge over the Salginatobel, hollow-box, three-hinged arch[b]	Schiers	Kanton Graubünden
	Apartment house	Geneva	S. I. Riant-Logis
	Factory and tanks[b]	Maroggia	S. A. Tannini Ticinesi
	Commercial building	Montreux	S. A. Passage du Kursaal

Date	Structure Title and/or Description	Location	Client
	Commercial building	Zürich	Gesellschaft Seefeldegg
	Distillation building[b]	Vernier	L. Givaudan & Cie.
	Gymnasium by the Sihl and Music pavillion[b]	Zürich	Stadt Zürich
1929–31	Bank building[b]	Geneva	Schweizerische Kreditanstalt
1930	Bridge for railway over the Landquart River, deck-stiffened arch[b]	Klosters	Rhätische Bahn
	Apartment houses	Geneva	S. I. La Colombière
	Warehouse, Gieshübel[b]	Zürich	Zürcher Lagerhaus-Gesellschaft
1930–1	Transformer station[b]	Zürich	Stadt Zürich
	Main hall for trade school[b]	Zürich	Stadt Zürich
	Small dwellings	Geneva	Société Coopérative d'Habitation
	Bridge over the Engstligen (Spital), deck-stiffened arch[b]	Adelboden	Kanton Bern
	Swimming pool building	Wengen	Losinger & Cie. Gemeinde Wengen
	Hall	Geneva	Paroisse de St. Jeanne
	Machine building	Vernier	Boccard Frères
1930–2	Enlargement and rebuilding of the church	Geneva	Paroisse du Sacré-Coeur
1931	Reservoir on the Käferberg	Zürich	Stadt Zürich
	Gravel silos	Geneva	S. A. Arva
	Administration building[b]	Zürich	Société d'Assurances "Vita"
	Dwelling house	Albisrieden	Hr. Bretscher
	Waiting room	Zürich	Strassenbahn
	Footbridge over the Engstligen, deck-stiffened arch	Frutigen (Ladholz)	Gemeinde Frutigen
	Hombach Bridge, deck-stiffened arch	Schangnau	Kanton Bern
	Apartment house	Geneva	S. I. Clarte
	Luterstalden Bridge, deck-stiffened arch	Schangnau	Kanton Bern
	Apartment house	Geneva	S. I. Bertrand Athénée
	Warehouse[b]	Vernier	Benzin- und Petroleum A.-G.
1931–2	Apartment house with large garage	Geneva	S. I. LeLoriot
1932	Bridge over the Triftwasser Brook, beam	Nessental	Kanton Bern

Date	Structure Title and/or Description	Location	Client
	New factory for screw manufacturing[b]	Gerlafingen	L.v. Roll'sche Eisenwerke
	Reservoir on the Mannenberg of 15,000 m³ volume[b]	Ostermundigen (Bern)	Stadt Bern
	Rebuilding of the sunken Quai[b]	Vevey	Ville de Vevey
	Harbor works at Vevey	Vevey	Ville de Vevey
	School house with gymnasium[b]	Wipkingen	Stadt Zürich
	Bridge over the Rossgraben, hollow-box, three-hinged arch[b]	Schwarzenburg (Bern)	Gemeinde Schwarzenburg
	Two bridges crossing the Traubach and Bohlbach, both deck-stiffened arches[b]	Habkern	Gemeinde Habkern
	Apartment houses	Geneva	S. I. Chemin Vermont D.
	Apartment houses	Geneva	S. I. Rue Caroline 19–25
	Transformer groups (rebuilding), foundations for the generator	Mühleberg	Bernische Kraftwerk
	Parish house	Zürich	Kirchgemeinde Unterstrass
1933	Bridge over the Thur, hollow-box, three-hinged arch[b]	Felsegg (Henau)	Kanton St. Gallen
	Apartment houses	Zürich	Baugenossenschaft Talwies
	Correction of River Aire: Bridge Déversoir	Lully	Canton de Genève
	Bridge over the Schwandbach, deck-stiffened arch[b]	Hinterfultigen (Bern)	Gemeinde Schwarzenburg
	Apartment house	Geneva	S. I. Boulevard du Pont d'Arve
	Apartment house	Geneva	S. I. Rue Jean-Charles
	Municipal building	Albisrieden	Gemeinde Albisrieden
	Institute	Geneva	S. I. Chemin Dumas
	Apartment houses and garage	Geneva	S. I. Rue du Roveray
	Apartment house	Geneva	Stè Coop. d'Habitations salubres, Rue des Allobroges
	Transformer building	Geneva	Service Industriels
	School museum	Bern	Stiftung Schulmuseum
	Farming apprenticeship boarding school	Albisbrunn	Stiftung Schulwarte
	Enlargement of the main building	Zofingen	Ringier & Company

Date	Structure Title and/or Description	Location	Client
	New church, reinforced-concrete nave and steeple	Lyss	Kirchgemeinde
	Apartment houses	Zürich	Gemeinde Baugenossenschaft Limmattal
	Omnibus covering or shed, steel construction of flat roof	Bern	Stadt Bern
1934	Bridge over the Aare, hollow-box, three-hinged arch[b]	Innertkirchen	Kanton Bern
	Bridge over the Töss, deck-stiffened arch[c]	Wülflingen-Winterthur	Gemeinde Winterthur
	Covering against falling stones for water-pressure system	Kandergrund	Bernische Kraftwerke
	Apartment house	Geneva	S. I. L'Acajou
	Apartment house	Geneva	S. I. L'Ebène
	Apartment house	Geneva	S. I. L'Amarante
	Apartment houses	Geneva	S. I. Délia
	Reservoirs on the Länggasse	Bern	Stadt Bern
	Enlargement of the toll house	Perly (Geneva)	Schweizerische Eidgenossenschaft
	Retaining walls and enlargement of the restaurant "Vieux Bois"	Geneva	Canton de Genève
	Pressure pipes, water gates, reservoir reinforcement piles	Kanton Schwyz	Etzelwerk A.-G.
	Administration building[a]	Zürich	Stadt Zürich
	Apartment house	Geneva	S. I. Rue Caroline A
	Apartment house	Geneva	S. I. Malgnou-Ermitage
1935	Apartment house	Geneva	S. I. Rue du Vieux Collège
	Bridge for railway over the Birs, continuous beam[b]	Liesberg (Bern)	Portlandzementfabrik Laufen
	Construction	Bern	Heim "Favorite"
	Automobile garage	Vernier (Geneva)	L. Givaudan & Cie.
	Bridge over the Huttwil-Wolhusen-Bahn, continuous beam[a]	Huttwil (Bern)	Gemeinde Huttwil
	Bridge over the Hämelbach	Langnau	Kanton Bern (listed as 1932 in Bern catalog)
	Automobile elevator in the Sihlpost	Zürich	Schweizerische Bundesbahnen und Postverwaltung

Date	Structure Title and/or Description	Location	Client
	Footbridge over the Maggia	Someo (Tessin)	Gemeinde Someo
	Factory	Vernier (Geneva)	L. Givaudan & Cie.
	Factory annex addition	Geneva	Laboratorium Sauter A.-G.
	Bridge over the Wohlei, rebuilding; reinforced track and steel construction	Wohlensee	Bernische Kraftwerke A.-G.
	Aqueduct over the Kander	Wimmis	Bernische Kraftwerke A.-G.
	Villa	Zofingen	Ringier-Brack
	Garage and auto repair shop	Geneva	A.-G. Garage Mon Repos
	Bridge over the Twannbach, hollow-box, three-hinged arch[a]	Twann (Bern)	Gemeinde Twann
1936	Dwelling house	Zürich	Knapp
	Bridge over the Arve, hollow-box, three-hinged arch[a]	Vessy (Geneva)	Bewirtschaftungsyndikat Vessy und Umgebung
	Air raid shelter and retaining walls	Bern	Stadt Bern
	Quai Turrettini[a]	Geneva	Ville de Genève
	Telephone building	Zürich	Schweizerische Eidgenossenschaft
	Rebuilding of the Poste du Stande	Geneva	Schweizerische Eidgenossenschaft
	Apartment house	Geneva	S. I. La Tour de Rive
	Swimming pool	Klosters	Strandbadgenossenschaft
	Vegetable cellar	Kühlewil	Armenanstalt Kühlewil
	Transformer house	Geneva	Services Industriels
	Factory	Geneva	Chocolats Croisier
	Gymnasium	Versoix (Geneva)	Commune de Versoix
	Powder storage shelter	Wimmis	Schweizerische Eidgenossenschaft
1937	Rebuilding of a church[a]	Delsberg (Geneva)	Paroisse de St. Joseph
	Warehouse	Bern	Galenica A.-G.
	Air raid shelter for the Wimmis powder factory	Wimmis	Pulverfabrik Wimmis
	Large kitchen in the Congress house	Zürich	Tonhalle-Stiftung
	Bridge over the Lütschine, continuous beam[a]	Gündlischwand	Kanton Bern

Date	Structure Title and/or Description	Location	Client
	Factory building	Geneva	S.A.D.E.D., A.-G.
	Apartment houses	Geneva	S. I. Sésia
	School complex in Marzili	Bern	Stadt Bern
	Lithography building superstructure	Zofingen	Ringier & Cie., A.-G.
	Apartment houses	Geneva	S. I. "Sig"
	Villa	Onex (Geneva)	Dumarest & Eckert
	Welfare house	Neuenegg (Bern)	Dr. Wander A.-G.
1938	Reservoir	Cortébert (Bern)	Trinkwasserversorg. -Syndikat der "Franches Montagnes"
	Storehouse reconstruction	Bern	Bürgi & Cie.
	Apartment houses	Geneva	Immobilien-Gesellschaft L'Amarante
	Swimming pool building with air raid shelter[a]	Zürich	Stadt Zürich
	Apartment houses	Geneva	Versicherungskasse der städtischen Angestellten
	Apartment house	Colombo (Ceylon)	Baur & Co.
	Bridge over Weissenstein Street, continuous beam[a]	Bern	Stadt Bern
	Cement Hall[a]	Zürich	Landesausstellung, 1939
	Strengthening of a foundation	Bern	Hr. Arm
	Reinforcing understructure	Bern	Bureau für Befestigungsbauten
	Bridge over the Unterwasser by the Sustenstrasse, hollow-box, three-hinged arch	Wiler	Kanton Bern
1939	Apartment house on Anemonenstrasse	Zürich	Architect W. Müller
	Air filter construction	Vernier (Geneva)	Ciment Portland S.A.
	Bridge over the Simme, hollow-box, three-hinged arch[a]	Garstatt	Kanton Bern
	Bridge over the Simme, beam	Laubegg	Kanton Bern
	Apartment houses	Geneva	S. I. "Le Charme"
	Renovation	Geneva	Paroisse du Sacré-Coeur
	Factory	Langenthal	Gugelmann & Co. A.-G.
	Silo	Chippis	Aluminum A.-G., Neuhausen

Date	Structure Title and/or Description	Location	Client
1940	Bridge over the rail line, skewed, continuous beam[a]	Altendorf	Schweizerische Bundesbahnen
	Bridge over the rail line, hollow-box, three-hinged arch[a]	Lachen	Schweizerische Bundesbahnen

[a] Designated by the author as Maillart's major works either before 1902 or after mid-1934.
[b] Designated by Maillart as major works in a 1934 office brochure, which did not include works before 1902.
[c] Not listed by Maillart in his 1934 brochure, but one of his major consulting projects.

Maillart's Writings

1901 "Das Hennebique-System und seine Anwendungen," *SBZ* 37(21) (May 25, 1901): 225–6.

"Armierte Betonbauten," *Schweizerischer Bau- und Ingenieur-Kalender* 3: 108–10.

1904 "Neuere Anwendungen des Eisenbetons," *Protokoll der ordentlichen Generalversammlung, am 16. und 17. September 1904* (Basel: Verein schweizerischer Zement- Kalk- und Gipsfabrikanten), App. III, 16–23.

1907 "Versuche über die Schubwirkungen bei Eisenbetonträgern," *SBZ* 49(16) (April 20, 1907): 198–202.

"Belastungsprobe eines Eisenbetonkanals," *SBZ* 50(10) (September 7, 1907): 125–7.

1908 "Korrespondenz [Guggersbach Bridge]," *SBZ* 51(12) (March 21, 1908): 157; and 51(13) (March 28, 1908): 169.

1909 "Die Sicherheit der Eisenbetonbauten," *SBZ* 53(9) (February 27, 1909): 119–20.

"Armierte Betonbauten," *Schweizerischer Ingenieur-Kalender* 1: 268–70.

1911 "Die Eisenbeton-Konstruktionen im Neuen Kulissenmagazin des Zürcher Stadttheaters," *SBZ* 57(1) (January 7, 1911): 12–13.

"Diskussion: Würfelprobe oder Kontrollbalken?" *Armierter Beton* 4 (May 1911): 159–60.

1912 "Zur Berechnung der Deckenkonstruktionen," *SBZ* 59(22) (June 1, 1912): 295–9.

1913 "Die Wasserkraftanlage Augst-Wyhlen: II Das Kraftwerk Whylen, Die Kabelbrücke," *SBZ* 62(8) (August 23, 1913): 97–9.

1918 "Die Grundwasser-Vorkommnisse in der Schweiz," *SBZ* 72(5) (August 3, 1918): 40–2.

1920 Review of "Versuche mit eingespannten Eisenbetonbalken" (by C. Bach and O. Graf), *SBZ* 76(23) (December 4, 1920): 267.

1921 "Zur Frage der Biegung," *SBZ* 77(18) (April 30, 1921): 195–7.

"Bemerkungen zur Frage der Biegung," *SBZ* 78(2) (July 9, 1921): 18–19.

"Le béton armé à la galerie du Ritom," *Bulletin technique de la suisse romande* 47(17) (August 20, 1921): 198–201.

Review of *Elastizität und Festigkeit* (by C. Bach), *SBZ* 77(18) (April 30, 1921): 203–4.

1922 "Ueber Drehung und Biegung," *SBZ* 79(20) (May 20, 1922): 254–7.

"De la construction de galeries sous pression intérieure," *Bulletin technique de la suisse romande* 48(22) (October 28, 1922): 256–60; 48(23) (November 11, 1922): 271–4; 48(25) (December 9, 1922): 290–3; 49(4) (February 17, 1923): 41–5; 49(5) (March 3, 1923): 53–8.

Miscellaneous Reviews, *SBZ* 79(14) (April 8, 1922): 186–7; 79(15) (April 15, 1922): 200; 79(17) (April 29, 1922): 227.

1923 "Ueber Gebirgsdruck," *SBZ* 81(14) (April 7, 1923): 168–71.

"Zum Vernietungs-Problem," *SBZ* 82(4) (July 28, 1923): 43–5.

"Dispositif de sécurité pour la circulation sur les ponts," *Schweizerische Zeitschrift für Strassenwesen* 9(5) (March 8, 1923): 70–1.

"Les autoroutes en Italie," *Schweizerische Zeitschrift für Strassenwesen* 9(8) (April 19, 1923): 109–12.

"Korrespondenz: zum Vernietungsproblem," *SBZ* 82(6) (August 11, 1923): 79–80.

"Korrespondenz: Nochmals zum Vernietungsproblem," *SBZ* 82(11) (September 15, 1923): 144–5.

"Nochmals zum Vernietungsproblem," *SBZ* 82(23) (December 8, 1923): 304–5.

1924 "Der Schubmittelpunkt," *SBZ* 83(10) (March 8, 1924): 109–11.

"Zur Frage des Schubmittelpunktes," *SBZ* 83(15) (April 12, 1924): 176–7.

"Zur Frage des Schubmittelpunktes," *SBZ* 83(22) (May 31, 1924): 261–2.

"Le centre de glissement," *Bulletin technique de la suisse romande* 50(13) (June 21, 1924): 158–62.

1925 "Questions relatives à l'exportation d'énergie électrique et à la mise en valeur de nos forces hydrauliques," *Bulletin technique de la suisse romande* 51(4) (February 14, 1925): 41–5.

"Die Brücke in Villeneuve-sur-Lot, nebst Betrachtungen zum Gewölbebau," *SBZ* 85(12) (March 21, 1925): 151–4; 85(13) (March 28, 1925): 169–70.

1926 "Eine schweizerische Ausführungsform der unterzuglosen Decke – Pilzdecke," *Schweizerische ingenieurbauten in Theorie und Praxis*, Internationaler Kongress für Brückenbau and Hochbau (Zurich 1926): 21 pages.

"Zur Entwicklung der unterzuglosen Decke in der Schweiz und in Amerika," *SBZ* 87(21) (May 22, 1926): 263–7.

"Le centre de glissement" (correspondance), *Le Génie Civil* 89(14) (October 2, 1926): 284.

1928 "Druckbeanspruchung bei Biegung" *First International Congress for Testing Materials*, Amsterdam, 1927 (The Hague, 1928), 2:13–17.

"Gewölbe-Staumauern mit abgestuften Druckhöhen," *SBZ* 91(15) (April 14, 1928): 183–5.

"Die Wahl der Gewölbestärke bei Bogenstaumauern," *SBZ* 92(5) (August 4, 1928): 55–6.

"Report on Sessions about Reinforced Concrete," *SBZ* 92(21) (November 24, 1928): 262–3.

"Diskussion: Weitgespannte Wölbbrücken," *Bericht über die II. Internationale Tagung für Brückenbau und Hochbau* Vienna, September 24–8, 1928 (1929): 417–19.

1929 "Die Material-Beanspruchung der Betonstrassen," *Schweizerische Zeitschrift für Strassenwesen* 15(16) (August 1, 1929): 215–18; 15(17) (August 15, 1929): 219–24.

1930 "Note sur les ponts voûtés en Suisse," *Premier congrès international du béton et du béton armé*, Liège, September 1930, III–4, 7 pages.

"Masse oder Qualität im Betonbau," *Beitrag zur Denkschrift anlässlich des 50-Jährigen Bestehens der Eidg. Material-prüfungsanstalt an der Eidg. Technischen Hochschule*, Zurich. Reprinted in 1931, *SBZ* 98(12) (September 19, 1931): 149–50.

1931 "Der Brand eines Fabrikgebäudes der Gummifabrik 'Prowodnik' in Riga," *Beton und Eisen* 30(11) (June 5, 1931): 206–7.

"Die Lorraine-Brücke über die Aare in Bern," *SBZ* 97(1) (January 3, 1931): 1–3; 97(2) (January 10, 1931): 17–20; 97(3) (January 17, 1931): 23–7; 97(5) (January 31, 1931): 47–9.

"Die Erhaltung des schiefen Turmes in St. Moritz," *SBZ* 98(3) (July 18, 1931): 29–31.

"Leichte Eisenbeton-Brücken in der Schweiz," *Der Bauingenieur* 12(10) (December 1931): 165–71.

1932 "Zum Entwurf der neuen schweizerischen Vorschriften für Eisenbetonbauten," *SBZ* 99(5) (January 30, 1932): 55–9.

"Diskussion," *Beton und Eisen* 31(1) (January 1932): 10.

"Zum Entwurf der neuen schweizerischen Vorschriften für Eisenbeton" (correspondence), *SBZ* 99(10) (March 5, 1932): 125–6.

"Ueber Erdbebenwirkung auf Hochbauten," *SBZ* 100(24) (December 10, 1932): 309–11.

"Die Wandlung in der Baukonstruktion seit 1883," *SBZ* 100(27) (December 31, 1932): 360–4.

"Die Verhältniszahl n = 15 und die zulässigen Biegungsspannungen" (discussion), *Beton und Eisen* 31(1) (January 5, 1932): 10.

1933 (With M. Turrettini) "Neues Bankgebäude der Schweizer. Kreditanstalt an der Place Bel Air in Genf," *SBZ* 101(4) (January 28, 1933): 47.

"Die Zürcher Sport-und Grünanlagen im neuen 'Sihlhölzli': Konstruktives," *SBZ* 101(9) (March 4, 1933): 103–5.

"Discussion: Théorie des dalles à champignons," *Premier congrès international des ponts et charpentes de Paris*, Final Report, Paris 1932 (Zurich, December 1933): 197–208.

1934 "Ponts-Voûtes en béton armé," *Bulletin du ciment* 2(8) (August 1934): 2–6

"Gekrümmte Eisenbeton-Bogenbrücken," *SBZ* 103(11) (March 17, 1934): 132–3.

"Robert Maillart, Ing. ETH," *Schweizer Echo* 14(2) (February 1934): 5. An account of Maillart's Russian experience.

1935   "Flachdächer ohne Gefälle," *SBZ* 105(15) (April 13, 1935): 175–6.

"Die Schweizerischen Normen für Eisenbeton von 1935," *Der Bauingenieur* 16(47/48) (November 22, 1935): 481–5.

"Verstärkung einer Eisenbetonkonstruktion," *SBZ* 105(11) (March 16, 1935): 130–2.

"La construction des ponts en béton armé, envisagée au point de vue esthétique," *Le Génie Civil* 106(11) (March 16, 1935): 262–3.

"Ponts-voûtes en béton armé; de leur développement et de quelques constructions spéciales exécutées en Suisse," *Travaux: Architecture, Construction, Travaux-Publics* 19(26) (February 1935): 64–71.

"The Construction and Aesthetic of Bridges," *The Concrete Way* 7(6) (May/June 1935): 303–9 (translation with some modifications of article in *Le Génie Civil* 106[11]: 262).

1936   "Einige neuere Eisenbetonbrücken," *SBZ* 107(15) (April 11, 1936): 157–63.

"Der Ausbau des Quai Perdonnet in Vevey," *SBZ* 108(15) (October 10, 1936): 159–61.

1937   "Lichtbilder Vortrag Basel," unpublished manuscript, Zurich Maillart Archive.

"The Modular Ratio," *Concrete and Constructional Engineering* (September 1937): 517–21.

1938   "Aktuelle Fragen des Eisenbetonbaues," *SBZ* 111(1) (January 1, 1938): 1–5.

"Ueber Eisenbeton-Brücken mit Rippenbögen unter Mitwirkung des Aufbaues," *SBZ* 112(24) (December 10, 1938): 287–93.

1939   "Evolution de la construction des ponts en béton armé," *Bulletin technique de la suisse romande* 65(7) (April 8, 1939): 85–91 (partial reprint of 1935 article in *Le Génie Civil* with some remarks by S. Giedion).

1940   "L'ingénieur et les authorités," *Vie, art et cité* (January–February, 1940): 3 pages.

1943   "Kongressaal," *SBZ* 121(23) (June 5, 1943): 282–3.

General Note on Cost Conversions

Costs in public works are always uncertain because they depend on the time and the place of construction. In addition, we need the exchange rate to convert costs in Switzerland to those in the United States. Between 1900 and 1940 that rate was mostly stable at the rate of five Swiss francs to one United States dollar. In the United States, the trend of construction costs over time during that same period can be approximated by a cost index. For conversions in this text, I assume that labor and material costs in Switzerland and the United States were comparable even though it is usually thought that in Europe materials were more costly and labor cheaper than in the United States. The conversions I have made in the text must be seen, therefore, as rough estimates only; the results for overpass bridges, however, do correspond reasonably well to costs in the early 1990s for comparable works in the United States.

To adjust costs for their equivalency in the United States, I use an American index. The *Engineering News Record* publishes a yearly Construction Cost Index, which began in 1906 and has been tied to costs in 1913. Thus, the 1913 index is 100 and the others are as shown on the table from *ENR*, March 25, 1996, p. 72.

The Tavanasa Bridge officially opened in 1906. The cost index for that year was 95, and the bridge cost was 28,000 Swiss francs, or $2.36 per square foot. With an exchange rate then of 5 francs = $1, the cost in the United States would have been $5,600. The cost index for 1996 (March) was 5537, and hence the ratio of 1906 to 1996 would be 58, so that the total 1996 cost would be about 5,600 × 58 = $324,800. The cost per square foot of bridge surface would be $2.36 × 58 = $138 per square foot.

To demonstrate why Maillart shifted his emphasis to buildings, we take Maillart's contract of 206,000 francs, or $41,200, for the concrete structure of the Zurich University building in 1910. The index is 96 for that year, so that the ratio between 1910 and 1996 is about 57.5. The adjusted 1996 cost would be about $2,370,000, a substantial contract – enough to build over seven Tavanasa bridges.

As a simple rule of conversion 100 Swiss francs before 1914 would be about $1,000 in 1996. By 1921, the cost index had risen to 202 and it remained close to 200 until 1937 when it rose to 236. Thus, roughly we can say that Maillart works for 1921 to 1936 would convert from Swiss francs by a factor of 5 to 1996 dollars and by a factor of 4 for 1937 to 1940 dollars. In the text, I use these approximate conversions when it seems useful to have a modern dollar figure.

Index

Note: *Page numbers in bold italics refer to illustrations.*